CISM COURSES AND LECTURES

The series presents lecture notes, monographs, edited works and proceedings in the field of Mechanics, Engineering, Computer Science and Applied Mathematics.
Purpose of the series in to make known in the international scientific and technical community results obtained in some of the activities organized by CISM, the International Centre for Mechanical Sciences.

INTERNATIONAL CENTRE FOR MECHANICAL SCIENCES

COURSES AND LECTURES - No. 334

STOCHASTIC APPROACH TO FATIGUE:
EXPERIMENTS, MODELLING AND RELIABILITY ESTIMATION

EDITED BY

K. SOBCZYK
POLISH ACADEMY OF SCIENCES, WARSAW

SPRINGER-VERLAG WIEN GMBH

Le spese di stampa di questo volume sono in parte coperte da
contributi del Consiglio Nazionale delle Ricerche.

This volume contains 124 illustrations.

Originally published by Springer Verlag Wien-New York in 1993

In order to make this volume available as economically and as rapidly as possible the authors' typescripts have been reproduced in their original forms. This method unfortunately has its typographical limitations but it is hoped that they in no way distract the reader.

ISBN 978-3-211-82452-8 ISBN 978-3-7091-2622-6 (eBook)
DOI 10.1007/978-3-7091-2622-6

PREFACE

Fatigue fracture of engineering materials has been recognized as an important cause of failure of such engineering structures as: aircrafts, vessels, road vehicles, cranes etc. However, fatigue is a very complicated process that is difficult for accurate description and prediction. Because of the inherent randomness found in fatigue data it is no doubt nowadays that a fatigue of real materials should be regarded as a random phenomenon and analysed by use of the stochastic theory. Today, the stochastic approach to fatigue is perceived as important and challenging task in mechanics and structural/mechanical reliability.

The CISM course and the presented volume of the lecture notes have been intended to summarize the latest achievements in stochastic modelling and analysis of fatigue. The lectures cover all main aspects of fatigue, namely: methodology of stochastic modelling of fatigue, tools for characterization of random fatigue loads, physical and mechanical aspects of stochastic fatigue crack growth, basic stochastic models for fatigue and the estimation of fatigue reliability of specific structural systems. The proposed lecture notes should be of interest to engineers and researches in industry and at universities who want first-hand information on the present status and trends in this rapidly developing field of engineering science.

Finally, I wish to express my thanks to the authors for their efforts in presenting the lectures and preparing the manuscripts for publication. In this context I additionally would like to thank Dr. I. Rychlik and Mr. Mats Frendahl from the University of Lund (Sweden) for their kindness in undertaking the arduous task of preparing the last, camera-ready, version of the lectures of Prof. Bolotin. Of course, many thanks are also due to CISM for its hospitality and excellent organization of the course.

Kazimierz Sobczyk

CONTENTS

STOCHASTIC MODELLING OF FATIGUE: METHODICAL BACKGROUND

K. Sobczyk
Polish Academy of Sciences, Warsaw, Poland

I. MODELLING — METHODICAL PROBLEMS

1. INTRODUCTORY REMARKS

Fatigue of engineering materials is a complicated and intriguing phenomenon that takes place in components and structures subjected to time - varying external loadings and that manifests itself in the deterioration of the material's ability to carry the intended loadings.

It would be extremely important to formulate a theory of fatigue fracture of materials that deals with all physical and chemical processes on micro – scale and provides the observed macroscopic characteristics of the fatigue process. Such modelling – truly rooted in the physics of fatigue – does not seem to be possible at present. However, fatigue damage assessment must be made for the purpose of engineering design (especially of such structures as: aircraft, bridges, pressure vessels etc.) In such a situation it is rational and important to formulate macroscopic/phenomenological mathematical models in order to recognize regularities in dispersed fatigue data and to provide a consistent basis for fatigue reliability estimation.

Constructing the mathematical models for fatigue – similarly as modelling other real phenomena – is a complicated intellectual process whose objective is to represent consistently an information contained in data in the language of mathematics. This process includes however many subtle methodical questions. The very first question is: *what is really meant by the statement that a certain system of mathematical symbols and relations is a representation of a real phenomenon?* In what follows we will try to shed some light on the methodology of modelling.

2. MATHEMATICAL MODELS OF REAL PHENOMENA

2.1. Reality and models

The concept of "model" is not univocal in the literature. For example, a material counterpart of a real system (made with preservation of the basic features of a real system) is often called a laboratory model. Among theoretical (or abstract) representations (descriptions) of real phenomena the basic role is played by the mathematical models in which the most essential aspects of real phenomenon are represented in the language of mathematics. Mathematical modelling proved to be very powerful method of science, and constitutes an important step in building scientific theories (cf. E. Wigner [1], R. Schlegel [2]).

Let us denote symbolically all information (data) which we have about the phenomenon (e.g. fatigue) by D. There exists a class of possible (hypothetical) models of the phenomenon:

$$\mathcal{M} = \{M(\theta), \quad \theta \in \Theta_{\mathcal{M}}\}$$

where $\Theta_{\mathcal{M}}$ is a set of parameters. Modelling consists in representation of the information contained in D in the form of one of the members of family $\mathcal{M}$. Formally, one should construct a mapping

$$D \longrightarrow M(\Theta_D).$$

This mapping should preserve a structure of D (that is, the mutual relations between elements of D). Mathematical "system" $M(\Theta_D)$ is regarded as a model of a real phenomenon (characterized by data D) if - generally speaking - it reflects the structure of the phenomenon in question; this means that there exists a relationship between "internal" (empirical) relations in D and mathematical relations in $M(\Theta_D)$. Science is less interested in facts themselves, but rather in systematic relationships between them. History of physics (where mathematical modelling has led to fundamentally significant results) indicates that a way toward most valuable theories goes through mutual interaction of rich empirical data and appropriate mathematical structures. General scheme of mathematical modelling of a real phenomenon can be represented as follows:

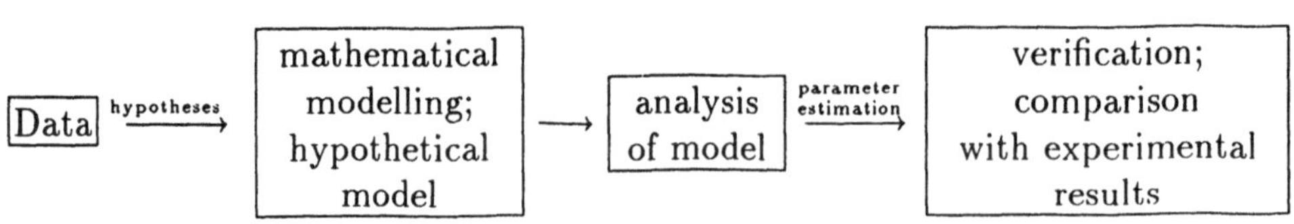

In the process of constructing a mathematical model one has to introduce some hypotheses (postulates) about the phenomenon. Some hypotheses result from the universal physical principles, such as principle of energy preservation, Newton law etc. But these hypotheses are usually insufficient and one has to adopt some phenomenological or empirical relations (models). Such empirical relations result from elaboration of experimental data and dimension analysis, and provide a relationships between basic characteristics of the phenomenon. It should be kept in mind that usually the applicability range of empirical models is not clearly determined; as a rule it is specified by the experimental conditions under which a considered "model" is obtained. Another group of hypotheses (in addition to those associated with model itself) is introduced to link quantities in the model and the observables. It often happens that the meaning of theoretical quantities (in the model) and the corresponding observables are not the same (e.g. test of the model for fatigue crack length can be realized by measurement of striations on the fracture surface). Let us characterize briefly requirements associated with models of real phenomena.

2.2. Adequacy of a model

It is clear that a mathematical model should be as adequate (to a real phenomenon) as possible. This means that a model should provide a good qualitative and quantitative description of the phenomenon (with respect to a selected class of features and external excitations). Adequacy of a model depends both on the correctness of primary hypotheses and on the possibility of a rigorous and exact analysis of a model. Data are usually not complete; should one require a model to be completely adequate? An adequacy of a model should be treated comparatively. Various models of the same phenomenon can have different degrees of adequacy depending, for example, on the number of parameters introduced. An appropriate selection of the "best" parameters (and rational restriction of their number) constitutes, undoubtely, one of the basic problems in mathematical modelling. There is a common conviction (cf. Blekhman, Mushkis, Panovko [3]) that the most efficient models of mechanics and physics are mainly concerned with the phenomena for which a number of empirical and theoretical parameters is not large.

2.3. Simplicity of a model

Most often the abstract or mathematical models of physical phenomena characterize these phenomena in a simplified way; a degree of this simplification depends upon the particular objective in modelling or, more often, on our insufficient knowledge of the phenomenon in question. A model should provide information about the phenomenon in a usable form. If the model is too complex, its usefulness becomes questionable. Therefore, relative simplicity seems to be an important feature of model construction. In the model construction a balance must be established between adequacy of a model

(which usually requirtes complex relations and many parameters) and its simplicity. Although adequacy of the model most often causes its complexity, it is not a general rule. It is clear that of two acceptable models M_1 and M_2 of the same simplicity, the model M_2 may have a higher degree of adequacy. Similarly, of two models of the same adequacy, one model may be simpler than other. The "optimal" models in which adequacy and simplicity are in good balance should be the aim of modeling.

3. ANALYSIS OF A MODEL

When a mathematical model is built, the next step is the analysis whose objective is to provide the appropriate qualitative and quantitative information about the basic model predictions. For example, if model has the form of the differential equation, one performs the analysis of this equation, which in particular can provide its solution. In many cases, the mathematical analysis of a model can create serious difficulties. The difficulty in the model analysis usually says much about the model itself; therefore it should generate some questions. For example, is the specific difficulty a result of complicated physics of a phenomenon, or - perhaps - it results from wrong model construction ?. If the difficulty stems from the complexity of the phenomenon (and there is no possibility to simplify the model) one is forced to perform an *approximate analysis*. In such a case, however, one should bear in mind that the approximations, when performed without appropriate care, can lead to distorted conclusions, or to wrong predictions. The literature provides many such cases (cf. [3]). Although the analysis of a mathematical model of a real phenomenon should be rigorous and belongs to mathematics, the approximations necessary to obtain the model predictions can be suggested by our information on the physical features of the phenomenon. Such information usually gives a possibility to estimate a relative significance of the individual terms in equations, and consequently leads to rational simplification of the model. Of course, all approximations which are introduced to perform analysis should not contradict underlying hypotheses of the model.

4. VERIFICATION OF A MODEL

The problem of verification (or,validation) of a mathematical model constitutes, in general, a complex issue, which includes very essential questions of methodological nature. Here, we wish to give only some comments. There are a number of important conditions which should be satisfied in order that a certain mathematical structure (or, "pre-model") could be regarded as a true model of a real phenomenon. In general, every body of scientific ideas (including mathematical models) is evaluated (verified) in the light of the result of (at least) three batteries of tests (cf. Bunge [4]): *metatheoretical*, *intertheoretical* and *empirical.* The first kind of tests (metatheoretical) means that the form and content of the model must be internally consistent (i.e. logically well built, empirically testable, etc.) An interteoretical test consist in examination whether a new model is compatible with other previously accepted theories (i.e. the model predictions can not go against justified, verified, and commonly accepted beliefs). Technically, the above two (non - empirical) tests include the following elementary tests.

a) Control of dimensions consist in checking whether both sides of mathematical equation have the same physical dimensions.

b) Control of orders consists in rough estimation of orders of magnitude of quantities in equation.

c) Control of a character of mathematical structure consisting in verification of the basic inter - relations between variables of mathematical model. For example, let us assume that the model has the structure: $D = A^2B - C, A, B, C > 0$. This means that if A and B increase the quantity D should increase as well, whereas increase of C should result in decrease of D . In addition, rate of increase of D is greater with respect to A than with respect to B .

d) Control of extremal situations provides an information about the form of a model, when some parameters and variables approach to certain (specified) values (for example, when parameters take their "critical" values, and variables (e.g. time) tend to zero or infinity). In some limiting cases one should obtain known models.

e) Control of sensivity of a model consist in checking what is the effect of variations in parameters on the model predictions.

The third kind of test is an empirical one. An empirical test of the mathematical model consists in confrontation of some logical consequences or predictions of the model with information obtained via observations or experiments. It seems that according to common opinion, agreement with empirical data is not only necessary, but also sufficient for the acceptance of a model. Especially for practitioners, agreement with data seems to be the highest court of appeal.

Of course, the agreement with experimental data must play a principal role in building a theory of any real phenomenon, but one also should keep in mind that the mutual relationship between experiment and theory is very complex issue (cf. Sobczyk, Spencer [5]). To consider an empirical test valid (and truly informative), the experiments have to be properly designed and performed and the results should be carefully and correctly elaborated and interpreted. Improper data handling can significantly affect the reliability of the final results.

An important issue is the fact that usually the subject of experimental measurement is not all population of objects, but its sample. The judgments about all population on the basis of investigation of a sample can, in general, be right or wrong. The possibility of obtaining correct information depends on representativeness of a sample, and proper (or, optimal) experiment design. According to Goodwin and Payne [6] "Experiment design is concerned with the selection of conditions during the experiment so that it is maximally informative about these variables of a system that are pertinent to a particular application". As far as experiments on fatigue process are conserned they are very seldom performed according to requirements of mathematical statistics and the optimal experiment design.

5. SCATTER IN FATIGUE DATA; STOCHASTIC MODELLING

Experimental data obtained from testing specimens under various loading conditions constitute the main source of information about fatigue of engineering materials. However, these data, regardless of how carefully they are generated, show significant scatter which depends on various uncontrolled factors, such as material properties and metallurgical structure, type of loading, environment etc. This scatter of fatigue data is today commonly regarded as an inherent feature of fatigue process in real materials and rises a conviction that deterministic methods can not be adequately used in constructing mathematical models of fatigue.

The basic question which arises is : what kind of mathematical language should we use in order to describe unpredictable fatigue process?. It turns out to be most rational to regard fatigue as a random phenomenon and to use the tools of contemporary probability theory, including stochastic processes. This theory proved to be very effective in studying regularities of various complicated and uncertain phenomena in physics and in many other fields. Therefore, although the concept of randomness - its origin and manifestation in real world - can be interpreted and quantified in various ways (cf. Littlewood [7], Nalimov [8], Fine [9]) stochastic modeling of fatigue is perceived today as important and challenging task in mechanics and applied stochastics. All what we said in preceding sections about mathematical modeling applies entirely to the stochastic case. In spite of that the stochastic modelling has its specific features.

Above all, one should distinguish situations which are stochastic from those which ought to be regarded as undetermined. The point is that in probability theory the characteristics of random variables (their probability distributions) are regarded as a priori given, whereas in analysis of real phenomena the probability distributions are seldom known ; in such cases a situation should rather be regarded as undetermined. Hence, in stochastic modelling an attention should be focused on determination of the probabilistic characteristics of the "primary" random quantities entering the model (e.g. random initial conditions, excitations etc.). Often the probability distributions are introduced as hypotheses; so, they should be a subject to subsequent verification. Also, estimation of parameters in stochastic models as well as design of experiments (for verification of the models) are strongly connected with the methods of mathematical statistics.

6. COMPARISONS AND DISCRIMINATING AMONG STOCHASTIC MODELS

The validity of a model can only be specified in relative terms. To validate a model we have to compare its predictions with the corresponding characteristics of empirical data. Also, comparisons between various models, and discrimination among them are of interest. In all such situations there is a need for appropriate measures of divergence (or, diversity) between a model and observation, and between two models. Since predictions of stochastic models take usually a form of probability distributions, we are interested in comparisons between and discrimination among various competing distributions. Various divergence measures between two probability distributions have been proposed (cf. Csiszar [10]). One of the most popular one is the following Kullback - Leibler distance measure (between probability distributions with densities $f_1(x)$ and $f_2(x)$)

$$J(f_1, f_2) = \int_{-\infty}^{+\infty} f_1(x) log \frac{f_1(x)}{f_2(x)} dx$$

There is a relationship between the Shannon mutual information $I(X_1, X_2)$ between random variables X_1 and X_1 (with densities f_1 and f_1) and the Kullback - Leibler measure. It turns out (cf. [5]) that the Shannon amount of information regarding X_1 provided by observation of X_2 is equal to the Kullback - Leibler divergence between $f_1 = f_{x_1 x_2}(X_1, X_2)$ and $f_2 = f_{x_1}(X_1) f_{x_2}(X_2)$. In modelling of empirical phenomena, i.e. particularly in modeling of the fatigue process, one wish to find a model which is the best among all admissible models. This requires introducing an appropriate *quality criteria* and *discriminating quantities.* For example, in fatigue analysis a variety of probability distributions have been proposed to describe the random life time (or, random time for a crack to increase to a specific critical value). Among them are: Weibull distribution, inverse - Gaussian distribution, log - normal distribution and others. In order to discriminate among such competing distributions Kozin and Bogdanoff [11] compared the coefficient of stewness of theoretical distributions and their empirical estimate. With this discriminating parameter and with the Wirkler empirical data, they found that the first asymptotic extreme value distribution (of maximum values) seems to be the most suitable. However, the problems of discriminating among various stochastic models can be formulated and analyzed in a more general way.

II. PROBABILITY DISTRIBUTIONS RELEVANT TO FATIGUE - SHORT RESUME

1. BASIC CONCEPTS

In order to quantitatively investigate the regularities of random phenomena, one has to first introduce a mathematical model for randomness together an appropriate measure of the possibility of occurrence of various "uncertain outcomes of an experiment. Such a model forms a basic scheme for probability theory in which the main notions are: sample space (i.e. a collection of all possible outcomes of an experiment), random event and probability. A sample space will be denoted by Γ and its elements, the possible or elementary outcomes of an observation, by $\gamma (i.e., \gamma \in \Gamma)$. Let $\mathcal{F}$ denote the family of subsets of Γ , termed the family of random events, on which a probability P is defined. The probability P is a function whose arguments are random events (i.e. elements of $\mathcal{F}$), such that

1. $0 \leq P(A) \leq 1$, for each $A \in \mathcal{F}$,
2. $P(\Gamma) = 1$,.
3. for a countable collection of mutually disjoint events $A_1, A_2, \ldots, A_n, \ldots$ in $\mathcal{F}$: $P\{\cup_j A_j\} = \sum_j P(A_j)$.

These three statements constitute the axioms of probability. It is clear that in experiments on random phenomena, various outcomes, or elementary events, can occur. In most situations they are represented by real numbers; other types of outcomes, while not originally numerical can be made to correspond to numbers by choice of a suitable mapping. Thus, it is commonly assumed that one can assign a real number $X(\gamma)$ for each elementary event $\gamma \in \Gamma$. This leads to the concept of a random variable. The definition is the following.

A *random variable* is a real-valued function $X = X(\gamma),\ \gamma \in \Gamma$, defined on a sample space Γ such that for every real number x there exists a probability $P\{\gamma : X(\gamma) \leq x\}$. The existence of a probability of event $\{\gamma : X(\gamma) \leq x\}$ ensures us (via set operations) that the probability of any finite or countably infinite combination of such events is well defined (e.g. $P\{x_1 \geq: X(\gamma) \leq x_2\}$). The probabilistic behaviour of a random variable $X(\gamma)$ is completely and uniquely specified by its (cumulative) *distribution function* $F_X(x)$ defined as

$$F_X(x) = P\{X(\gamma) \leq x\} \tag{2.1}$$

By the definition, the distribution function always exists and is a non-negative and non-decreasing function of the real variable x ; furthermore, it is continuous from the right. From the properties of probability it follows that

$$F_X(-\infty) = 0 \qquad F_X(+\infty) = 1, \tag{2.2}$$

and for any two real numbers a, b such that $a < b$

$$P\{a < X \leq b\} = F_X(b) - F_X(a) \tag{2.3}$$

There are two basic classes of random variables: *discrete random variables* - taking on only a finite or countably infinite number of distinct values, and *continuous random variables* - taking on a non-countable number of values. A discrete random variable is specified by the values it assumes, x_i , and the corresponding probabilities $p_i = P(X = x_i)$. Of course

$$\begin{aligned} F_X(x) &= \sum_{\substack{i \\ x_i \leq x}} p_i \\ \sum_i p_i &= 1 \end{aligned} \tag{2.4}$$

Strictly speaking, a random variable $X(\gamma)$ is termed a *continuous* random variable if its probability distribution function $F_X(x)$ has a density function, i.e. for each x there exists a nonnegative and integrable (e.g. continuous) function $f_X(x)$ such that

$$F_X(x) = \int_{-\infty}^{x} f_X(\xi)d\xi \tag{2.5}$$

The function $f_X(x)$ is called the *probability density function* of the random variable $X(\gamma)$. Clearly

$$\frac{dF_X(x)}{dx} = f_X(x) \tag{2.6}$$

and

$$\begin{gathered} f_X(x) \geq 0 \\ \int_a^b f_X(x)dx = F_X(b) - F_X(a) \\ \int_{-\infty}^{\infty} f(x)dx = 1 \end{gathered} \tag{2.7}$$

The next important notion is the *mean* or *average value* a random variable. The mean or average value of a random variable $X(\gamma)$ is defined as

$$
\begin{aligned}
m_X &= E(X) = \langle X \rangle = \int_{-\infty}^{\infty} x dF_X(x) \\
&= \begin{cases} \sum_i x_i p_i, & \text{for a discrete random variable} \\ \int_{-\infty}^{\infty} x f_X(x) dx, & \text{for a continuous random variable.} \end{cases}
\end{aligned}
\tag{2.8}
$$

An important class of average values is formed by the average of the powers of a random variable. These quantities are called *moments* . The kth order moment of $X(\gamma)$ is defined as

$$
m_k = \langle X^k \rangle = \begin{cases} \sum_i x_i^k p_i, & \text{for a discrete random variable} \\ \int_{-\infty}^{\infty} x^k f_X(x) dx, & \text{for a continuous random variable.} \end{cases}
\tag{2.9}
$$

The *central moment* of order k of a random variable $X(\gamma)$ is the quantity

$$
\mu_k = \langle [X(\gamma) - m_X]^k \rangle
\tag{2.10}
$$

In particular, the second central moment is

$$
\sigma_X^2 = \mu_2 = \begin{cases} \sum_i (x_i - m_X)^2 p_i, & \text{for a discrete random variable} \\ \int_{-\infty}^{\infty} (x - m_x)^2 f(x) dx, & \text{for a continuous random variable.} \end{cases}
\tag{2.11}
$$

which is known as the *variance* of $X(\gamma)$. It measures the spread or dispersion of the random variable $X(\gamma)$ about its mean. The positive square root of the variance, namely σ_X , is called the *standard deviation* of $x(\gamma)$. The ratio

$$
\nu_X = \frac{\sigma_X}{| m_X |} \qquad m_X \neq 0
\tag{2.12}
$$

normalizes the spread of $X(\gamma)$ with respect to the mean, and is called the *coefficient of variation* .

Other central descriptors of a random variable are the mode and the median. The *mode* $\tilde{x}$ is the most probable value of a random variable. For continuous random variables, $\tilde{x}$ is the value for $f_X(x)$ assumes the maximum value. The *median* x_m is the value of $X(\gamma)$ for which $F_X(x_m) = 1/2$.

In order to characterize the symmetry (or lack of symmetry) of a probability distribution of random variable $X(\gamma)$, the *skewness* or *asymmetry coefficient* is defined as

$$\zeta_X^{(1)} = \frac{1}{\sigma_X^3}\langle (X - m_X)^3 \rangle = \frac{\mu_3}{\sigma_X^3} \tag{2.13}$$

A probability distribution which is symmetric about m_X has a zero skewness coefficient (i.e. $\zeta_X^{(1)} = 0$).

The concentration of the values of $X(\gamma)$ around the mean is characterized by the *excess coefficient*

$$\zeta_X^{(2)} = \frac{1}{\sigma_X^4}\langle (X - m_X)^4 \rangle - 3 = \frac{\mu_4}{\sigma_X^4} - 3 \tag{2.14}$$

The *kurtosis coefficient* is defined as μ_4/σ_X^4 . The coefficients of skewness and excess are zero for a normal, or Gaussian, distribution (i.e. $\zeta_X^{(1)} = \zeta_X^{(2)} = 0$).

Often the joint behaviour of two or more random variables is of interest. In such cases, multiple random variables and their joint probability distributions are defined and analyzed (cf. Ang and Tang [12], Soong [13]).

2. PROBABILITY DISTRIBUTIONS RELEVANT TO FATIGUE

The specific probability distributions described analytically by the distribution function $F_X(x)$, or by the probability density function $f_X(x)$ can take a wide variety of forms. However not all distributions are of equal importance. Several particular distributions have proven to be extremely useful both in theory and applications. Here, we briefly characterize the distributions which are directly relevant to fatigue life prediction.

2.1. Poisson distribution

Among the distributions of discrete random variable, the Poisson distribution plays a dominant role. A discrete random variable $X(\gamma)$ taking no-negative integer values $k = 0, 1, 2, \ldots$ is said to have a Poisson distribution if

$$P\{X(\gamma) = k\} = \frac{\lambda^k}{k!} e^{-\lambda} \quad k = 0, 1, 2, \ldots \tag{2.15}$$

where λ is a positive constant characterizing the mean occurrence rate of events (per unit time). If the number of occurrences of an event in time interval of length t (e.g. in $[0, t]$) is of interest, than

$$P\{X_t = k\} = \frac{(\lambda t)^k}{k!} e^{-\lambda t} \quad k = 0, 1, 2, \ldots \tag{2.16}$$

The mean and variance are given by

$$\begin{aligned} \langle X_t \rangle &= m_X = \lambda t \\ \sigma^2_{X_t} &= \lambda t \end{aligned} \tag{2.17}$$

2.2. Uniform Distribution

The simplest distributions of continuous random variables is the uniform distribution. A random variable $X(\gamma)$ has an uniform distribution in the interval $[a, b]$ if

$$f_X(x) = \begin{cases} \dfrac{1}{b-a} , & a \leq x \leq b \\ 0 , & \text{otherwise} \end{cases} \tag{2.18}$$

The mean and variance are given by

$$\begin{aligned} m_X &= \frac{a+b}{2} \\ \sigma^2_X &= \frac{(b-a)^2}{12} \end{aligned} \tag{2.19}$$

2.3. Gaussian (normal) Distribution

The most important probability distribution in the theory as well as in applications is the Gaussian or normal distribution. A random variable $X(\gamma)$ with m_X and standard deviation σ_X has a Gaussian or normal distribution $N(m_X, \sigma_X)$ if its probability density function is given by

$$f_X(x) = \frac{1}{\sigma_X \sqrt{2\pi}} \exp\left[-\frac{1}{2}\left(\frac{x - m_X}{\sigma_X}\right)^2\right] , \qquad -\infty < x < \infty \tag{2.20}$$

The cumulative distribution function $F_X(x)$ is

$$F_X(x) = \frac{1}{\sigma_X \sqrt{2\pi}} \int_{-\infty}^{x} \exp\left[-\frac{1}{2}\left(\frac{\xi - m_X}{\sigma_X}\right)^2\right] d\xi = \Phi\left(\frac{x - m_X}{\sigma_X}\right) \tag{2.21}$$

where the function $\Phi(u)$, called the standard normal integral, is defined as

$$\Phi(x) = \frac{1}{\sqrt{2\pi}} \int_{-\infty}^{x} e^{-\frac{\xi^2}{2}} d\xi \tag{2.22}$$

Tabulated values of $\Phi(u)$ are readily available (e.g. see[3], [4]).

It is seen that the density function (2.20) is symmetric with respect to its mean value m_X. Therefore, all odd central moments are equal to zero. In general,

$$\begin{aligned} \mu_{2k+1} &= 0 \\ \mu_{2k} &= 1 \cdot 3 \cdot \ldots \cdot (2k-1)\sigma_X^{2k} \end{aligned} \tag{2.23}$$

The skewness and excess coefficient defined by equations (2.13) and (2.14) are equal to zero for the Gaussian distribution.

2.4. Lognormal Distribution

A random variable $X(\gamma)$ has a lognormal probability distribution if lnX (the natural logarithm of X) is normal. The density function of $X(\gamma)$ is then

$$f_X(x) = \frac{1}{\zeta x\sqrt{2\pi}} \exp\left[-\frac{1}{2}\left(\frac{\ln x - \lambda}{\zeta}\right)^2\right], \qquad 0 < x < \infty, \quad \zeta > 0 \tag{2.24}$$

where λ and ζ^2 are the mean and variance of the normal random variable $\ln X$, respectively. The mean and variance of a lognormal distribution are given in terms of the distribution parameters λ and ζ

$$\begin{aligned} m_X &= \exp\left(\lambda + \frac{\zeta^2}{2}\right) \\ \sigma_X^2 &= m_X^2(e^{\zeta^2} - 1) \end{aligned} \tag{2.25}$$

The cumulative distribution $F_X(x)$ is directly related to the standard normal distribution, namely

$$F_X(x) = \Phi\left(\frac{\ln x - \lambda}{\zeta}\right) \tag{2.26}$$

From (2.25) we have

$$\begin{aligned} \lambda &= \ln(m_X) - \frac{1}{2}\zeta^2 \\ \zeta^2 &= \ln\left(1 + \frac{\sigma_X^2}{m_X^2}\right) = \ln(1 + \nu_X^2) \end{aligned} \tag{2.27}$$

2.5. Gamma Distribution

A random variable $X(\gamma)$ has a gamma distribution $G(a,p)$ if its probability density function takes the form

$$f_X(x) = \frac{a^p}{\Gamma(p)}\, x^{p-1}e^{-ax}, \qquad x \geq 0,\ a > 0,\ p > 0 \tag{2.28}$$

where $\Gamma(p)$ is the well known gamma function

$$\Gamma(p) = \int_0^\infty x^{p-1} e^{-x}, \qquad p > 0 \tag{2.29}$$

When p is positive integer $\Gamma(p) = (p-1)!$.

The cumulative distribution function is

$$F_X(x) = \int_0^x f_X(u) du = \frac{\Gamma(p, ax)}{\Gamma(p)} , \qquad x \geq 0 \tag{2.30}$$

where $\Gamma(p,u)$ is the socalled incomplete gamma function given by

$$\Gamma(p,u) = \int_0^u \xi^{p-1} e^{-\xi} d\xi \tag{2.31}$$

Then mean and variance are

$$m_X = \frac{p}{a} , \quad \sigma_X^2 = \frac{p}{a^2} \tag{2.32}$$

The gamma distribution serves as a useful model because of its versality in the sense that a wide variety of shapes of the density function can be obtained by varying the values of p and a . The parameter p characterizes the shape of the distribution, whereas a is a scale parameter of the distribution. In general, the gamma density function is unimodal with its peak at $x = 0$ for $p \leq 1$ and at $x = (p-1)/a$ for $p > 1$.

If the occurrences of an event in time are characterized by Poisson distribution, then the time until the kth occurrence of the events is described by the gamma distribution. A special case of the gamma distribution, when p is integer, is known as the *Erlang distribution.*

2.6. Exponential Distribution

When $p = 1$, the gamma distribution reduces to an exponential form

$$f_X(x) = \begin{cases} ae^{-ax} , & x \geq 0, \; a > 0 \\ 0 , & \text{elsewhere} \end{cases} \tag{2.33}$$

The associated cumulative distribution function, mean and variance are determined from equations (2.30) and (2.32) by setting $p = 1$ to obtain

$$F_X(x) = 1 - e^{-ax} , \quad x \geq 0$$
$$m_X = \frac{1}{a} , \quad \sigma_X^2 = \frac{1}{a^2} \tag{2.34}$$

If the occurrence of events obey a Poisson distribution, then the exponential distribution describes the associated recurrence time (i.e. the time between two consecutive occurences of the event, cf. [12]).

2.7. Chi-Square (χ^2) Distribution

Another particular case of the gamma distribution is the χ^2 distribution obtained by setting $a = 1/2)$ and $p = n/2)$ in (2.28), where n is positive integer. Therefore, the (χ^2) distribution has the form

$$f_X(x) = \frac{1}{2^{n/2}\Gamma(\frac{n}{2})} x^{(\frac{n}{2}-1)} e^{-\frac{x}{2}} \tag{2.35}$$

The parameter n is termed as the *number of degrees of freedom* of the χ^2 distribution. This distribution is best known as a distribution of a sum of the squares of n independent standard normal variables. Because of this fact, the (chi^2 distribution plays an important role in statistical inference and hypothesis testing (cf. (χ^2) – test [13]).

2.8. Inverse-Gaussian Distribution

Another distribution which might be appropriate in fatigue problems is the inverse-Gaussian distribution. Although originally related to particular problems in physics (Schrödinger and Smoluchowski, 1915) and statistics (Wald, 1944) this distribution has recently become useful in reliability theory and in modelling of random fatigue.

A positive variable $X(\gamma)$ has an inverse-Gaussian distribution if for $x > 0$

$$f_X(x) = f_X(x; a, b) = \sqrt{\frac{b}{2\pi}}\, x^{-3/2} \exp\left[-\frac{b(x-a)^2}{2a^2x}\right], \quad a > 0,\ b > 0 \tag{2.36}$$

If we introduce parameter $\xi = \frac{b}{a}$, then the density function (2.36) can be written in the following equivalent forms

$$\begin{aligned} f_X(x; a, b) &= a^{-1} f_X(\frac{x}{a}; 1, \xi) \\ &= b^{-1} f_X(\frac{x}{b}; \xi, 1) \end{aligned} \tag{2.37}$$

The case in which $a = 1$ is often regarded as a standard form (sometimes referred to as the standard Wald distribution), because the shape of the distribution depends only on ξ . Thus, ξ is termed the shape parameter.

The above distribution (2.37) is unimodal and skewed. However, the density function in (2.36) represents a wide class of distributions running from highly skewed distributions to symmetric ones as ξ varies from zero to infinity.

The cumulative distribution function $F_X(x)$ can be expressed in terms of the standard normal distribution function Φ as follows

$$F_X(x) = \Phi\left[\sqrt{\frac{b}{x}}\,(\frac{x}{a} - 1)\right] + e^{2b/a}\Phi\left[-\sqrt{\frac{b}{x}}\,(1 + \frac{x}{a})\right] \tag{2.38}$$

The first order moments are

$$\begin{aligned} m_X &= \langle X \rangle = a \\ m_2 &= \langle X^2 \rangle = a^2 + \frac{a^3}{b} \\ m_3 &= \langle X^3 \rangle = a^3 + 3\frac{a^4}{b} + 3\frac{a^5}{b^2} \\ m_4 &= \langle X^4 \rangle = a^4 + 6\frac{a^5}{b} + 15\frac{a^6}{b^2} + 15\frac{a^7}{b^3} \end{aligned} \tag{2.39}$$

The central moments are

$$\mu_2 = \sigma_X^2 = \frac{a^3}{b}, \quad \mu_3 = 3\frac{a^5}{b^2}, \quad \mu_4 = 15\frac{a^7}{b^3} + 3\frac{a^6}{b^2} \tag{2.40}$$

The inverse Gaussian distribution is the first passage time distribution of the Brownian motion process (cf. Cox and Miller [14]); it seems that the name inverse Gaussian was used for the first time just in this context by Tweede in 1956.

2.9. Extreme Value Distributions

In many applications, especially in those associated with reliability of engineering systems there is a necessity to determine probability distribution of maximum or minimum values of a sequence of random variables, $X_1, X_2, \ldots, X_n$ (cf. Gumbel [15]).

Let us assume that $X_1, X_2, \ldots, X_n$ are independent and identically distributed with the same distribution $F_X(x)$. The following two random variables are of interest

$$Y_n = \max(X_1, X_2, \ldots, X_n) \tag{2.41}$$

$$Z_n = \min(X_1, X_2, \ldots, X_n) \tag{2.42}$$

It is clear that distribution functions of the variables are

$$F_{Y_n}(n) = P(Y_n \le y) = P(\text{all} X_k \le y) = [F_X(y)]^n \tag{2.43}$$

$$\begin{aligned} F_{Z_n}(z) &= P(Z_n \leq z) = P(\text{at least one}(X_k \leq z) \\ &= P(X_1 \leq z \cup X_2 \leq z \cup \ldots \cup X_n \leq z) \\ &= 1 - P(X_1 > z \cap X_1 > z \cap \ldots \cap X_n > z) \\ &= 1 - [1 - F_{X_1}(z)][1 - F_{X_2}(z)] \ldots [1 - F_{X_n}(z)] \\ &= 1 - [1 - F_X(z)]^n \end{aligned} \tag{2.44}$$

If the random variables $X_i, i = 1, 2, \ldots, n$, are continuous, then the corresponding density functions are given by

$$f_{Y_n}(y) = n[F_X(y)]^{n-1} f_X(y) \tag{2.45}$$

$$f_{Z_n}(z) = n[1 - F_X(z)]^{n-1} f_X(z) \tag{2.46}$$

In many situation the number of random variables X_i is very large. Moreover, the distribution $F_X(x)$ of each X_i is often unavailable. Thus, the problem of great importance is: what are the limiting (or, asymptotic) distributions of Y_n and Z_n when $n \to \infty$ and what is a possible class of distributions $F_X(x)$ for which such limiting distributions exist, are unique and can be effectively constructed?

There are three basic types of "parent" distributions $F_X(x)$ for which the asymptotic (limiting) distribution exist.

a. Type I Asymptotic Distribution (of maximum values)

If the distribution $F_X(x)$ has the form

$$F_X(x) = 1 - e^{-g(x)} \tag{2.47}$$

where $g(x)$ is an increasing function of x , then it is a so called exponential type distribution (i.e. $F_X(x)$ approaches one at least as fast as an exponential distribution). The normal, log-normal, gamma and exponential distributions are of this type.

For exponential type distributions, the distribution for the largest of n random variates X_i, $i = 1, 2, \ldots, n$ as $n \to \infty$ are given by

$$F_Y(y) = \exp[-e^{-\alpha(y-u)}], \qquad -\infty < y < +\infty \tag{2.48}$$

$$f_Y(y) = \alpha \exp[-\alpha(y - u) - e^{-\alpha(y-u)}], \qquad -\infty < y < +\infty \tag{2.49}$$

The distribution parameters are the mode u of the distribution and α which is an inverse measure of the dispersion of the distribution. The moments are

$$\begin{aligned} m_Y &= u + \frac{\delta}{\alpha} \quad \delta \approx 0.577, \text{ Euler's constants} \\ \sigma_Y^2 &= \frac{\pi^2}{6\alpha^2} \end{aligned} \tag{2.50}$$

The above distribution is also termed the *"double exponential"*, *Gumbel,* or *Fisher-Tippet Type I* distribution. The Type I distribution for minimum value can be derive in a similar manner (cf. [15]).

b. Type II Asymptotic Distribution (of maximum values)

A random variate X_i is of the Pareto type if its distribution function $F_X(x)$ is given by

$$F_X(x) = 1 - ax^{-k}, \qquad a > 0,\ k > 0,\ x \geq 0 \tag{2.51}$$

For random variates of the Pareto type, the asymptotic distribution of the *largest value* of n random variates X_i, $i = 1, 2, \dots, n$ as $n \to \infty$ has the form

$$F_Y(y) = \exp\left[-(\frac{u}{y})^k\right]; \qquad y \geq 0,\ k > 0,\ u > 0 \tag{2.52}$$

$$f_Y(y) = \frac{k}{y}(\frac{u}{y})^k \exp\left[(\frac{u}{y})^k\right]; \qquad y \geq 0 \tag{2.53}$$

The distribution parameters are the characteristic distribution value u (mode $< u <$ median) and k which is a dimensionless inverse measure of the dispersion. The moments are

$$\begin{aligned} m_Y &= u\Gamma(1 - \frac{1}{k}), \quad k > 1 \\ \sigma_Y^2 &= u^2\left[\Gamma(1 - \frac{2}{k}) - \Gamma^2(1 - \frac{1}{k})\right], \quad k > 2 \end{aligned} \tag{2.54}$$

The moments of order l of Y do not exist for $l \geq k$; this fact complicates the estimation of the parameters u and k . It should be noticed that the "parent" distribution (2.51) has moments only up to the order r , whre r is the largest integer less than k .

The type II asymptotic distribution of minimum values can be derived under analogues conditions (cf. [15]).

c. Type III Asymptotic Distribution (of minimum values)

Because the type III asymptotic distribution for the maximum values is of limited practical interest, we present here the minimum - value distribution which has proven to be very useful in applications.

Consider the distribution $F_X(x)$ of the random variable s X_i which is of the form

$$F_X(x) = c(x-\epsilon)^k, \qquad x \geq \epsilon, c > 0, \ k > 0 \tag{2.55}$$

The uniform $(k = 1)$, triangular $(k = 2)$ and the gamma distribution $(\epsilon = 1)$ belong to this class.

For variates distributed as in (2.55), the distribution of the *smallest value* of n random variates $X_i, \ i = 1, 2, \ldots, n$ as $n \to \infty$ has the form

$$F_Z(z) = 1 - \exp\left[-\left(\frac{z-\epsilon}{w-\epsilon}\right)^k\right], \qquad z \geq \epsilon, \ k > 0, \ w > \epsilon \tag{2.56}$$

$$f_Z(z) = \frac{k}{w-\epsilon}\left(\frac{z-\epsilon}{w-\epsilon}\right)^{k-1} \exp\left[\left(\frac{z-\epsilon}{w-\epsilon}\right)^k\right] \tag{2.57}$$

The mean and variance are

$$\begin{aligned} m_Z &= \epsilon + (w-\epsilon)\Gamma(1+\frac{1}{k}) \\ \sigma_Z^2 &= (w-\epsilon)^2\left[\Gamma(1+\frac{2}{k}) - \Gamma^2(1+\frac{1}{k})\right] \end{aligned} \tag{2.58}$$

The distribution (2.56), (2.57) is widely known as the three-parameter *Weibull distribution*; if $\epsilon = 0$, it is termed the two-parameter Weilbull distribution. For $\epsilon = 0$ and $k = 2$, the distribution (2.56), (2.57) becomes the *Rayleigh distribution*, which has the form $(w = \sigma\sqrt{2})$

$$f(z) = \frac{z}{\sigma^2}\exp\left(-\frac{z^2}{2\sigma^2}\right) \tag{2.59}$$

The Weilbull distribution has been widely used in characterization of random fatigue life time. It is worth noting that the two-parameter Weilbull distribution fit experimental data quite well at high stress levels. When stress levels are lower, the three-parameter distribution (with an appropriate threshold) provides a better fit of the data.

III. STOCHASTIC PROCESSES FOR MODELLING OF FATIGUE

1. BASIC CONCEPTS

In many applications one has to deal with random quantities which depend upon a certain deterministic parameter. One of the well known examples of such phenomena is Brownian motion. Each coordinate of the Brownian particle is a random variable which depends on time. Other examples are the ground motion of a fixed spatial point during an earthquake (at each instant of time it is a random variable), wind pressure acting on structure, sea waves excitation, etc.

To create a mathematical model of phenomena similar to those mentioned above, the concept of a random function has been introduced. Most often, the parameter (or dependent variable) in the random functions is time, and the concepts of stochastic processes are commonly used.

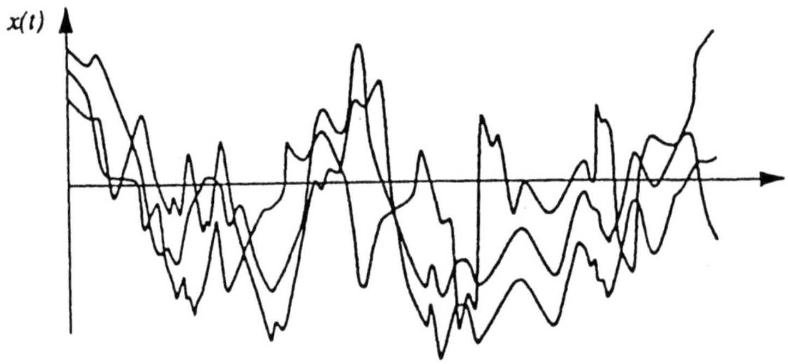

Figure 3.1. Realizations of a stochastic process.

Let T be a set of real number and let $t \in T$, where t has the meaning of time. A *stochastic process* $X(t)$ is a family of random variables $X_t(\gamma), t \in T, \gamma \in \Gamma$ depending upon the parameter t and defined on the probability space $(\Gamma, \mathcal{F}, \mathcal{P})$. One can also say that a stochastic process $X(t)$ is a function which maps the index set T into the space S of random variables defined on $\Gamma, \mathcal{F}, \mathcal{P})$.

Since a stochastic process is a family of random variables, its specification is similar to that for random vectors; the differences are associated with the fact that the number of random variables may now be (countably or uncountably) infinite. It turns out, however, that an infinite number of random variables can be described by means of finite-dimensional distributions, For an arbitrary finite set of t-values, say $\{t_1, t_2, \ldots, t_n\}$, the random variables $X(t_1), X(t_2), \ldots, X(t_n)$ have a joint dimensional distribution with the distribution function

$$F_{t_1,t_2,\dots,t_n}(x_1,x_2,\dots,x_n) = P\{X(t_1) \le x_1, X(t_2) \le x_2, \dots, X(t_n) \le x_n\} \tag{3.1}$$

The function $F_{t_1,t_2,\dots,t_n}(x_1,x_2,\dots,x_n)$ is called the *n-dimensional distribution function* of random process $X(t)$. The collection of all joint distribution functions for $n = 1, 2, \dots$ at all possible time instants t_i constitutes the family of finite-dimesional distributions of the process $X(t)$; this family of distributions (for all finite sets $t_1, t_2, \dots, t_n$ of $t \in T$) characterizes the process $X(t)$. The associated joint density function $f_{t_1,t_2,\dots,t_n}(x_1,x_2,\dots,x_n)$ is called the *n-dimensional probability density function* of the process $X(t)$.

A stochastic process can also be defined in another way. For every fixed elementary event γ in the given probability space $\Gamma, X(t)$, or more explicitly $X(t,\gamma)$, becomes a function of t, defined for all $t \in T$. This function $x(t) = X(t,\gamma)$, for a fixed $\gamma \in \Gamma$), describe a particular realization of the stochastic process and is called a *realization, trajectory* or a *sample function* of the process. The sample function $x(t)$ may be regarded as a "point" in the space $\mathcal{X}$ of all finite realvalued functions of $t \in T; \mathcal{X}$ is real-valued called the *sample function space* of the process $X(t)$ (cf. Figure 3.1). Thus, the stochastic process $X(t)$, or $X(t,\gamma)$, can be defined as a *generalized random variable (or random element)*, i.e., as a function mapping every point $\gamma \in \Gamma$ into a point of space $\mathcal{X}$. According to such an interpretation, a stochastic process is specified by characterizing the probability (or probability measure) defined on the sample function space $\mathcal{X}$. A possible way to construct such a probability measure is associated with the concept of the characteristic functionals of the process.

Like random variables, stochastic processes can be conveniently described by their moments; the simplest characteristic is the *mean* or *average value*, $m_X(t)$. It is defined as a function which, for each t, is equal to the mean value of the corresponding random variable, i.e.

$$m_X(t) = \langle X(t) \rangle \tag{3.2}$$

where $\langle\ \rangle$ denotes the ensemble average. The average value of a real stochastic process $X(t)$ is expressed using the one-dimensional density function of the process $f_t(t)$ as follows

$$m_X(t) = \int_{-\infty}^{\infty} x dF_t(x) = \int_{-\infty}^{\infty} x f_t(x) dx \tag{3.3}$$

The variance of the process $X(t)$ is defined by

$$\sigma_X^2(t) = \langle [X(t) - m_X(t)]^2 \rangle \tag{3.4}$$

One of the basic quantities of interest both in theory and in applications is the covariance function of the stochastic process. The *covariance function* of real stochastic process $X(t)$ is defined as

$$K_X(t_1,t_2) = \langle [X(t_1) - m_X(t_1)][X(t_2) - m_X(t_2)]\rangle \tag{3.5}$$

It is easily seen that $\sigma_X^2(t) = K_X(t,t)$. The function $\langle [X(t_1)X(t_2)]\rangle$ is often called the *correlation function* . A more complete characterization of the stochastic process can be provided by the higher order moments, for example by the function

$$K_X(t_1,\ldots,t_2) = \langle [X(t_1) - m_X(t_1)]\ldots[X(t_n) - m_X(t_n)]\rangle \tag{3.6}$$

which is sometimes called the *moment function of the* n^{th} *order* .

Often, the simultaneous behavior of n stochastic process $X_1(t),\ldots,X_n(t)$ is of interest. One can then introduce and n -dimensional stochastic vector process $\mathbf{X}(t) = [X_1(t),\ldots,X_n(t)]$. The mean value of such a process is defined as: $\mathbf{m}_X(t) = [m_{X_1}(t), m_{X_2}(t)\ldots,m_{X_n}(t)]$. The component processes $X_i(t)$ may be statistically related to each other, their mutual dependence is conveniently characterized by the covariance matrix $\mathbf{R}(t_1,t_2) = R_{X_i,X_j}(t_1,t_2), i,j,=1,2,\ldots,n$, where

$$\mathbf{R}_{X_i,X_j}(t_1,t_2) = \langle [X_i(t_1) - m_{X_i}(t_1)][X_j(t_2) - m_{X_j}(t_2)]\rangle \tag{3.7}$$

is said to be the *cross-covariance function* of the process $X_i(t)$ and $X_j(t)$. Obviously, $R_{X_i,X_j}(t_1,t_2) = K_{X_i}(t_1,t_2)$.

A stochastic process $X(t), t \in T$, is called a *Gaussian* or *normal stochastic process* if all of its finite-dimensional distributions are Gaussian, i.e., joint probability density function has the form

$$\begin{aligned} &f_{t_1,\ldots,t_n}(x_1,\ldots,x_n) \\ &= \frac{1}{(2\pi)^{n/2}\sqrt{|\mathbf{K}|}}\exp\{-\frac{1}{2\sqrt{|\mathbf{K}|}}\sum_{i,j=1}^{n}|K_{ij}|[x_i - m_X(t_i)][x_j - m_X(t_j)]\} \end{aligned} \tag{3.8}$$

where $x_i = X(t_i)$, $|\mathbf{K}| \neq 0$ is the determinant of the covariance matrix $\mathbf{K}_X(t_i,t_j)$, $i,j = 1,2,\ldots,n$, of the process $X(t)$, and $|K_{i,j}|$ is the cofactor of the element of the matrix $\mathbf{K}_X(t_i,t_j)$. Hence n -dimensional distributions of a Gaussian process are completely specified by the mean are $m_X(t)$ and the covariance function $K_X(t_1,t_2)$. The moment functions of higher order can be expressed in terms of $K_X(t_1,t_2)$. For example

$$K_X(t_1,t_2,t_3,t_4) = K_X(t_1,t_2)K_X(t_3,t_4) + K_X(t_1,t_3)K_X(t_2,t_4) + K_X(t_1,t_4)K_X(t_2,t_34) \tag{3.9}$$

2. STATIONARY PROCESSES

In many fields of application we encounter random processes which are invariant under a translation of time. A useful model of such phenomena is commonly called a *stationary stochastic process.*

A stochastic process $X(t), t \in T$ is said to be *strictly stationary* if the family of all finite-dimensional distributions (3.1) remains invariant under an arbitrary translation of the time parameter, i.e., for each n and for an arbitrary τ such that $t_i + \tau \in T$, we have

$$F_{t_1,t_2,\ldots,t_n}(x_1,x_2,\ldots,x_n) = F_{t_1+\tau,t_2+\tau,\ldots,t_n+\tau}(x_1,x_2,\ldots,x_n) \tag{3.10}$$

Given a physical process, it is often quite difficult to determine whether it is strictly stationary since the equality in (3.1) has to hold for all n. To make things easier, a wider class of stationary processes is introduced.

A stochastic process $X(t), t \in T$ is called a *weakly stationary* process (or, a *wide-sense* stationary process) if

$$\begin{aligned} \langle X(t)\rangle &= m_X(t) = m_X = \text{constant} < \infty \\ K_X(t_1,t_2) &= K_X(t_2-t_1) = K_X(\tau), \quad \tau = t_2 - t_1 \end{aligned} \tag{3.11}$$

Clearly, a strictly stationary process whose second moment is finite is also weakly stationary. The converse statement is not true in general. An exception is the Gaussian process, since it is completely characterized by its moments of first and second order. Therefore, a weakly stationary Gaussian process is also strictly stationary.

A stochastic vector process $\mathbf{X}(t) = [X_1(t), X_2(t), \ldots, X_n(t)]$ is defined as *weakly stationary* if its mean is constant and all elements of its covariance matrix $\mathbf{R}(t_1,t_2)$ depend only on the difference $\tau = t_2 - t_1$. From this definition it follows that the stationarity of a vector stochastic process is not assured by the stationary of its components; the component-processes have also to be stationarily correlated.

Stationary processes are of great practical importance. This is mainly due to their regularity. The estimation of the statistical characteristics of such processes from experimental data is much simpler than in the case of nonstationary processes. Furthermore, for stationary processes there exists the apparatus of spectral analysis, analogous to harmonic analysis of deterministic functions. The spectral method follows from the

Bochner-Khincin theorem (c.f. [16]) which states that a function $K_X(\tau)$ can be a correlation function of a stationary (and mean square continuous) stochastic process if and only if it has the representation

$$K_X(\tau) = \int_{-\infty}^{+\infty} e^{i\omega\tau}\, dG_X(\omega) \tag{3.12}$$

where $G_X(\omega)$ is termed the spectral distribution function and is real, nondecreasing, bounded and continuous from the left.

If $G_X(\omega)$ is absolutely continuous, i.e. there exists a function $g_X(\omega)$ such that

$$G_X(\omega) = \int_{-\infty}^{\omega} g_X(\xi) d\xi \tag{3.13}$$

$$g_X(\omega) = \frac{dG_X(\omega}{d\omega} \tag{3.14}$$

The function $g_X(\omega)$ is called *spectral density* of the process $X(t)$. Equation (3.12) can be written in terms of the spectral density function as

$$K_X(\tau) = \int_{-\infty}^{\infty} e^{i\omega\tau}\, g_X(\omega) d\omega \tag{3.15}$$

and (provided $\int_{-\infty}^{\infty} |K_X(\tau)| d|\tau < \infty$) we have

$$g_X(\omega) = \frac{1}{2\pi} \int_{-\infty}^{\infty} e^{-i\omega\tau}\, K_X(\tau) d\tau \tag{3.16}$$

It is easily seen that

a) $g_X(\omega) \geq 0$,

b) $\sigma_X^2 = K_X(0) = \int_{-\infty}^{\infty} g_X(\omega) d\omega$,

c) for real processes: $g_X(\omega) = g_X(-\omega)$.

Figure 3.2 shows a typical correlation function and the corresponding spectral density of a stochastic process $X(t)$.

A useful method of specifying stochastic processes is to characterize them in terms of a deterministic function of time with random variables as parameters; that is

$$X(t) = g(t; X_1(\gamma), \ldots, X_n(\gamma)) \tag{3.17}$$

where the functional form of g is given and the random variables $X_1(\gamma), \ldots, X_n(\gamma)$ have specified probabilistic properties. A common example is a random harmonic oscillation

$$X(t) = A\cos\omega t + B\sin\omega t \tag{3.18}$$

where A and B are random variables and ω is constant. If A and B are uncorrelated random variables with zero mean and equal variance then the process in (3.18) is weakly stationary. If additionally A and B are Gaussian, then this process is also Gaussian. The stochastic process in (3.18) can be equivalently represented as

$$X(t) = R\cos(\omega t + \Phi) \tag{3.19}$$

where $\sqrt{A^2 + B^2}$, and $\Phi =$ arctan (A/B) . In Gaussian case R , and Φ are independent, R having the Rayleigh distribution and Φ being uniformly distributed on the interval $(0, 2\pi]$.

It is worth noting that the process $Y(t)$ termed a "periodic signal in noise" and defined by

$$Y(t) = r\cos(\omega t + \varphi) + X(t) \tag{3.20}$$

where r, ω and φ are deterministic constant and $X(t)$ is a stationary Gaussian stochastic process is Gaussian but nonstationary.

Figure 3.2. Correlation function and corresponding spectral density of a stationary process.

In the analysis of various practical problems employing stationary stochastic processes, it is important to know the range of frequencies which contain the most significant power in a random signal. This leads to the concepts of a broad-band and narrow-band stationary processes.

Usually, one says that a stationary process $X(t)$ is *broad (or wide) band* if its spectral density function has significant values over a wide range of frequencies. If the process $X(t)$ play the rôle of an excitation to a dynamical system, and its band of "significant" frequencies is large in comparison with the frequency band of the system in question, then the process $X(t)$ is said to be a broad-band excitation. A process with the reverse property (i.e. $g_X(\omega)$ has significant values over a narrow band of frequencies) is called a *narrow-band process.* Figure 3.3 depicts trajectories and spectral densities of narrow and broad-band processes. We see that the narrow band process is characterized by its approximate constant period or frequency; the frequency content of the wide band process is more complex.

A common idealization of a broad-band process is to assume that $g_X(\omega) = g_o =$ *constant* for all frequencies. Such a process is called a *white noise* , which will be denoted by $\xi(t)$. This implies that

$$g_\xi(\omega) = \frac{1}{2\pi}\int_{-\infty}^{\infty} K_\xi(\tau)e^{-i\omega\tau} = g_o = \frac{1}{2\pi}\nu \tag{3.21}$$

where the constant ν is called the intensity of a white noise. Such a process does not exist in the classical sense because (3.21) is only compatible with the following correlation function

$$K_\xi(\tau) = \nu\delta(\tau) \tag{3.22}$$

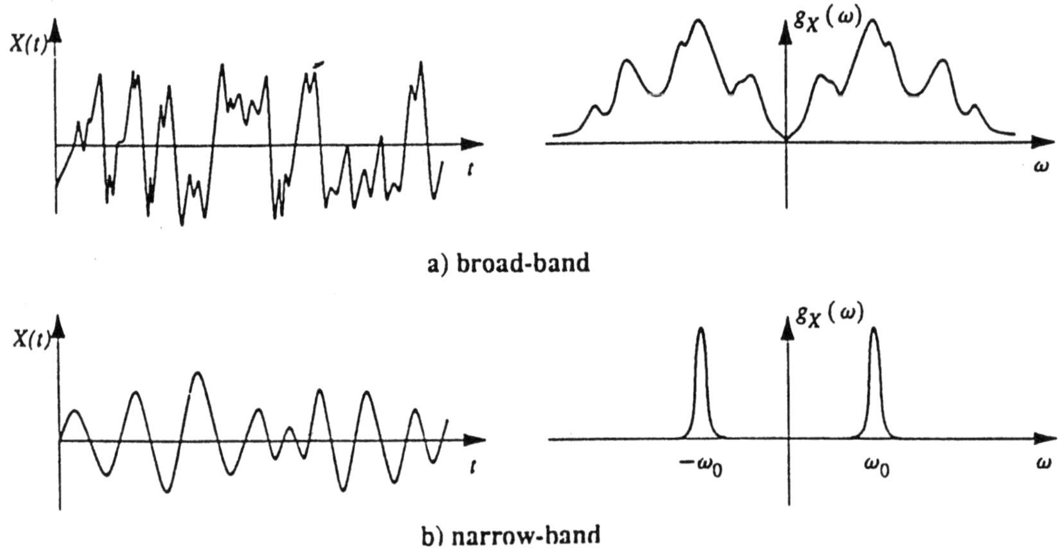

Figure 3.3. Trajectories and spectral densities of a) broad-band process; b) narrow-band process.

where $\delta(\tau)$ is the Dirac delta function. Equation (3.22) implies that a white noise process has an infinite variance (i.e. $K_\xi(0) = \sigma_\xi^2 = \infty$) . Thus, white noise is physically unrealizable. Despite these defects, white noise provides a useful approximation of real processes with very short correlation time and is widely used in applications. The use of a white noise in stochastic analysis resembles in many ways the use of Dirac delta function in deterministic theory of dynamical systems. A rigorous definition of white noise can be given within the theory of generalized stochastic process.

Spectral characteristics are commonly used in the analysis of stochastic processes. The spectral content of the process is often characterized by several parameters. The most important of them - the *spectral moments* λ_k and the *regularity factor* α - are defined as follows

$$\lambda_k = \int_{-\infty}^{+\infty} \omega^k g_X(\omega) d\omega \tag{3.23}$$

$$\alpha = \frac{\lambda_2}{\sqrt{\lambda_0 \lambda_4}} \tag{3.24}$$

We see that spectral moments, defined by an integral over infinite range, may or may not be finite. The moment λ_{2k} is finite if and only if $K_X(\tau)$ possesses a derivative of order $2k$ at $\tau = 0$. It can be easily shown that (cf. [16])

$$\lambda_{2k} = (-1)^k \frac{d^{2k} K_X(\tau)}{d\tau^{2k}} |_{\tau=0} \tag{3.25}$$

where it is clear that $\lambda_0 = K_X(0) = \sigma_X^2$.

The parameter α characterizes the bandwidth of the process. If the process is narrow-band, $\alpha \approx 1$. The parameter λ_k and α are especially useful in characterizing the extreme values of a stochastic process. The rate of positive zero-crossings as well as the rate of maxima above the zero level are shown to be functions of spectral moments.

3. STOCHASTIC DERIVATIVES AND INTEGRALS

In application of stochastic process theory to the analysis of dynamic phenomena one often has to deal with differential equations for random functions. To be able to interpret such equation correctly, it is necessary to define the concepts of a derivative and integral of a stochastic process. This two concepts are of primary interest in applications.

If almost all sample functions of a process $X(t)$ (i.e., all except events with zero probability) are differentiable (integrable), then one usually says that the process is *differentiable (integrable) with probability one.* Knowledge of the analytic properties of sample functions of the stochastic process under consideration is of considerable interest in many applied problems. It turns out, however, that in correlational analysis of stochastic processes (dealing only with the first and second moments of a process) the concept of a derivative and integral in the mean-square sense is very natural. These concepts are defined in the following way. A stochastic process $X(t)$ with finite second order moments is said to have a *mean-square derivative* $\dot{X}$ at $t \in T$ if

$$\lim_{h\to 0}\left\langle\left|\frac{X(t+h)-X(t)}{h}-\dot{X}(t)\right|^2\right\rangle = 0, \quad t+h \in T \tag{3.26}$$

Analogously, the *mean-square integral* of a stochastic process $X(t)$ is defined by the relation

$$\lim_{n\to\infty}\left\langle\left|\sum_{i=1}^{n} X(\tau_i)(t_{i+1}-t_i)-\int_a^b X(t)dt\right|^2\right\rangle = 0 \tag{3.27}$$

for each partition of the interval $[a,b]$ and for arbitrary $\tau_i \in [t_i, t_{i+1}]$.

It is possible to establish the properties of mean-square derivatives and integrals and the rules of their operation which prove to be analogous to those in calculus of ordinary functions (cf. [17]). Here we shall only give those properties which occur most commonly in applications.

If the mixed derivative $\partial^2 K_X/\partial t_1 \partial t_2$ of the correlation function $K_X(t_1,t_2)$ exists, then $X(t)$ has a mean-square derivative $X(t)$ and

$$K_{\dot{X}}(t_1,t_2) = \frac{\partial^2}{\partial t_1 \partial t_2} K_X(t_1,t_2) \tag{3.28}$$

If a stochastic process $X(t)$ is stationary, then its derivatives are also stationary and

$$K_{\dot{X}}(\tau) = -\frac{d^2 K_X(\tau)}{d\tau^2} \tag{3.29}$$

The spectral density of the derivative process $\dot{X}(t)$ is given by

$$g_{\dot{X}}(\omega) = \omega^2 g_X(\omega) \tag{3.30}$$

and in general

$$g_{\dot{X}^n}(\omega) = \omega^{2n} g_X(\omega) \tag{3.31}$$

A stochastic process $X(t)$ with correlation $K_X(t_1,t_2)$ is mean-square integrable over the interval $[a,b]$ if and only if the ordinary double Riemann integral

$$\int_a^b\int_a^b K_X(t_1,t_2)dt_1 dt_2 \tag{3.32}$$

exists (and is finite). Let $\varphi(t)$ be an ordinary function defined and integrable on the interval $[a, b]$ and let

$$Y(t) = \int_{t_0}^{t} \varphi(s) X(s) ds \tag{3.33}$$

Then

$$K_Y(t_1, t_2) = \int_{t_0}^{t_1} \int_{t_0}^{t_2} \varphi(s_1) \varphi(s_2) K_X(s_1, s_2) ds_1 ds_2 \tag{3.34}$$

Remarks

It is well known that differential equation provide very convenient models for numerous physical processes. Additionally, the fatigue processes can be modelled by means of differential equations. However, if we wish to take into account the complexity and uncertainty associated with the process considered, then we come to model expressed by *stochastic differential equations*; that is - by differential equations for stochastic processes.

The differential equation for a stochastic process $Y(t)$

$$\frac{dY(t)}{dt} = F[Y(t), X(t, \gamma), t], \quad Y(t_0) = Y_0 \tag{3.35}$$

where F is a deterministic function of its argument, and $X(t, \gamma)$ is a given stochastic process, represents a stochastic differential equation. The initial conditions for (3.35) Y_0 can be random or deterministic. If the random processes occurring in (3.35) are sufficiently regular, than the main problems are similar to those in the classical theory of differential equations; in this case, equation (3.354) can be interpreted as a family of equation for the individual sample functions. Of course, introducing random elements into differential equation also generates new problems; of most interest are the questions concerned with the probabilistic properties of the solution process.

If equation (3.35) involves highly irregular stochastic processes like a white noise or Brownian motion process, then we have a special class of stochastic differential equations which require a specific theory. The most common representation of such equations is the *Ito stochastic differential equation* whose usual form is

$$dY(t) = m[Y(t), t]dt + \sigma[y(t), t]dW(t) \tag{3.36}$$

where $Y(t)$ is the unknown stochastic process, $W(t)$ is the Wiener process (defined in next sub-Section) and $m(y,t)$ and $\sigma(y,t)$ are given functions defined for $t \in [t_0,T]$, $y \in (-\infty,\infty)$.

Since the realizations of the Wiener-process are non-differentiable (and are, with probability one, of unbounded variation in any interval), the integration of equation (3.36) requires a special definition of the Stieltjes-type integral with respect to the Wiener process $W(t)$. Such integral is known as the *Ito stochastic integral* and plays a fundamental role in the analysis of equations of the form (3.36). The Ito stochastic differential equations serve as model of dynamical systems with rapidly varying random excitations, for example - with excitations being with noise. The important feature of Ito stochastic equations is that , under quite general conditions, the solutions are diffusion Markov processes. For further information the reader is referred to reference [17]. Stochastic differential equation models of fatigue will be discussed by Spencer in this volume.

4. MARKOV DIFFUSION PROCESSES

An important class of stochastic processes possessing a well-developed theory as well as a very broad range of applications, is the class of Markov processes. One can say that in Markov process the "future" is independent of the "past" when we know the "present". More exactly, a stochastic process $X(t), t \in T$, is said to be a *Markov process* if for an arbitrary set of times $t_1 < t_2 < \ldots < t_n$, the conditional distribution of $X(t_n)$ when $X(t_1),\ldots,X(t_{n-1})$ are given depends only on $X(t_{n-1})$, i.e.

$$\begin{aligned} &P\{X(t_n) \geq x_n | X(t_{n-1}) = x_{n-1};\ldots;X(t_2) = x_2;\; X(t_1) = x_1\} \\ &\qquad = P\{X(t_n) \geq x_n | X(t_{n-1}) = x_{n-1}\} \end{aligned} \tag{3.37}$$

If the distribution of the initial state is known, then the Markov process is completely specified by the following conditional distribution function

$$F(s,x;t,y) = P\{X(t) < y | X(s) = x\} \tag{3.38}$$

which is commonly known as the *transition probability function* of the Markov process $X(t)$.

Let us restrict our attention to a *continuously valued Markov processes.* The basic class of such processes is formed by diffusion Markov processes. Without going into details, one can say that *diffusion processes* are the processes for which the probability of any substantial change within a small time interval Δt is very small (of order less than Δt) and there exist functions $a(t,x)$ and $b(t,x)$ characterizing, respectively, the average tendency in the evolution of the process $X(t)$ within an infinitesimal time interval $(t, t+\Delta t)$ and the mean square of the change in the process within the time interval $(t, t+\Delta t)$. For a diffusion Markov process, its transition probability function $F(s,x;t,y)$ and the corresponding density $p(s,x;t,y)$ satisfy certain partial differential equations of the probabilistic type. One of these equations is used particularly often in applications; it is the *Fokker-Planck-Kolmogorov* equation

$$\frac{\partial p}{\partial t} + \frac{\partial}{\partial y}[a(t,y)p(s,x;t,y)] - \frac{1}{2}\frac{\partial^2}{\partial y^2}[b(t,y)p(s,x;t,y)] = 0 \tag{3.39}$$

For a vector Markov diffusion process $\mathbf{X}(t) = [X_1(t), \ldots, X_n(t)]$, the Fokker-Planck-Kolgomorov equation has the form

$$\begin{aligned} \frac{\partial p(s,\mathbf{x};t,\mathbf{y})}{\partial t} &+ \sum_{i=1}^{n} \frac{\partial}{\partial y_i}[a_i(t,\mathbf{y})p(s,\mathbf{x};t,\mathbf{y})] \\ &- \frac{1}{2}\sum_{i,j=1}^{n} \frac{\partial^2}{\partial y_i \partial y_j}[b_{ij}(t,\mathbf{y})p(s,\mathbf{x};t,\mathbf{y})] = 0 \end{aligned} \tag{3.40}$$

where $\mathbf{x} = (x_1, \ldots, x_n), \mathbf{y} = (y_1, \ldots, y_n)$ denote the states of the process, and $a_i(t,\mathbf{y})$, $b_{ij}(t,\mathbf{y})$ are the components of the vector-function $\mathbf{A}(t,\mathbf{y})$ and the matrix-function $\mathbf{B}(t,\mathbf{y})$, respectively; like $a(t,y), b(t,y)$ in the one-dimensional case, the functions $\mathbf{A}(t,\mathbf{y})$ and $\mathbf{B}(t,\mathbf{y})$ characterize the infinitesimal properties of the process $\mathbf{X}(t)$.

An important example of a Markov diffusion is the *Brownian motion process* or the *Wiener process* - a mathematical model of the Brownian motion of a free particle immersed in liquid. Thorough understanding of the Wiener process proves particularly important since many other diffusion Markov processes can be constructed using the Wiener process (cf. [17]).

A stochastic process $X(t), t \in [0, \infty]$, is said be a *Wiener process* and is denoted by $W(t)$ if it has the following properties:

a) $P\{W(0) = 0\} = 1$

b) the increments $W(t) - W(s), s < t$ are independent and Gaussian,

c) $\langle W(t) - W(s) \rangle = 0$,

d) $\langle [W(t) - W(s)]^2 \rangle = \sigma^2(t-s)$, where σ^2 is a positive constant.

The correlation function of a Wiener process is

$$K_W(t_1, t_2) = \sigma^2 \min(t_1, t_2). \tag{3.41}$$

The density of the probability of a transition from state x to state y is given by

$$p(s,x;t,y)] = \frac{1}{\sigma\sqrt{2\pi(t-s)}} \exp\left[-\frac{(y-x)^2}{2(t-s)\sigma^2}\right]. \tag{3.42}$$

References

1. Wigner E., The Unreasonable Effectiveness of Mathematics in the Natural Sciences, Comm. Pure Appl. Math. Vol 13, 1, 1960,1-14.

2. Schlegel R. "Do We Know The World Through Science" in : Mind, Science and History (Eds. H. E. Kiefer, M. K. Munitz), State Univ. of N. York Press, Albany, 1970.

3. Blekhman I. I, Myshkis A. D., Panovoko J.G., Mechanics and Applied Mathematics; Logic and Features of Mathematical Applications (in Russian), Moskow, Nauka 1983.

4. Bunge M., Theory Meets Experience in: Mind, Science and History (Eds. Kiefer H. E., Munitz M.K.), State Univ. of N. York, Albany, New York, 1970.

5. Sobczyk K., Spencer B. F., Random Fatigue: Form Data to Theory, Academic Press, Boston, N. York, 1992.

6. Goodwin G. C., Payne R. I., Dynamic System Identification: Experiment Design and Data Analysis, Acad. Press, N. York, 1977.

7. Littlewood J. E., Dillema of Probability Theory, in: A Mathematical Miscellany, London 1957.

8. Nalimov V., Faces of Science, ISI Press, 1981.

9. Fine T. L., Theories of Probability. An Examination of Foundations, Acad. Press, 1973.

10. Csiszar I., Information - Type Distance Measures and Indirect Observations, Studia Scient. Math. Hungarica, Vol. 2, 299 - 318, 1967.

11. Kozin F., Bogdanoff J. L., Role of Third Order Statistics in Discriminating among Models of Fatigue Crack Growth, Metallurgical Trans. A, Vol. 18 A, N. 11, Nov. 1987.

12. Ang A.H-S, Tang W.H., Probability Concepts in Engineering Planning and Design, vol. I. Basic Principles, J. Willey & Sons, New York, 1975.

13. Soong T.T., Probabilistic Modelling and Analysis in Science and Engineering, J. Willey & Sons, New York, 1981.

14. Cox D.R., Miller H.D., The Theory of Stochastic Processes, Chapman & Hall, London, 1977.

15. Gumbel E.J., Statistic of Extremes, Columbia Univ. Press, New York, 1958.

16. Cramer H., Leadbetter M.R., Stationary and Related Stochastic Processes, Willey, New York, 1967.

17. Sobczyk K., Stochastic Differential Equations with Applications to Physics and Engineering, Kluwer Acad. Publ., Dordrecht, Boston, 1991.

CHARACTERIZATION OF RANDOM FATIGUE LOADS

I. Rychlik
University of Lund, Lund, Sweden

1. RANDOM FATIGUE LOADS AND CYCLE COUNTS

1.1. Introduction

The general approach in fatigue life prediction is to relate the fatigue life of a construction, subjected to a random load, to laboratory fatigue experiments of simple specimens subjected to constant amplitude load, so called S-N data. Therefore, it is necessary to define amplitudes of equivalent "load cycles" S_k, which are functions of the sequence of maxima and minima in the load, and assume a damage rule, i.e. a method to measure the damage caused by each simple cycle.

A commonly used damage rule, due to Palmgren and Miner, postulates that the total damage caused by a stress history $\{S_k\}$ in a load function $x(t)$ is

$$D(T) = \sum_{k=1}^{k(T)} \frac{1}{N_{S_k}}, \tag{1}$$

where the sum is extended over all cycles completed at time T and N_s is the cycle life obtained from S-N data from tests with constant amplitude s. Fatigue failure occurs when $D(T)$ exceeds one. Observe that often $x(t)$ is an elongation function, then we are dealing with displacement cycles instead of load cycles. The two approaches are equivalent and we shall use the same notation $x(t)$ for both load or elongation processes.

Real loads are complicated functions of time, and their variation can not be predicted in advance. In order to describe loads we shall use a mathematical model of continuous time random processes.

In most situations properties of the load vary with time, e.g. loading conditions can change between the different operations of an aircraft or a car. However we shall assume that locally the load process is statistically stationary, and in these lectures we shall only be concerned with stationary processes.

1.2. Gaussian processes

The most commonly used model for a random load is the Gaussian process. The common use of the Gaussian model is due to its nice stability properties, e.g. if the input to a linear system is a Gaussian process the output is also a Gaussian process. Furthermore the mathematical tools to analyze Gaussian loads are very well developed. Loosely speaking, the Gaussian assumption is approximatively satisfied, if the global load is a sum of a large number of small contributions. We turn now to a short introduction to Gaussian variables, vectors and processes.

A random variable X with mean $E[X] = m$ and variance $V[X] = E[(X-m)^2] = \sigma^2 > 0$ is a Gaussian variable if its density is given by

$$f_X(x) = \frac{1}{\sigma\sqrt{2\pi}} e^{-\frac{(x-m)^2}{2\sigma^2}}.$$

The vector $\mathbf{X} = (X_1, \ldots, X_n)$ with mean $\mathbf{m} = (m_1, \ldots, m_n)$ and covariance matrix $\mathbf{C} = [c_{i,j}]$, $c_{i,j} = C[X_i, X_j] = E[(X_i - m_i)(X_j - m_j)]$, has a joint nonsingular Gaussian distribution if the determinant $\det \mathbf{C} \neq 0$ and its joint density is given by

$$f_{\mathbf{X}}(\mathbf{x}) = (2\pi)^{-n/2} \frac{1}{\sqrt{\det \mathbf{C}}} e^{-\frac{1}{2}(\mathbf{x}-\mathbf{m})\mathbf{C}^{-1}(\mathbf{x}-\mathbf{m})^T}.$$

Further, for any constants a_0, $\mathbf{a} = (a_1, \ldots, a_n)$ the random variable $Y = a_0 + \sum_{i=1}^n a_i X_i$ has a Gaussian distribution with mean $m = a_0 + \sum_{i=1}^n a_i m_i$ and variance $\sigma^2 = \mathbf{a}\mathbf{C}\mathbf{a}^T$.

The set of random variables $X(t)$, $t > 0$, is called a Gaussian process if for every finite linear combination of the form $Y = a_0 + \sum_{i=1}^n a_i X_i(t_i)$ is a Gaussian random variable.

The important property of Gaussian processes is that they are defined by the mean function $m(t) = E[X(t)]$ and the covariance function $r_X(t_1, t_2) = C[X(t_1), X(t_2)]$. The process $X(t)$ is stationary if the mean function is constant, $m(t) = m$, and the covariance function depends only on time difference, i.e. $r_X(t_1, t_2) = r(t_2 - t_1)$. Further, if the covariance function $r(\tau)$, $\tau = t_2 - t_1$, is a continuous function of τ then, by Bochner's theorem, it can be represented as follows

$$r_X(\tau) = \int_{-\infty}^{+\infty} e^{i\lambda\tau} S_X(\lambda)\, d\lambda,$$

where λ is an angular frequency and S_X is a spectral density function. (We shall equivalently use the term spectral density and spectrum.) The spectral density S_X is a positive and symmetric generalized function, which can contain a countable number of δ-Dirac functions. Note that, if $S_X(\lambda)$ contains a continuous component, e.g. $S_X(\lambda)$

is not exclusively a sum of δ-dirac functions, then the vector $X(t_1), X(t_2), \ldots, X(t_n)$ has a joint nonsingular Gaussian density function for any sequence of time points $0 \le t_1 \le \cdots \le t_n$.

Since loads are continuous functions of time, we shall always assume that the random processes have continuous sample paths, see [1] for a detailed discussion of the properties of sample paths of random processes.

An example of a Gaussian process which we shall use in several sections is taken from oceanography.

EXAMPLE 1. (Gaussian process with JONSWAP spectrum.) The JONSWAP spectrum is designed to fit empirical wave data from the North Sea, i.e. $x(t)$ is a horizontal movement in time of a fixed point of the sea surface under stationary storm conditions. The spectrum has the explicit form

$$S(\lambda) = \frac{g^2\alpha}{\lambda^5} e^{-\beta(\lambda_m/\lambda)^4} \gamma^{\psi(\lambda)}, \quad \psi(\lambda) = e^{-(\lambda-\lambda_m)^2/(2\sigma^2\lambda_m^2)}, \quad \sigma = \begin{cases} \sigma_a, & \text{for } |\lambda| \le \lambda_m, \\ \sigma_b, & \text{otherwise.} \end{cases}$$

We choose $g = 9.81$, $\alpha = 0.0029$, $\beta = 1.25$, $\lambda_m = 0.075$, $\gamma = 1$, $\sigma_a = 0.07$, $\sigma_b = 0.09$. Note that a process with JONSWAP spectrum has no second derivative and it has infinite intensity of local extremes. In order to obtain a smoother process we truncate the spectrum at 0.2, e.g. $S(\lambda) = 0$ for $|\lambda| > 0.2$. The second derivative of the Gaussian process with truncated JONSWAP spectrum exists (it will be needed in the definition of Morison force, see Example 2). □

The JONSWAP spectrum, in the case when $\gamma = 1$ is the so called Pierson Moskowitz spectrum defined as follows

$$S(\lambda) = \frac{g^2\alpha}{\lambda^5} e^{-\beta(g/v_0\lambda)^4},$$

where g is the acceleration of gravity, α, β dimensionless constants and v_0 is the mean wind velocity at a height of 19.5 m above the still sea surface.

In many fields of applications one has developed standard spectra, e.g. the surface of the road is often assumed as a Gaussian process with the following spectrum

$$S(\lambda) = \frac{C}{\lambda^a}, \qquad \lambda_{\min} \le \lambda \le \lambda_{\max},$$

where C is a constant, $2 \le a \le 3$ and $\lambda_{\min}, \lambda_{\max}$ are suitable cut off frequences, see [27]. In the analysis of aircraft structures one often assumes that the longitudinal component of stationary wind velocity $V(t)$, say, has the following spectrum

$$S(\lambda) = \frac{\sigma_Z^2 c_V}{\pi v_0} \frac{1 + 3\hat{\lambda}^2}{(1 + \hat{\lambda})^2}, \qquad \hat{\lambda} = \frac{c_V}{v_0}\lambda,$$

where v_0 is the mean wind speed, σ_Z^2 is the variance of the vertical component and c_V denotes the scale of the turbulance in the longitudinal direction. (Note that $\int_0^{+\infty} S(\lambda)\, d\lambda = \infty$, i.e. the process has infinite variance. The process can be rigorously defined using the theory of generalized random process. Another example of a generalized process is a Gaussian white noise with a constant spectrum.) In the last section we shall be concerned only with Gaussian loads.

1.3. Memoryless transformations of Gaussian processes

Another large class of processes used to model loads or elongations are obtained by memoryless transformations of Gaussian vector valued processes. Here we shall consider two such models.

EXAMPLE 2. (Morison force.) Let $y(t)$ be the velocity of a fluid around a circular pile, and $F(t)$ be the force acting on the pile. By Morison's equation the force F is the following function of the process $y(t)$,

$$F(t) = K_M y'(t) + K_D y(t)|y(t)|, \tag{2}$$

where K_M, K_D are positive constants. The properties of $F(t)$ are often discussed in ocean engineering literature, e.g. see [2]. Usually the velocity $y(t)$ is assumed to be a stationary Gaussian process, e.g. the process with truncated JONSWAP spectrum from Example 1. □

Another simple example is related to wind pressure load. If $V(t)$ denotes the instantaneous wind velocity then the wind pressure $P(t)$ is given by

$$P(t) = \frac{1}{2}\rho C V^2(t),$$

where ρ is the density of air (about 1.2 kq/m^3) and C is the wind pressure coefficient, which depends on the geometry of the structure. If $V(t)$ is a Gaussian process then $P(t)$ is the simplest example of the so called χ^2-load.

EXAMPLE 3. (χ^2-load.) Let $x(t)$ be a χ^2-load, i.e. $x(t)$ is a sum of squares of stationary, continuously differentiable, Gaussian processes

$$x(t) = \sum_{i=1}^{n} x_i^2(t), \tag{3}$$

where $x_1(t), \ldots, x_n(t)$ are independent random variables with mean zero and variance one. For simplicity, we assume that the processes x_i are independent. □

1.4. Markov processes

The loads discussed up to now are all related to Gaussian processes. Another general class of processes are Markov processes. Markov models have well developed mathematical tools for analysis of the first passage problem both for linear and for non-linear systems, see [3]. A simple example of a Markov process is the output of a damped harmonic oscillator driven by white noise.

EXAMPLE 4. (Oscillator filter driven by white noise.) In many technical applications one uses the motion of a damped harmonic oscillator driven by a stationary excitation as a simple model for an elongation process. The governing equation is

$$x''(t) + 2\zeta\omega x'(t) + \omega^2 x(t) = \sigma X(t), \qquad t \geq 0, \quad 0 < \zeta < 1, \tag{4}$$

where ζ, ω are system parameters and X is a suitable random process, e.g. a Gaussian load, Morison force or wind pressure. A particularly important special case is when X is a Gaussian white noise process. In that case $x(t), x'(t)$ form a Markov process. Assume that the initial conditions are $x(0) = x'(0) = 0$. Then x is given by

$$x(t) = \int_0^t h(t-s)\, dB(s), \qquad t \in [0, T],$$

where $B(s)$ is a Brownian motion and the kernel h is given by

$$h(t) = \frac{\sigma}{\omega\sqrt{1-\zeta^2}}\, e^{-\zeta\omega t} \sin(\omega\sqrt{1-\zeta^2}\, t).$$

We choose $\omega = 1$ and $\sigma^2 = 2\zeta\omega$; hence the asymptotic variances of x and x', $\sigma_x^2 = \sigma_{x'}^2 = 1$. □

1.5. Markov chain of extremes

Usually cycles are defined as a function of a sequence of local extremes of x. Since the distribution of cycle characteristics can be expressed in terms of absorption probabilities in the sequence of extremes, the Markov chain model for the dependence in the sequence gives both a broad class of loads and effective algorithms to compute the distribution of cycle amplitudes.

A sequence of random variables X_n, $n \geq 1$, is called a Markov chain if for any $n > 0$, $m > 0$ the conditional distribution of X_n given m previous values of the sequence is equal to the conditional distribution of X_n given the last value, i.e. the following equality holds

$$P[X_n \leq x \mid X_{n-1} = x_1, \ldots, X_{n-m} = x_m] = P[X_n \leq x \mid X_{n-1} = x_1],$$

for all $x \in R$ and all vectors $(x_1, \ldots, x_m)$.

A generalization of the Markov chain models are the so called m-step Markov chains, which means that the distribution of X_n depends only on m previous values of the sequence, i.e. for all n and $k > 0$

$$\begin{aligned} P[X_n \leq x \mid X_{n-1} = x_1, \ldots, X_{n-m} = x_m, X_{n-m-1} = x_{m+1}, \ldots, X_{n-m-k} = x_{m+k}] \\ = P[X_n \leq x \mid X_{n-1} = x_1, \ldots, X_{n-m} = x_m] \end{aligned}$$

for all x and all vectors $(x_1, \ldots, x_{m+k})$.

Section 3 will be devoted to Markov chain models. We shall show by numerous examples that approximation of the sequence of local extremes in the load by a Markov chain gives an excellent agreement between computed expected damages and simulated damages for Gaussian loads with various spectral densities, as well as for some non-Gaussian loads, as discussed in Examples 2–3. The method can be applied to smooth as well as to very irregular loads with an infinite number of cycles in finite time. Obviously there are several other types of models for random loads but in these lectures we shall confine ourselves for models mentioned.

1.6. Cycle counting methods

Generally, a cycle is defined by specifying its highest and lowest points, x, y, respectively, and the time at which the cycle is counted t, say, i.e. a cycle is a pair $(x, y)_t \in \mathbf{R}^2$, $x \geq y$. (We consider also cycles with zero amplitude, although such cycles are redundant since they do not cause any damage.) A cycle count is a procedure transforming a function $x(t)$, $0 \leq t \leq T$ into a set of cycles $\{(x, y)_t\}$.

In the literature one can find several definitions of cycle counting procedures, see [4]. Dowling, [5] has studied the accuracy of the predictors of the fatigue life based on the eight most commonly used counting methods, and finds that only the rainflow count leads to prediction agreeing with actual lives.

The rainflow method identifies the ranges of strain (stress) of the closed hysteresis loops in the stress-strain response of a material subjected to cyclic loading. The definition proposed first by Endo et al. [6], and other modified algorithms, see e.g. [7], has a complicated sequential structure in order to "on-line" identify ranges of closed hysteresis loops. Although writing the program implementing the algorithm is an easy task, the original definition is not suitable for determination of the probabilistic characteristics of the rainflow count. We turn therefore to a general definition of the rainflow count, first given in [8], which is more convenient for mathematical analysis of the rainflow method.

The rainflow count defines a one to one correspondence between the sequences of local maxima and minima. In the case of smooth functions there exists only a finite number of local extremes in $[0, T]$, and hence the number of cycles is finite. However, in many applications one uses mathematical models for elongation which are less regular, e.g. the second derivative may not exists and hence there can be infinitely many extremes in any finite interval. Consequently, it is not obvious how to define cycles from the sequence of local extremes. (For example, the output of an oscillator filter driven by a white noise process, see Example 4, is only once continuously differentiable.)

This problem can be resolved by approximating the continuous load by a sequence of sampled values of the process and defining cycles using standard counting methods. However the dependence of the damage $D(t)$, defined by (1), on sampling frequency is complicated and hard to investigate.

In the following general definition of rainflow count we shall define a continuum of cycles, i.e. we attache a cycle to all points $t \in [0,T]$; then the rainflow count is defined as a subset of cycles with nonzero amplitudes.

DEFINITION 1. (RAINFLOW CYCLE.) *Let $x(t)$, $0 \le t \le T$, be a continuous function. For any fixed $t \in [0,T]$, let t^-, t^+ be the following time points,*

$$t^- = \begin{cases} \sup\{s \in [0,t);\ x(s) \ge x(t)\}, & \\ 0, & \text{if } x(s) < x(t) \text{ for all } s \in [0,t), \end{cases}$$

$$t^+ = \begin{cases} \inf\{s \in (t,T];\ x(s) > x(t)\}, & \\ T, & \text{if } x(s) \le x(t) \text{ for all } s \in (t,T], \end{cases} \tag{5}$$

and let $h^-(t), h^+(t)$ be the lowest points in the intervals $[t^-,t]$, $[t,t^+]$, respectively, i.e.

$$h^-(t) = \inf\{x(s);\ t^- \le s \le t\},$$
$$h^+(t) = \inf\{x(s);\ t \le s \le t^+\}.$$

The rainflow cycle originating at t, $0 \le t \le T$, is defined as a pair $(x,y)_t$, where $x = x(t)$ and

$$y = \begin{cases} \max(h^-(t), h^+(t)) & \text{if } t^- > 0, \\ h^+(t) & \text{if } t^- = 0, \end{cases} \tag{6}$$

see Figure 1.

Observe that, in Definition 1, rainflow cycles are defined for all $t \in [0,T]$. For most points the rainflow cycle has zero amplitude, i.e. $x = y$. Note, that for smooth loads (second derivative is continuous) the nonzero amplitude loads $x > y$ occur only at the points of local maximima.

The algorithm for the rainflow count given in Definition 1 is equivalent to the original method proposed by Endo. The local extremes of $x(t)$ which are the endpoints of closed hysteresis loops are identified by both methods and paired into cycles. The remaining local extremes form the so called *residual* (or a set of *half cycles*), see Figure 2.

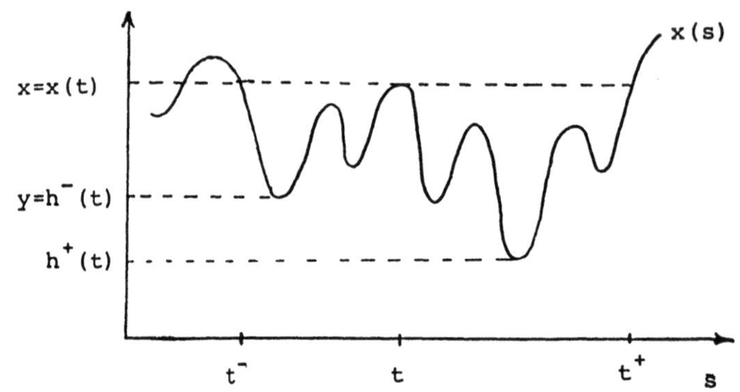

Figure 1: The definition of rainflow cycle.

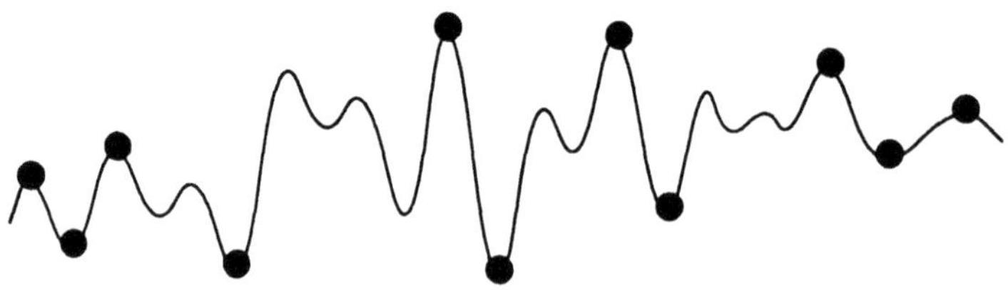

Figure 2: Example of a load with local extremes which do not form closed hysteresis loops marked by •.

The difference between the original rainflow method and the new by Rychlik is the way how half cycles are treated. In Endo's definition the set of half cycles is found while in the new definition the maxima in the residual are paired with the following minima (in the residual) to form cycles.

Although we shall be mostly concerned with properties of the rainflow method, we give here definitions of other frequently used counting methods.

DEFINITION 2. (PEAK-VALLEY CYCLE.) *Let $x(t), 0 \leq t \leq T$, be a continuous function and denote by M_i, m_i the local maxima and the following minima of x. Let t_i be the time for the occurrence of the local maximum M_i. Then the peak-valley cycle is defined as $(M_i, m_i)_{t_i}$.*

DEFINITION 3. (MEAN-CROSSINGS CYCLE.) *Let $x(t)$, $0 \leq t \leq T$, be a continuous function with a mean m_x, where $m_x = \frac{1}{T}\int_0^T x(t)\,dt$. Denote by t_i the times of the upcrossings of the mean level m_x. The mean-crossing cycle $(x,y)_{t_i}$ originating at t_i is defined by*

$$x = \sup\{x(t);\ t_i \leq t \leq t_{i+1}\}, \qquad y = \inf\{x(t);\ t_i \leq t \leq t_{i+1}\}.$$

DEFINITION 4. (PEAK CYCLE.) *Let $x(t)$, $0 \leq t \leq T$, be a continuous function with mean m_x. Denote by M_i the local maxima of x at time t_i which are higher than the mean m_x. The peak cycle originating at t_i is defined by $(M_i, 2m_x - M_i)_{t_i}$*

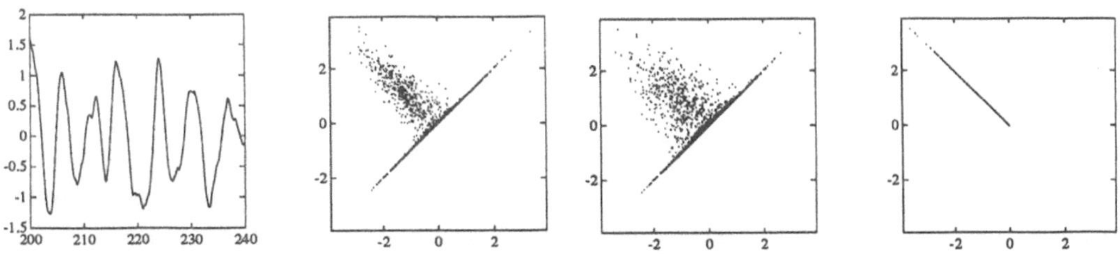

Figure 3: The comparison of different counting methods; (a) part of a load; (b) rainflow count; (c) peak-valley count; (d) peak count.

EXAMPLE 5. As we have mentioned before a cycle count transforms a function $x(t)$ into a set of cycles $\{(x, y)_t\}$. Now each cycle $(x, y)_t$ can be represented as a point in the plane so a cycle count is seen as a point process in the plane. In Figure 3 we present part of a load containing 2,047 extremes together with pictures showing the three different cycle counts; rainflow, peak-valley and peak-count. □

2. TOTAL DAMAGE $D(T)$

2.1. Linear damage rules

Let $x(t)$, $0 \leq t \leq T$, be a continuous function, and let $\{(x, y)_t\}$ be a cycle count, for example some of the counts of Definitions 1-4. Denote by $f(x, y)$ a damage caused by a cycle (x, y). Assume $f(x, y) \geq 0$ is increasing in x and decreasing in y. Usually one assumes $f(x, y) = N_{x-y}^{-1}$, where N_s is obtained from constant amplitude fatigue data. In applications one often finds an explicit expression for N_{x-y},

$$N_{x-y} = \alpha(x - y)^{\beta}, \qquad \beta \geq 1. \tag{7}$$

By (1), if the set of cycles with positive amplitudes is countable, the total damage $D(T)$ is given by

$$D(T) = \sum_{t \in [0,T]} f((x, y)_t). \tag{8}$$

EXAMPLE 5, (contd.) In Figure 4 we present the total damage $D(T)$ for the load based on the different cycle counts. The damage function is $f(x, y) = (x - y)^{\beta}$, $1 \leq \beta \leq 5$. □

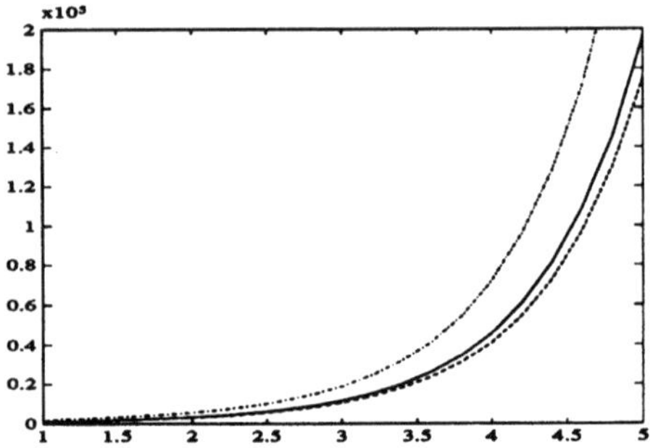

Figure 4: Total damage obtained using the rainflow count (solid), the peak-valley count (dashed) and the peak count (dash-dotted) obtained for the load of Example 5.

Obviously, given a cycle count $\{(x,y)_t\}$ formula (8) is computationally most convenient. However, for theoretical considerations and computation of expected damage, if the function f is known explicitly, it is more convenient to rewrite equation (8) using integration by parts. We begin with some general properties of cycle counts.

DEFINITION 5. *Let $x(t)$, $0 \le t \le T$, be a continuous function with a countable set of cycles $\{(x,y)_s\}$. Let $N_T(u,v)$, $u \ge v$, denote a counting distribution function of cycle count $\{(x,y)_s\}$, i.e.*

$$N_T(u,v) = \#\{(x,y)_s;\ s \in [0,T];\ x > u \ge v > y\}, \tag{9}$$

where $\#\{\cdot\}$ is a number of elements in the set $\{\cdot\}$.

Note that N_T is not a standard distribution function. Many applications of the counting distribution are consequences of the following simple properties of N_T:

$$\begin{aligned}&\#\{(x,y)_s;\ (x,y)_s \in (u_1,u_2] \times [v_1,v_2)\}\\ &\quad = N_T(u_2,v_1) - N_T(u_1,v_1) - N_T(u_2,v_2) + N_T(u_1,v_2),\\ &\qquad N_T(u,v) \le \min_{v \le x \le u} N_T(x,x).\end{aligned}$$

Consequently the counting distribution N_T is bounded by its diagonal values. In the following we shall consider only cycle counts such that the diagonal $N_T(u,u)$ is bounded. This implies the countability of cycle counts, see the following lemma.

LEMMA 1. *Let $x(t)$, $0 \le t \le T$, be a continuous function and $\{(x,y)_t\}$ a cycle count such that for a finite constant c, $N_T(u,u) < c$, for all u. For any $\epsilon \ge 0$, if*

$$A_\epsilon = \{(x,y)_t;\ x - y > \epsilon\}$$

then the number of cycles in A_ϵ is bounded by $c\epsilon^{-1}(\sup_{0\le s\le T} x(s) - \inf_{0\le s\le T} x(s))$ and the set A_0 is countable. Further, the "total variation" of a cycle count is finite and

$$\sum_{(x,y)_s \in A_0} (x - y) = \int_{-\infty}^{\infty} N_T(u,u)\, du.$$

PROOF. A simple proof can be found in [9]. □

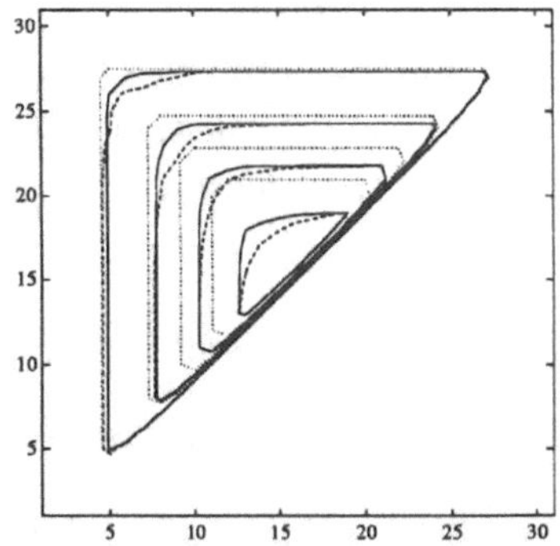

Figure 5: Contour levels of the counting distribution $N_T(u, v)$ for the rainflow (solid line); peak-valley (dashed line) and peak count (dotted line); for the counts of Figure 2. The contour levels are plotted at the levels 10, 60, 170 and 350.

EXAMPLE 5, (contd.) In Figure 5 we present contour levels of the counting distribution $N_T(u, v)$ for the rainflow-, peak-valley- and peak count. □

We turn now to the alternative formula for the total damage $D(T)$.

THEOREM 2. *Let $\{(x, y)_t\}$, $0 \leq t \leq T$, be a cycle count, $N_T(u, v)$ its counting distribution and $f(x, y)$ a damage caused by a cycle (x, y). Assume $f(x, y) \geq 0$ is increasing in x and decreasing in y and is twice continuously differentiable function. If $f(x, x) = 0$ and $N_T(u, u)$ is a bounded function of u, the total damage $D(T)$ is finite and given by*

$$D(T) = -\int_{-\infty}^{+\infty}\int_{-\infty}^{x} N_T(x, y)\frac{\partial^2 f(x, y)}{\partial x \partial y}\,dy\,dx - \int_{-\infty}^{+\infty} N_T(x, x)\left.\frac{\partial f(x, y)}{\partial y}\right|_{y=x} dx. \quad (10)$$

PROOF. The elementary but rather lengthy proof can be found in [9]. □

Note that for the function f given by (7) the second integral disappears if $\beta > 1$. Further in a special case when $f(x, y) = g(x - y)$ then

$$D(T) = \int_{-\infty}^{+\infty}\int_{-\infty}^{x} N_T(x, y) g''(x - y)\,dy\,dx + g'(0)\int_{-\infty}^{+\infty} N_T(x, x)\,dx. \quad (11)$$

Finally, if $\partial^2 f/\partial x \partial y \leq 0$ then the upper bound for the damage $D(T)$ is obtained by replacing $N_T(x, y)$ in (10) by

$$K_T(x, y) = \min_{x \leq u \leq y} N_T(u, u),$$

$$D(T) \leq -\int_{-\infty}^{+\infty}\int_{-\infty}^{x} K_T(x, y)\frac{\partial^2 f(x, y)}{\partial x \partial y}\,dy\,dx - \int_{-\infty}^{+\infty} K_T(x, x)\left.\frac{\partial f(x, y)}{\partial y}\right|_{y=x} dx. \quad (12)$$

2.2. Comparison of damage due to different counting methods

Here we shall compare damages due to different counting method. We begin with some additional properties of counting distributions.

By Definition 5, $N_T(u,u)$ is a number of cycles $(x,y)_t$ such that $x > u > y$. For many counting methods, e.g. rainflow and peak-valley count, $N_T(u,u)$ is equal to the number of downcrossings of level u of the function $x(t)$. This is not true for peak and mean-crossings cycle counts.

DEFINITION 6. *A cycle counting method is called "crossing consistent" if*

$$N_T(u,u) = \#\{t \in [0,T];\ t \textit{ is a } u\textit{-downcrossing of } x\}. \tag{13}$$

We turn now to the counting distribution of the rainflow count.

LEMMA 3. *Let $x(t)$, $0 \le t \le T$, be a continuous function and $\{(x,y)_t\}$ a set of rainflow cycles. Then*

$$N_T(u,v) = \#\{t \in [0,T];\ t \textit{ is a } u\textit{-downcrossing and } h^+(t) < v\}. \tag{14}$$

PROOF. For the proof see [10]. □

Now, if t is u-downcrossing point, then $h^+(t) < u$ and hence the rainflow count is crossing consistent, i.e. $N_T(u,u)$ is the number of the downcrossing of the level u by $x(t)$, $t \in [0,T]$.

Observe that the formula (14) for the distribution $N_T(u,v)$ of the rainflow count can be written in the following more convenient way:

$$N_T(u,v) = \#\{t \in [0,T];\ t \text{ is a } u\text{-downcrossing and } x(t+\cdot) \text{ crosses } v \text{ before } u\}. \tag{15}$$

The distribution of peak-valley count $N_T^*(u,v)$, say, can be defined by imposing an additional condition on u-downcrossings in (15). More exactly one has to require that $x(t+\cdot)$ is nonincreasing between t and the first v-downcrossing point. Obviously, for $u \ge v$, $N_T(u,v) \ge N_T^*(u,v)$ and $N_T(u,u) = N_T^*(u,u)$. Consequently, the damage obtained using peak-valley count is always bounded by the damage based on rainflow count.

The mean-crossings cycle count, see Definition 3, is related to the rainflow count, since the counting distributions of rainflow- and mean-crossing amplitude counts are identical for $v = m_x$ or $u = m_x$, where m_x is a mean of the load, and for all other values of (u,v) the rainflow counting distribution bounds the mean-crossing amplitude counting distribution. Again the damage based on rainflow count bounds the damage obtained using the mean-crossing cycle count.

Finally, we consider the simplest count, the peak cycle count, see Definition 4. One can give examples of functions $x(t)$ when damage based on rainflow count is greater or lower than damage based on peak count. However for stationary symmetric processes the expected damage caused by peak count is greater than the expected damage obtained using any crossing consistent cycle count, e.g. any count discussed in this subsection; see [11] for more details.

2.3. Expected damage $E[D(T)]$

Let $x(t)$, $0 \le t \le T$, be a random process, and let $\{(x,y)_t\}$ be a cycle count with bounded counting distribution $N_T(u,v)$. The total damage $D(t)$ is an increasing random process and $D(T)$ a.s. finite. Since the fatigue failure time T^f is defined by $D(T^f) = 1$, the failure probability is given by

$$P[T^f \le T] = P[D(T) \ge 1],$$

and hence the distribution of $D(T)$ is of great importance. However, in our setting of the problem, we define a probabilistic structure of the process $x(t)$ and since $D(T)$ is a complicated nonlinear functional defined on x the problem of computing $P[D(T) \ge 1]$ is extremely hard. One possibility is to obtain estimates of $P[D(T) \ge 1]$ by simulations. (This is not very simple, since one needs, in some cases, 10^7–10^9 cycles before $D(T) > 1$. Further, since we are interested in small probabilities, a large number of sample paths of x are needed.) The other possibility is to use some asymptotic results for cycles with very large amplitudes and neglect all the cycles with smaller amplitudes, what can be applied to low cycle fatigue. We shall not discuss this approximation here.

However, the fatigue life T^f, defined by $D(T^f) = 1$, is only a crude approximation of the "real" fatigue life. Even in well controlled laboratory experiments the failure of specimen can occur when the total damage $D(T)$ considerably differs from 1. Consequently, relatively fast methods to estimate T^f are of interest.

The simplest estimator $\hat{T}^f$, say, of T^f is obtained by using the moment methods, i.e. $\hat{T}^f$ is the solution of the moment equation

$$E[D(\hat{T}^f)] = 1$$

and for stationary loads

$$\hat{T}^f = \frac{1}{E[D(1)]}.$$

In the following we shall call $E[D(1)]$ the damage intensity. In this subsection we shall discuss some general properties of expected damage, and then, in Section 3, we give a Markov method to approximate $E[D(T)]$. Finally, in Section 5, we consider $E[D(T)]$ for Gaussian processes.

In the previous section the total damage $D(T)$ has been defined in two alternative ways. Using (8) the expected damage is given by

$$E[D(T)] = \sum_{t \in [0,T]} E[f((x,y)_t)]. \tag{16}$$

In general it is not obvious how to compute $E[f((x,y)_t)]$. Even for a stationary processes $x(t)$ with a finite number of cycles in a finite time interval the sequence $\{(x,y)_t\}$ is non-stationary. Consequently, for the n'th cycle counted at time $t = t_n$, $E[f((x,y)_{t_n})] \neq E[f((x,y)_{t_{n+1}})]$. However, in most important special case when the intensity of cycles c_t, say, is finite the equation (16) can be written in the following way

$$E[D(T)] = \int_0^T c_t E^{\text{int}}[f((x,y)_t)]\,dt.$$

The index "int" indicates that we are considering a "typical" cycle at time t. Another interpretation is that $E^{\text{int}}[f((x,y)_t)]$ is a conditional expected damage given that a cycle is counted at the time t. (For stationary loads both c_t and $E^{\text{int}}[f((x,y)_t)]$ are independent of time and equal to the expected number of cycles in the interval $[0,1]$ and the average damage of a typical cycle.)

In the case of ergodic loads the "intensity" distribution of a cycle $(x,y)_t$ is equal to the empirical ("long-run") distribution of cycles. The expectation $E^{\text{int}}[f((x,y)_t)]$ can be expressed in terms of the distribution of the process $x(t)$ using the concept of Slepian model processes. Slepian models have many other important engineering applications related to fatigue and degradation of components; we turn back to the definition and properties of Slepian models in Section 4.

We turn now to the second definition of $D(T)$, i.e. Formula (10). The expected damage is obtained by computing the expectations of the integrals in (10). Assume that we can change the order of expectation and integration, e.g. $\frac{\partial^2 f(x,y)}{\partial x \partial y} \leq 0$, for all $u > v$. Then

$$E[D(T)] = \int_{-\infty}^{+\infty}\int_{-\infty}^{x} E[N_T(x,y)]\frac{\partial^2 f(x,y)}{\partial x \partial y}\,dy\,dx - \int_{-\infty}^{+\infty} E[N_T(x,x)]\frac{\partial f(x,y)}{\partial y}\bigg|_{y=x}\,dx. \tag{17}$$

Consequently, while the counting distribution $N_T(x,y)$ defines the total damage, the expected count

$$\mu_T(u,v) = E[N_T(u,v)] \tag{18}$$

will define the expected damage. Obviously, for stationary load

$$\mu_T(u,v) = T\mu_1(u,v).$$

In the following we shall often skip the subindex "1" and call the function $\mu(u,v)$ the count intensity.

Since $N_T(u,v)$ is bounded $D(T)$ is a.s. finite. However, we need some additional conditions to ensure the finiteness of the expected damage. (Obviously, even if $D(T)$ is finite for almost all samples of the load x, $E[D(T)]$ can be infinite.)

Similarly as in formula (12), let for $u \geq v$,

$$k_T(u,v) = \min_{v \leq x \leq u} \mu_T(x,x), \tag{19}$$

then

$$E[D(T)] \le \int_{-\infty}^{+\infty} \int_{-\infty}^{x} k_T(x,y) \frac{\partial^2 f(x,y)}{\partial x \partial y}^{-} dy\, dx - \int_{-\infty}^{+\infty} k_T(x,x) \frac{\partial f(x,y)}{\partial y}\bigg|_{y=x} dx$$
$$= D_T^+, \tag{20}$$

where $x^- = \max(0, -x)$. Note that $\partial f(x,y)/\partial y \le 0$ since f is nonincreasing in y.

In the most important special case of crossing consistent counts, e.g. rainflow-, peak-valley counts, $\mu_T(u,u) = E[N_T(u,u)]$ is equal to the expected number of u-downcrossings in $[0,T]$ of the process $x(t)$, denoted by $\mu_T(u)$, i.e.

$$\mu_T(u) = \mu_T(u,u) = E[N_T(u,v)],$$

and hence $k_T(u,v) = \min_{v \le x \le u} \mu_T(x)$. Consequently, the upper bound D_T^+ depends only on $\mu_T(u)$ and not the particular counting method. For many processes the function $\mu_T(u)$ is given by Rice's formula

$$\mu_T(u) = \int_0^T E[x'(t)^- | x(t) = u] f_{x(t)}(u)\, dt. \tag{21}$$

Further in [11] we have proved that for Gaussian loads the upper bound D_T^+ can be evaluated explicitly and coincides with the so called narrow band approximation of the expected damage for stationary Gaussian loads. We shall turn back to this in the last section.

We finish the section with three examples of the upper bound D_T^+ given by (21) for $T = 1$ and the damage function $f(x,y) = g(x-y)$, where

$$g(s) = s^{\beta}, \qquad \beta = 2, 2.5, 3, 3.5, 4, 4.5, 5. \tag{22}$$

EXAMPLE 1, (contd.) Let x be a standardized Gaussian load, i.e. variances $V[x(0)] = V[x'(0)] = 1$. Then

$$k(u,v) = \min_{v \le x \le u} \mu(x) = \frac{1}{2\pi} e^{-\frac{\max(u^2,v^2)}{2}}.$$

Using (20), with $g(s)$ given by (22), the upper bound for the damage intensity $E[D(1)]$ is given by

$$D_1^+ = \frac{1}{2\pi} 2^{3\beta/2} \Gamma(1 + \beta/2)$$

and is equal to 1.27, 2.43, 4.79, 9.74, 20.37, 43.67, 95.75, respectively. □

For non-Gaussian loads, if the joint density of the load and its derivative $x(0), x'(0)$ is known, the conservative estimate of the damage intensity can be computed numerically using (20) and (21), see the next two examples.

EXAMPLE 2, (contd.) Let $y(t)$ be a stationary, twice continuously differentiable Gaussian process with mean $v = E[y(t)] > 0$. Assume that $V[y(t)] = V[y'(t)] = 1$ and for a positive constant K define

$$x(t) = y'(t) + Ky(t)|y(t)|. \tag{23}$$

The Morison force F can always be transformed into this form by suitable choice of scales. Observe that for $K = 0$ the load x is a zero mean stationary Gaussian process.

The intensity of u-downcrossings $\mu(u)$ of x can be expressed in a form of a single integral which has to be computed numerically, namely

$$\begin{aligned}\mu(u) = \; & \tfrac{1}{2\pi}\sqrt{\tfrac{\lambda_4}{\lambda_2}(1-\alpha^2)}\int_{-\infty}^{+\infty} e^{-\frac{1}{2}(z-v)^2-\frac{1}{2}(u-Kz|z|)^2} \\ & \cdot\Psi\left(\tfrac{\alpha}{\sqrt{1-\alpha^2}}(z - v - 2K|z|u + 2K^2z^3)\right) dz, \end{aligned} \tag{24}$$

where $\Psi(z) = \phi(z) + z\Phi(z)$, ϕ and Φ being the standardized normal density and distribution functions. Here α is the regularity factor, i.e. $\alpha = \frac{\lambda_2}{\sqrt{\lambda_4\lambda_0}}$, and $\lambda_0 = V[y(t)]$, $\lambda_2 = V[y'(t)]$, $\lambda_4 = V[y''(t)]$ are the so called spectral moments, see Section 5 for definition.

Let $y(t)$ be a stationary Gaussian process with covariance function

$$r(\tau) = \frac{\sin(\sqrt{3}\tau)}{\sqrt{3}\tau},$$

and mean v. The y-process has $\lambda_0 = \lambda_2 = 1$ and $\lambda_4 = 1.8$. As before, let $g(s)$ be given by (22). Since for $K = 0$ the load $x(s) = y'(s)$ is stationary Gaussian with $V[x(0)] = 1$ and $V[x'(0)] = 1.8$ the conservative bounds for the damage intensity can be obtained by multiplying the values given in the first example by $\sqrt{1.8}$, and they are: 1.71, 3.24, 6.42, 13.08, 27.35, 58.63, 128.55, respectively. For $K = 1$, the numerical integration of (20), with $\mu(u)$ given by (24), leads to the following conservative bounds for damage intensity for $v = 0$; 6.27, 20.20, 71.96, 275.93, 1130.0, 4899.8, 22342.7, respectively, and for $v = 1$; 10.34, 34.97, 126.81, 486.10, 1963.9, 8323.7, 36828.9, respectively.

Observe, the big difference between the cases K=0 and K=1 which is a consequence of the non-linear term $y(t)|y(t)|$ in (23). □

EXAMPLE 3, (contd.) Following [12] the u-downcrossing intensity of the χ^2-process $x(s)$ is given by

$$\mu(u) = \frac{\pi^{-\frac{1}{2}}}{2^{\frac{n-1}{2}}\Gamma(\frac{n}{2})}u^{\frac{n-1}{2}}e^{-\frac{u^2}{2}}. \tag{25}$$

By computing numerically the integral in (20), the conservative bounds for the expected damage of the χ^2-loads can be easily obtained. For example, let $n = 2$ in (25) and $g(s)$ be given by (22). Then the the conservative bound D_1^+ for the expected damage is: 5.83, 15.58, 44.19, 132.21 415.5 1366.9, 4690.3, respectively. □

3. RAINFLOW COUNT – MARKOV CHAIN APPROACH

3.1. Rainflow count – smooth, ergodic loads

In this subsection we shall consider only rainflow and peak-valley counts. Further, we assume that the loads are stationary and ergodic processes with finite intensity of local maxima c_M, i.e.

$$c_M = \lim_{T\to\infty} \frac{\#\{M_i\}_{[0,T]}}{T} < \infty,$$

where $\#\{M_i\}_{[0,T]}$ is the number of local maxima in $[0,T]$. Let the rainflow count be denoted $\{M_i, m_i^{\mathrm{RFC}}\}_{[0,T]}$ and peak-valley count, (maxima and the following minima), by $\{M_i, m_i\}_{[0,T]}$, respectively. Denote by $N_T^{\mathrm{RFC}}(u,v)$, $N_T^{\mathrm{PV}}(u,v)$, the counting distributions of the rainflow and peak-valley counts, respectively, and let

$$\mu^{\mathrm{RFC}}(u,v) = \lim_{T\to\infty} \frac{N_T^{\mathrm{RFC}}(u,v)}{T}, \qquad \mu^{\mathrm{PV}}(u,v) = \lim_{T\to\infty} \frac{N_T^{\mathrm{PV}}(u,v)}{T}, \tag{26}$$

be rainflow and peak-valley intensities, respectively.

Using the ergodicity of the load we simply obtain that

$$\begin{aligned}\mu^{\mathrm{PV}}(u,v) &= \lim_{T\to\infty} \frac{\#\{M_i > u, m_i < v\}_{[0,T]}}{T} = \lim_{T\to\infty} \frac{\#\{M_i\}_{[0,T]}}{T} \frac{\#\{M_i > u, m_i < v\}_{[0,T]}}{\#\{M_i\}_{[0,T]}} \\ &= c_M \mathrm{Prob}[\, M_1 > u, m_1 < v\,] \\ &= c_M \int_u^\infty \mathrm{Prob}[\, m_1 < v \mid M_1 = x\,]\, f_{M_1}(x)\, dx \end{aligned} \tag{27}$$

where "Prob" and f_{M_1} are ergodic ("long-run") probability and density functions, respectively.

In the case of rainflow count, $\mu^{\mathrm{RFC}}(u,v)$ depends in a very complex way on the sequence of extremes (see the following lemma).

LEMMA 4. *If $x(t)$ is a stationary ergodic load with finite intensity of local maxima c_M, then*

$$\mu^{\mathrm{RFC}}(u,v) = c_M \int_u^\infty \mathit{Prob} \left[\begin{matrix} m_1 < u \text{ and } \{m_i\}_{i\geq 1} \text{ crosses} \\ v \text{ before } \{M_i\}_{i\geq 2} \text{ crosses } u \end{matrix} \,\middle|\, M_1 = x \right] f_{M_1}(x)\, dx, \tag{28}$$

or, in an alternative form,

$$\mu^{\mathrm{RFC}}(u,v) = \mu^{\mathrm{PV}}(u,v) + \kappa(u,v). \tag{29}$$

The counting intensity μ^{PV} is given by (27) and the positive correction term $\kappa(u,v)$ is given by

$$\begin{aligned}\kappa(u,v) &= c_M \int_u^\infty \mathit{Prob} \left[\begin{matrix} v \leq m_1 < u \text{ and } \{m_i\}_{i\geq 2} \text{ crosses} \\ v \text{ before } \{M_i\}_{i\geq 2} \text{ crosses } u \end{matrix} \,\middle|\, M_1 = x \right] f_{M_1}(x)\, dx \\ &= c_M \int_v^u \int_u^\infty \mathit{Prob} \left[\begin{matrix} \{m_i\}_{i\geq 2} \text{ crosses } v \text{ be-} \\ \text{fore } \{M_i\}_{i\geq 2} \text{ crosses } u \end{matrix} \,\middle|\, M_1 = x, m_1 = y \right] f_{M_1,m_1}(x,y)\, dx\, dy. \end{aligned} \tag{30}$$

PROOF. By ergodicity of x, we have

$$\mu^{\mathrm{RFC}}(u,v) = \lim_{T\to\infty} \frac{\#\{M_i > u,\ m_i^{\mathrm{RFC}} < v\}_{[0,T]}}{T}$$
$$= \lim_{T\to\infty} \frac{\#\{M_i\}_{[0,T]}}{T} \frac{1}{\#\{M_i\}_{[0,T]}} \# \left\{ \begin{array}{l} M_i > u,\ m_i < u \text{ and } \{m_{i+j}\}_{j\geq 1} \\ \text{crosses } v \text{ before } \{M_{i+j}\}_{j\geq 2} \text{ crosses} \\ u. \end{array} \right\}_{[0,T]}$$

giving the statement of the lemma. The second equality follows from Formula (15) which states that the number of rainflow cycles $(x,y)_t$ such that $x > u$ and $y < v$ is equal to the number of u-downcrossings by x followed by downcrossing of the level v. □

In the last section we discuss approximations of $\mu^{\mathrm{RFC}}(u,v)$, based on Formula (15), for continuously differentiable Gaussian loads, using barrier crossings of the so called Slepian model processes. The algorithms are complex and restricted to Gaussian loads and hence one still needs more efficient and general methods to compute the rainflow intensity $\mu^{\mathrm{RFC}}(u,v)$.

Since $\mu^{\mathrm{RFC}}(u,v)$ can be expressed using the absorption probabilities for the sequence of extremes (see the previous lemma), the Markov type models can simplify the computations.

As before, let $\{M_i, m_i\}$, $i = 1,2,\ldots$, be a sequence of pairs of local maxima and the following local minima of the load. If the sequence of extremes $M_1, m_1, M_2, m_2, \ldots$ is a Markov chain, then we shall say that the load has Markov extremes. Assuming that the transition densities of the Markov chain are known the problem is to compute the expected damage $E[D(T)]$ due to rainflow count.

3.2. Rainflow cycles in loads with Markov extremes

In this section we assume that the sequence of local extremes forms a Markov chain. Since in practical applications one measures the load in a discretized form we shall here assume that the local extremes can take only a finite number of discrete values. It is a remarkable fact that if the sequence of discrete extremes is a Markov chain the rainflow intensity μ^{RFC} is obtained by using simple matrix manipulations.

Let $\{M_k^D, m_k^D\}$, $k = 0,1,2,\ldots$, denote the sequence of (discrete) maxima and the following local minima. Let u_i, $u_{i+1} < u_i$, $i = 1,2,\ldots,n$, be the discrete levels (with u_1 as the largest), and let $\mathbf{P} = (p_{ij})$, $\widehat{\mathbf{P}} = (\hat{p}_{ij})$ be the transition probabilities from maximum to the following minimum and from minimum to the following maximum, respectively, i.e.

$$p_{ij} = P[\, m_k^D = u_j \mid M_k^D = u_i\,], \qquad \hat{p}_{ij} = P[\, M_{k+1}^D = u_j \mid m_k^D = u_i\,].$$

(Obviously the matrix $\mathbf{P}$ is right triangular and $\widehat{\mathbf{P}}$ is left triangular.) The sequence of maxima $\{M_i\}$ is also a Markov chain with transition matrix $\mathbf{P}\widehat{\mathbf{P}}$. Assume that the intensity of load maxima, c_M is given. We now turn to the computation of $\mu^{\mathrm{RFC}}(u,v)$.

Let (u_i, u_j), $u_i \geq u_j$, be fixed levels. Then for $u = u_i$ and $v = u_j$ the discrete version of Formula (28) from Lemma 4 is

$$\mu^{\mathrm{RFC}}(u,v) = c_M \sum_{u_k > u} \mathrm{Prob}\left[\begin{matrix} m_1^D < u \text{ and } \{m_i^D\}_{i\geq 1} \text{ crosses} \\ v \text{ before } \{M_i^D\}_{i\geq 2} \text{ crosses } u \end{matrix} \,\middle|\, M_1^D = u_k\right] p_{M^D}(u_k), \tag{31}$$

where

$$p_{M^D}(u_k) = \lim_{T\to\infty} \frac{\#\{M_i^D = u_k\}_{[0,T]}}{\#\{M_i^D\}_{[0,T]}}.$$

Assume that the Markov chain of maxima is ergodic. In Theorem 5 we give a formula for the rainflow intensity as a function of the transition matrices $\mathbf{P}$ and $\widehat{\mathbf{P}}$. More exactly, for fixed indices (i,j), $i > j$, define the following submatrices of $\mathbf{P}$ and $\widehat{\mathbf{P}}$:

$$\begin{array}{llll} \mathbf{A} = (p_{kl}), & i \quad \leq k \leq j-1, & i+1 \leq l \leq j, & \\ \mathbf{B} = (\hat{p}_{kl}), & i+1 \leq k \leq j, & i \quad \leq l \leq j-1, & \\ \mathbf{C} = (p_{kl}), & 1 \quad \leq k \leq i-1, & i+1 \leq l \leq j. & \end{array} \tag{32}$$

The matrix $\mathbf{A}$ contains the probabilities that maxima at levels u_i, u_{i+1}, ..., u_{j-1} are followed by minima at levels u_{i+1}, u_{i+2}, ..., u_j, $\mathbf{B}$ the probabilities that minima at u_{i+1}, u_{i+2}, ..., u_j are followed by maxima at levels u_i, u_{i+1}, ..., u_{j-1}, and $\mathbf{C}$ the probabilities that maxima at u_1, u_2, ..., u_{i-1} are followed by minima at levels u_{i+1}, u_{i+2}, ..., u_j.

Further, let for a fixed level u_j,

$$\mathbf{p} = [\, p_k \,] = \left[\sum_{l=j+1}^{n} p_{kl}\right], \tag{33}$$

e.g. the vector $\mathbf{p}$ contains the conditional probabilities that a maximum $M_1^D = u_k$ is followed by a minimum $m_1^D < u_j$.

THEOREM 5. *Under the general assumptions of this section, let $\boldsymbol{\pi} = [\,\pi_i\,]$ be a stationary distribution of a Markov chain of maxima, e.g. the unique solution to the equation system*

$$\begin{cases} \boldsymbol{\pi} = \boldsymbol{\pi}\mathbf{P}\widehat{\mathbf{P}} \\ \sum_i \pi_i = 1. \end{cases}$$

For fixed values $(u,v) = (u_i, u_j)$, the rainflow intensity is given by

$$\mu^{\mathrm{RFC}}(u,v) = c_M \tilde{\boldsymbol{\pi}} \left(\mathbf{d} + \sum_{r=1}^{\infty} \mathbf{CB}(\mathbf{AB})^{r-1}\mathbf{e}\right) = c_M \tilde{\boldsymbol{\pi}} \left(\mathbf{d} + \mathbf{CB}(\mathbf{I} - \mathbf{AB})^{-1}\mathbf{e}\right), \tag{34}$$

where $\tilde{\boldsymbol{\pi}} = [\,\pi_1 \;\pi_2\; \dots\; \pi_{i-1}\,]$, $\mathbf{d}^T = [\,p_1\; p_2\; \cdots\; p_{i-1}\,]$, $\mathbf{e}^T = [\,p_i\; p_{i+1}\; \cdots\; p_{j-1}\,]$. The submatrices $\mathbf{A}$, $\mathbf{B}$ and $\mathbf{C}$ are defined by (32), respectively.

PROOF. The conditional probability in (31) can be written as a sum of the following probabilities,

$$p_0(u_k) = P[\, m_1^D < u_j \mid M_1^D = u_k\,],$$
$$p_r(u_k) = P[\, m_{r+1}^D < u_j,\ u_j \le m_{l-1}^D < M_l^D < u_i \text{ for all } l = 2,3,\dots,r+1 \mid M_1^D = u_k\,],$$

i.e. given a start in a local maximum at u_k, r indicates the number of local maxima between the barriers $u_j < u_i$ before crossing u_j. Consequently the rainflow intensity $\mu^{\mathrm{RFC}}(u_i, u_j)$ writes

$$\mu^{\mathrm{RFC}}(u_i, u_j) = c_M \sum_{l=1}^{i-1} \pi_l \Big(\sum_{r=0}^{\infty} p_r(u_l) \Big).$$

Since $[p_0(u_k)]_{k=1}^{i-1} = \mathbf{d}$, $[p_1(u_k)]_{k=1}^{i-1} = \mathbf{CBe}$ and $[p_r(u_k)]_{k=1}^{i-1} = \mathbf{CB}(\mathbf{AB})^{r-1}\mathbf{e}$, $r > 1$, the theorem simply follows. □

Observe that by Lemma 4 the intensity of the peak-valley count is simply

$$\mu^{\mathrm{PV}}(u_i, u_j) = c_M \tilde{\pi} \mathbf{d}. \tag{35}$$

The discrete version of formula (31) for the correction term $\kappa(u,v)$, $(u,v) = (u_i, u_j)$, is

$$\begin{aligned}
&\kappa(u,v)\\
&= c_M \sum_{l=i+1}^{j} \sum_{k=1}^{i-1} \mathrm{Prob}\left[\begin{matrix} \{m_i\}_{i\ge 2} \text{ crosses } v \text{ before} \\ \{M_i\}_{i\ge 2} \text{ crosses } u \end{matrix} \,\middle|\, M_1 = u_k, m_1 = u_l \right] P[\, M_1 = u_k,\, m_1 = u_l\,] \\
&= \sum_{l=i+1}^{j} \underbrace{\mathrm{Prob}\left[\begin{matrix} \{m_i\}_{i\ge 2} \text{ crosses } v \text{ before} \\ \{M_i\}_{i\ge 2} \text{ crosses } u \end{matrix} \,\middle|\, m_1 = u_l \right]}_{[\mathbf{B}(\mathbf{I}-\mathbf{AB})^{-1}\mathbf{e}]_l} \underbrace{c_M \sum_{k=1}^{i-1} P[\, M_1 = u_k,\, m_1 = u_l\,]}_{\mu^{\mathrm{PV}}(u_i, u_{l-1}) - \mu^{\mathrm{PV}}(u_i, u_l)} \\
&= \mathbf{qB}(\mathbf{I}-\mathbf{AB})^{-1}\mathbf{e}.
\end{aligned}$$

where, for fixed (i,j),

$$\mathbf{q} = [\, \mu^{\mathrm{PV}}(u_i, u_{l-1}) - \mu^{\mathrm{PV}}(u_i, u_l)\,], \qquad l = i+1, i+2, \dots, j. \tag{36}$$

Consequently, if μ^{PV} is given the algorithm for μ^{RFC} simplifies and is given in the following theorem.

THEOREM 6. *Under the general assumptions of this section, if the peak-valley intensity $\mu^{\mathrm{PV}}(u,v)$ is known then for for fixed values $(u,v) = (u_i, u_j)$, the rainflow intensity is given by*

$$\mu^{\mathrm{RFC}}(u,v) = \mu^{\mathrm{PV}}(u,v) + \mathbf{qB}(\mathbf{I}-\mathbf{AB})^{-1}\mathbf{e}. \tag{37}$$

where $\mathbf{A}$, $\mathbf{B}$, $\mathbf{e}$ are defined as in Theorem 5 and $\mathbf{q}$ is given by (36).

Theorem 6 will be used in Section 5 to construct an approximation to μ^{RFC} for Gaussian loads.

3.3. The Markov chain approximation

Let $\{M_i, m_i\}$ denote the sequence of maxima and the following minimum of the load $x(t)$. The sequence is approximated by a discrete sequence of extremes $\{M_i^D, m_i^D\}$. Let $\mathbf{P}$ and $\widehat{\mathbf{P}}$ be the transition probabilities for the discrete sequence $\{M_i^D, m_i^D\}$ and $\pi = [\,\pi_i\,]$ the stationary distribution of the Markov chain of maxima generated by the transition matrix $\mathbf{P}\widehat{\mathbf{P}}$.

Using the algorithm presented in the previous section we obtain $\mu^{\mathrm{RFC}}_{\mathrm{Markov}}(u,v)$ for some given intensity c_M of the maxima. This intensity can be difficult to obtain and hence in our computer programs we assume that $c_M = 1$. Consequently, if a discrete sequence of extremes $\{M_i^D, m_i^D\}$ is a Markov chain, then

$$\mu^{\mathrm{RFC}}(u,v) \propto \mu^{\mathrm{RFC}}_{\mathrm{Markov}}(u,v),$$

where $\mu^{\mathrm{RFC}}_{\mathrm{Markov}}(u,v)$ is obtained from (34) with $c_M = 1$. The proportionality constant is then found from the crossing consistency of rainflow count, i.e. that $\mu^{\mathrm{RFC}}(u,u) = \mu(u)$, (the downcrossing intensity of the load). Consequently, before computing the Markov approximation of the expected damage $E[D(1)]$, the counting intensity $\mu^{\mathrm{RFC}}_{\mathrm{Markov}}(u,v)$ is renormalized in the following way:

$$\tilde{\mu}^{\mathrm{RFC}}_{\mathrm{Markov}}(u,v) = \frac{\mu(0)}{\mu^{\mathrm{RFC}}_{\mathrm{Markov}}(0,0)}\mu^{\mathrm{RFC}}_{\mathrm{Markov}}(u,v). \tag{38}$$

The zero-crossing intensity can be obtained from the Rice formula for the expected number of downcrossings (21). Note that the choice of the point $(u,u) = (0,0)$ for normalization is optimal for zero mean Gaussian loads, since the intensity of u-downcrossings has a maximum at zero. For non-Gaussian loads we shall use other normalization points.

In the case when a discrete sequence of extremes $\{M_i^D, m_i^D\}$ is only approximatively a Markov chain we shall still use $\tilde{\mu}^{\mathrm{RFC}}_{\mathrm{Markov}}$ as an approximation to μ^{RFC}.

3.4. Examples

In previous sections we have presented the Markov method to approximate the counting intensity for the rainflow count μ^{RFC}, see Theorem 5 and formula (38). The approximations can be used to approximate the damage intensity $E[D(T)]/T$ $(= E[D(1)])$ by replacing $E[N_T(u,v)]$ $(= \mu_T(u,v))$ in formula (17) by $\tilde{\mu}^{\mathrm{RFC}}_{\mathrm{Markov}}$.

In this section we shall use the Markov approximation to compute μ^{RFC} and $E[D(1)]$ for many different load processes, such as Gaussian loads, χ^2-loads, Morison forces. The

Markov approximation for Gaussian loads will be further discussed in Section 5, where some additional examples are given.

The examples are computed in the following way: First we simulate a long sample path of a random load. The rainflow and peak/valley counts are found for that sample path. The rainflow count is used to compute an estimate of the rainflow intensity

$$\mu^{\mathrm{RFC}}(u,v) \approx N_T^{\mathrm{RFC}}(u,v)/T = \hat{N}(u,v), \tag{39}$$

say. The damage intensity $E[D(1)]$ is estimated by using formulas (17) and (39), i.e.

$$\begin{aligned} E[\,D(1)\,] &\approx -\iint_{u\geq v} \hat{N}(u,v)\frac{\partial^2 f(u,v)}{\partial u \partial v}\,du\,dv - \int_{-\infty}^{+\infty} \hat{N}(u,u)\frac{\partial f(u,v)}{\partial v}\bigg|_{u=v} du \\ &= \frac{D(T)}{T}. \end{aligned} \tag{40}$$

The accuracy of the numerical integration in (40) is checked by comparing the results with an alternative way to compute $\frac{D(T)}{T}$, i.e. using formula (8) to calculate $D(T)$.

The Markov transition matrices $\mathbf{P}$, $\widehat{\mathbf{P}}$ are estimated from the sample path and $\tilde{\mu}^{\mathrm{RFC}}_{\mathrm{Markov}}$ computed. For simplicity of notation we introduce

$$\hat{\mu}(u,v) = \tilde{\mu}^{\mathrm{RFC}}_{\mathrm{Markov}}(u,v). \tag{41}$$

The first check for correctness of the Markov approach is to plot the downcrossing intensity of the Markov chain of extremes, i.e. $\hat{\mu}(u,u)$ against the theoretical downcrossing intensity obtained from Rice's formula in the case of Gaussian and χ^2-loads. Next the counting intensities $\hat{N}(u,v)$ and $\hat{\mu}(u,v)$ are compared. Finally we give a plot of the estimated damage intensity, $E[D(1)]$, given by (40) and the computed expected damage intensity, $E_{\mathrm{Markov}}[D(1)]$ using Markov method, i.e.

$$E_{\mathrm{Markov}}[D(1)] = -\iint_{u\geq v} \hat{\mu}(u,v)\frac{\partial^2 f(u,v)}{\partial u \partial v}\,du\,dv - \int_{-\infty}^{+\infty} \hat{\mu}(u,u)\frac{\partial f(u,v)}{\partial v}\bigg|_{u=v} du.$$

We will throughout this section assume that the damage function is $f(u,v) = (u-v)^\beta$, $u \geq v$, $1 \leq \beta \leq 5$.

Example 4, (contd.) We choose $(\zeta,\omega) = (0.01, 1)$ and $(\zeta,\omega) = (0.30, 1)$ and simulate two loads with 150,000 points each ($\Delta t = 0.02$). Figure 6 shows a part of the simulated load. Observe that even the narrow band load, $\zeta = 0.01$, has an infinite numbers of extremes. This can be seen from Figure 7 which shows the rainflow count. The dots on the diagonal $u = v$ represent the infinite set of cycles with very small amplitudes. However, for any $\epsilon > 0$, the number of cycles such that $M_i - m_i^{\mathrm{RFC}} > \epsilon$ is finite. In Figure 7 we compare the contour plots of $\hat{N}(u,v)$ and $\hat{\mu}(u,v)$ which are almost identical.

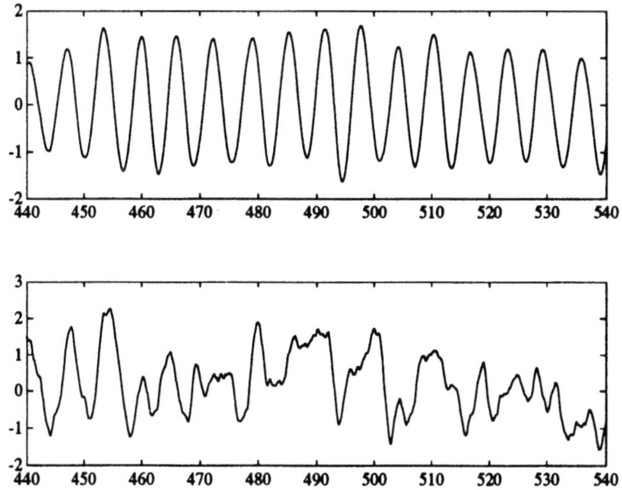

Figure 6: Filtered Gaussian white noise process; Load processes; upper part: $\zeta = 0.01$, lower part: $\zeta = 0.30$.

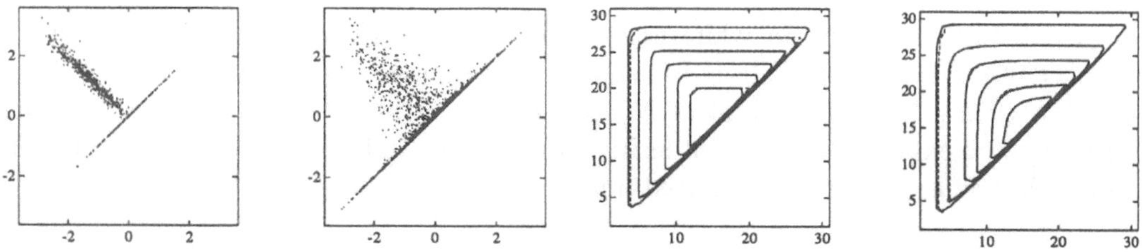

Figure 7: Filtered Gaussian white noise process; Rainflow count for a Gaussian process with $\zeta = 0.01$, 744 rainflow cycles (a) and $\zeta = 0.30$, 3,284 rainflow cycles (b). Contour plot of $\hat{N}(u,v)$ (dashed line) and $\hat{\mu}(u,v)$ (solid line), $\zeta = 0.01$ (c) and $\zeta = 0.30$ (d).

The estimated and computed damage intensity, obtained from $\hat{N}(u,v)$, $\hat{\mu}(u,v)$, respectively, are compared in Figure 8.

Finally, we add a deterministic mean, $y(t) = 3\sin(t/200)$, to the filtered Gaussian white noise process ($\zeta = 0.30$). As can be seen in Figure 9, the Markov approximation is not good. The computed expected damage intensity using Markov approximation (dotted line) is much larger than the estimated expected damage intensity (dashed line). □

EXAMPLE 3, (contd.) (χ^2-load.) Let $x_1(t)$ and $x_2(t)$ be two independent samples of the process $x(t)$ defined by (4) with $\zeta = 0.30$, $\omega = 1$. Then $x(t) = x_1^2(t) + x_2^2(t)$ is a sample path of a χ^2-process with $n = 2$. A part of the simulated path is shown in Figure 10.

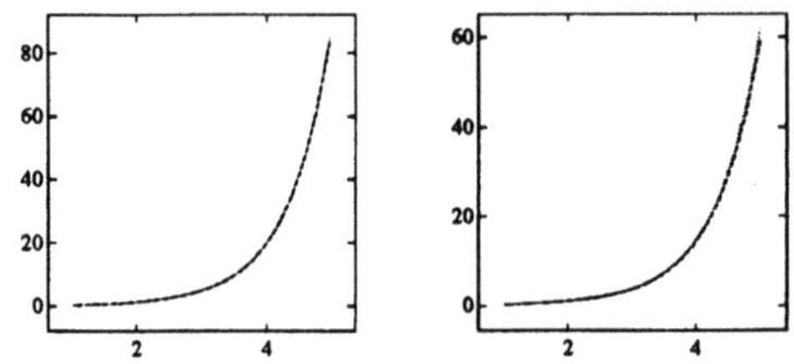

Figure 8: Filtered Gaussian white noise process; Estimated expected damage (dashed line) and computed damage using Markov approximation (dotted line), $\zeta = 0.01$ (a) and $\zeta = 0.30$ (b).

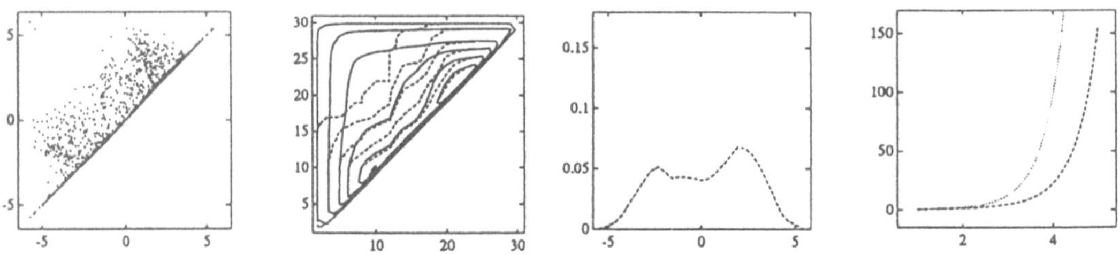

Figure 9: Filtered Gaussian white noise process; a) Rainflow count for $y(t) = x(t) + 3\sin(t/200)$, where $x(t)$ is filtered Gaussian white noise process with $\zeta = 0.30$, 3,277 rainflow cycles. b) Contour plot of $\hat{N}(u,v)$. c) The estimated downcrossing intensity. d) Estimated expected damage intensity (dashed line) and computed damage intensity using Markov approximation (dotted line).

Since $x(t)$ always is positive the contour plot of $\hat{N}(u,v)$ and $\hat{\mu}(u,v)$ are defined only for $u > v \geq 0$ (see Figure 11). The results presented in Figure 11 show the applicability of the Markov method for the load. □

EXAMPLE 2, (contd.) (Morison force when $y(t)$ is a Gaussian process with truncated JONSWAP spectrum.) Let $x(t)$ be a stationary, twice continuously differentiable Gaussian process with truncated JONSWAP spectrum described earlier and let $E[y(t)] = 0$,

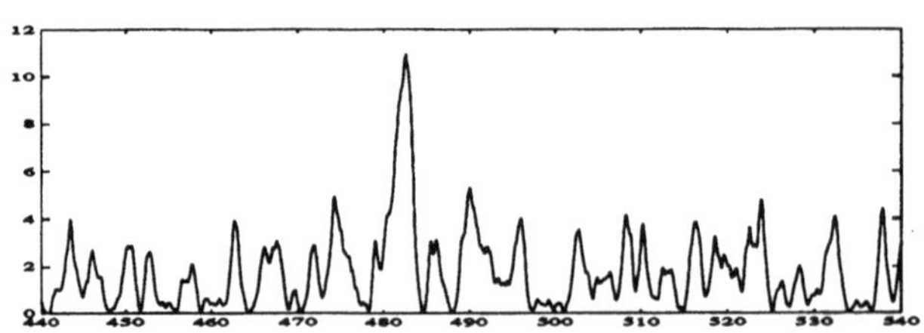

Figure 10: A part of the sample path of a χ^2-process.

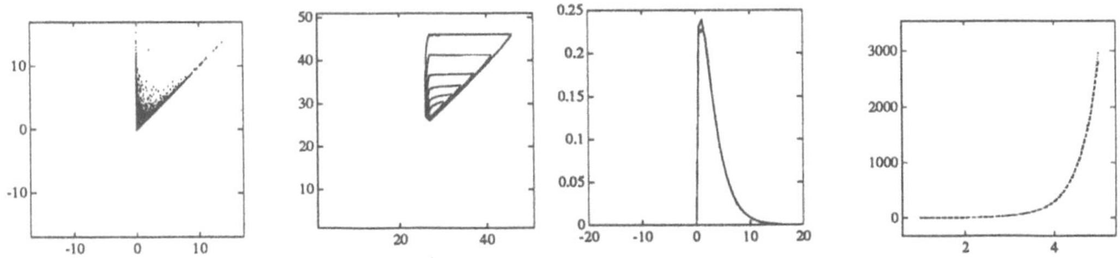

Figure 11: χ^2-load process; a) 5,028 rainflow cycles. b) Contour plots of $\hat{N}(u,v)$ (dashed line) and $\hat{\mu}(u,v)$ (solid line). c) Theoretical expected downcrossing intensity (solid line) and $\hat{\mu}(u,u)$ (dashed line). d) Estimated expected damage (dashed line) and computed damage using Markov approximation (dotted line).

$\sigma_y^2 = \sigma_{y'}^2 = 1$. For $K_1 = K_2 = 1$, define the Morison force by

$$x(t) = y'(t) + y(t)\,|y(t)|.$$

The results presented in Figure 12 show the applicability of the Markov method for the force. □

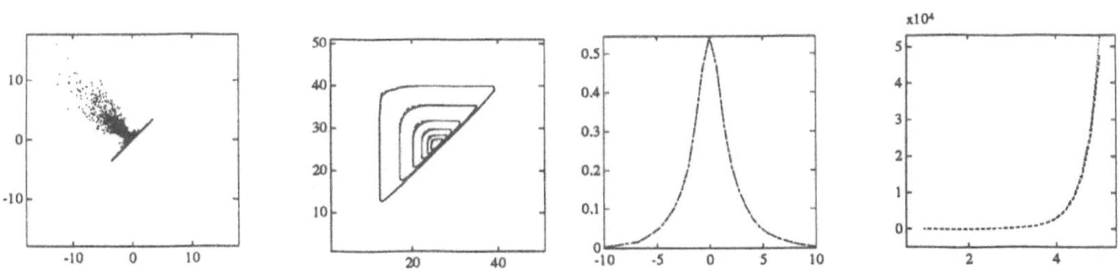

Figure 12: Morison force when $y(t)$ is a Gaussian process with truncated JONSWAP spectrum; a) 2,054 rainflow cycles. b) Contour plots of $\hat{N}(u,v)$ (dashed line) and $\hat{\mu}(u,v)$ (solid line). c) Estimated expected downcrossing intensity (dashed line) and $\hat{\mu}(u,u)$ (dotted line). d) Estimated expected damage (dashed line) and computed damage using Markov approximation (dotted line).

EXAMPLE 2, (contd.) Morison force when $y(t)$ is a Gaussian process with rectangular spectrum. Let $y(t)$ be a stationary, twice continuously differentiable Gaussian process with rectangular spectrum

$$S(\lambda) = \begin{cases} 1/2\sqrt{3}, & \lambda \in [-\sqrt{3}, \sqrt{3}], \\ 0, & \text{otherwise}. \end{cases}$$

For $K_1 = K_2 = 1$, define the Morison force as before by $x(t) = y'(t) + y(t)\,|y(t)|$. The results presented in Figure 13 show the applicability of the Markov method for the force. □

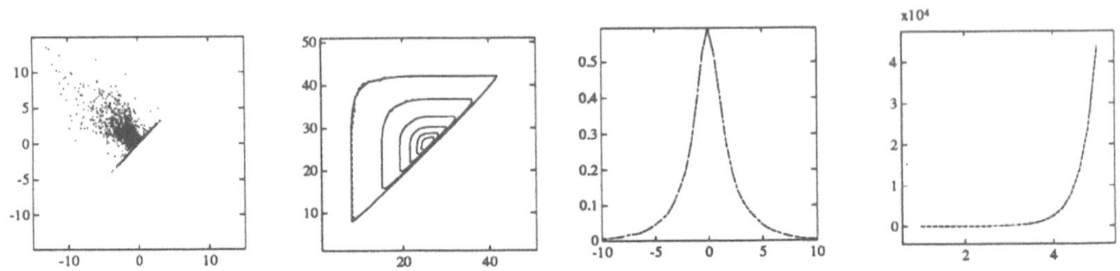

Figure 13: Morison force when $y(t)$ is a Gaussian process with rectangular spectrum; a) 1,958 rainflow cycles. b) Contour plots of $\hat{N}(u,v)$ (dashed line) and $\hat{\mu}(u,v)$ (solid line). c) Estimated expected downcrossing intensity (dashed line) and $\hat{\mu}(u,u)$ (dotted line). d) Estimated expected damage (dashed line) and computed damage using Markov approximation (dotted line).

4. MARKED CROSSINGS AND SLEPIAN MODEL PROCESSES

4.1. Counting distribution of marked crossings

In previous sections we have used the counting distribution $N_T(u,v)$ and its expectation $\mu_T(u,v)$ to compute the total damage $D(T)$ and the expected damage $E[D(T)]$ due to the cycle count $\{(x,y)_t\}$, respectively. For crossing consistent counts, e.g. rainflow and peak-valley, the counting distribution can be related to the properties of $x(t)$ after its downcrossings of a fixed level. For example, $N_T(u,u)$ is equal to the number of the downcrossings of the level u in the interval $[0,T]$ and in the case of rainflow count, for any $v < u$,

$$N_T(u,v) = \#\{t \in [0,T];\ t \text{ is a } u\text{-downcrossing of } x \text{ and } h^+(t) < v\},$$

see Lemma 3 in Section 2. These are examples of a general problem of counting the number of crossings of a fixed level by a process x satisfying some additional conditions, so called *marked crossings.*

The concept of marked crossings can be set up in a very general way and has found several engineering applications. Here we shall discuss only some aspects of the theory and defined Slepian model processes, first introduced in [13] and [14].

DEFINITION 7. *Let $x(t)$, $0 \le t \le T$, be a continuously differentiable random process with finite number of downcrossings of the level u and let t_n be the times of u-downcrossings of x. Consider, in addition to x, a sufficiently smooth, vector valued process $y(t) = (y_1(t), \ldots, y_n(t))$, and let H be a vector valued function defined on the sample pathes of y. Then*

$$\{(t_n, H_n)\} = \{(t_n, H(y(t_n + \cdot)))\}$$

is called a sequence of marked crossings. Define a counting distribution of marks H_n *by*

$$N_H(h) = \#\{t_n \in [0,T];\ H_n < h\}.$$

In this subsection we shall discuss the problem how to compute the expectation of N_H, i.e.

$$\mu_H(h) = E[\#\{t_n \in [0,T];\ H_n < h\}].$$

We begin with two examples of marked crossings sequences.

EXAMPLE 6. (Rainflow count.) Rainflow count with a finite number of cycles can be described using the concepts of marked crossings. Assume we are counting rainflow cycles in a process $y(t)$. Since the times t_n when cycles are counted are the times of local maxima of y, then, with $x(t) = y'(t)$, t_n is a sequence of zero-downcrossings of x. By Definition 1 the n'th rainflow cycle is given by $H_n = (y(t_n), \max(h^-(t_n), h^+(t_n)))$ and is a two dimensional mark attached to t_n. □

There are many other marked crossing sequences used in applications and we give only one example taken from oceanography.

EXAMPLE 7. (Wavelength and amplitude.) Let $\{(x,y)_{t_n}\}$ be a peak-valley count. Define wave-length and amplitude of a wave as $T_n = t_{n+1} - t_n$ and $H_n = x - y$, respectively. Obviously wave-length and amplitude T_n, H_n are the marks attached to the time t_n of the n'th wave. The question raised in oceanography is how to compute the distribution of the wave-length and amplitude of a typical wave, defined as

$$F_{T,H}(t,h) = \lim_{S\to\infty} \frac{\#\{t_n \in [0,S];\ T_n < t, H_n < h\}}{\#\{t_n \in [0,S]\}} = \frac{E[\#\{t_n \in [0,1];\ T_n < t, H_n < h\}]}{E[\#\{t_n \in [0,1]\}]},$$

for ergodic waves. (Note, $E[\#\{t_n \in [0,1]\}]$ is the intensity of waves.) □

We turn now to the explicit formula for $\mu_H(h)$. For a fixed u and T,

$$\mu_H(h) = E[\#\{t_n \in [0,T];\ H_n < h\}] = \int_0^T E[x'(t)^- 1_{\{H(y(t+\cdot))<h\}} \mid x(t) = u] f_{x(t)}(u)\, dt, \tag{42}$$

where $x^- = \max(0,-x)$ and $1_{\{\cdot\}}$ is an indicator function of the set $\{\cdot\}$, see [15] for details. (Note that μ_H depends on T, u; however for notational convenience we do not write the dependence explicitly.) For $h = +\infty$ the indicator is always equal to one and hence

$$E[\#\{t_n \in [0,T]\}] = \int_0^T E[x'(t)^- \mid x(t) = u] f_{x(t)}(u)\, dt, \tag{43}$$

which is a well known Rice's formula for the expected number of u-downcrossings by x.

4.2. Intensity distribution of marks

Let x, y be jointly stationary processes and let $\{(t_n, H_n)\}$ be a sequence of marked crossings with finite intensity of marks, i.e. $E[\#\{t_n \in [0,1]\}] < +\infty$, where t_n are downcrossings of a fixed level u by x, see Definition 7.

Since x, y are jointly stationary, the Formulas (42) and (43) simplify to

$$\frac{\mu_H(h)}{T} = E[x'(0)^- 1_{\{H(y)<h\}} \mid x(0)=u] f_{x(0)}(u), \tag{44}$$

$$\frac{E[\#\{t_n \in [0,T]\}]}{T} = E[x'(0)^- \mid x(0)=u] f_{x(0)}(u),$$

Now $\mu_H(h)/T$ and $E[\#\{t_n \in [0,T]\}]/T$ are called the intensity measure of mark H, and intensity of marks, respectively. Further, the intensity distribution of the mark H is defined as the normalized (to one) intensity measure of H, i.e.

$$F_H^{\text{int}}(h) = \frac{E[x'(0)^- 1_{\{H(y)<h\}} \mid x(0)=u] f_{x(0)}(u)}{E[x'(0)^- \mid x(0)=u] f_{x(0)}(u)} = \frac{E[x'(0)^- 1_{\{H(y)<h\}} \mid x(0)=u]}{E[x'(0)^- \mid x(0)=u]}. \tag{45}$$

Obviously, the distribution of wavelength and amplitude of typical waves, discussed in Example 7, is an intensity distribution of a mark T, H.

REMARK. Using (44) we obtain an explicit formula for the functions $\mu_T^{\text{RFC}}(u,v)$, $\mu_T^{\text{PV}}(u,v)$, in the case of stationary load x, i.e.

$$\mu_T^{\text{RFC}}(u,v) = T \cdot E[x'(0)^- 1_{\{x \text{ crosses } v \text{ before } u\}} \mid x(0)=u] f_{x(0)}(u), \tag{46}$$

and

$$\mu_T^{\text{PV}}(u,v) = T \cdot E[x'(0)^- 1_{\{x'(s) < 0 \text{ for all } 0<s<S(x)\}} \mid x(0)=u] f_{x(0)}(u), \tag{47}$$

where $S(x)$ is the first passage time of v by x, i.e. $S(x) = \inf\{s>0; x(s)=v\}$. □

We turn now to the problem of computation of the intensity measure, or equivalently the intensity distribution of a mark H, see (44), (45), respectively.

Obviously, by (44), if the joint density of the random vector $x(0), x'(0), H(y)$ is known then

$$E[x'(0)^- 1_{\{H(y)<h\}} | x(0)=u] f_{x(0)}(u) = \int_{-\infty}^{0} \int_{-\infty}^{h} |z| f_{x(0),x'(0),H}(u,z,r)\, dr\, dz \tag{48}$$

and

$$E[x'(0)^- \mid x(0)=u] f_{x(0)}(u) = \int_{-\infty}^{0} |z| f_{x(0),x'(0)}(u,z)\, dz. \tag{49}$$

Furthermore, the density of the intensity distribution of H is given by

$$f_H(h) = \frac{\int_{-\infty}^{0} |z| f_{x(0),x'(0),H}(u,z,h)\, dz}{\int_{-\infty}^{0} |z| f_{x(0),x'(0)}(u,z)\, dz}$$

$$= \frac{E[x'(0)^- \mid x(0)=u, H(y)=h] f_{x(0),H(y)}(u,h)}{E[x'(0)^- \mid x(0)=u] f_{x(0)}(u)}. \tag{50}$$

Thus the intensity distribution of H is simply obtained from (48)–(50). However, usually, the densities of $x(0), x'(0), H(y)$ as well as the integrals in (48)–(50), have to be computed numerically.

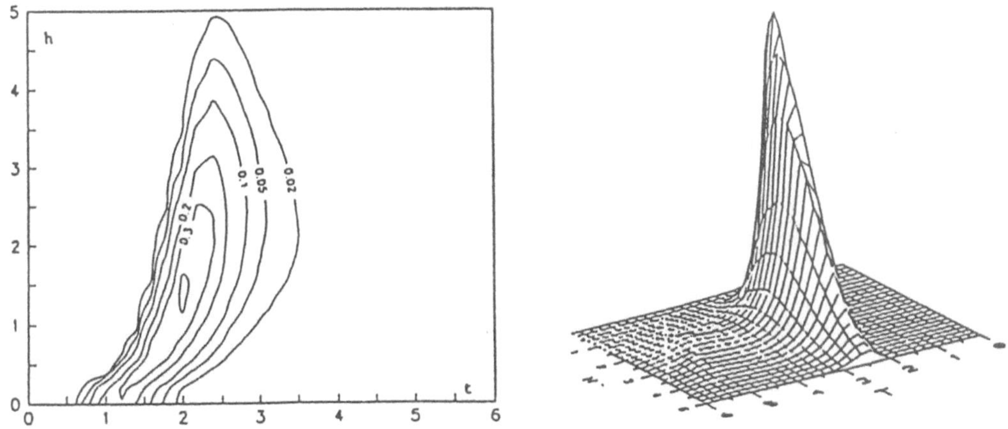

Figure 14: Contour plot of the density $f_{T,H}(t,h)$ of wavelength and amplitude of a typical wave for a zeromean Gaussian process with covariance function $r(\tau) = \sin(\sqrt{3}\,\tau)/(\sqrt{3}\,\tau)$.

EXAMPLE 7, (contd.) In Figure 14, we give isolines of the density $f_{T,H}(t,h)$ of wavelength and amplitude of a typical wave for a zeromean Gaussian process with covariance function $r(\tau) = \sin(\sqrt{3}\,\tau)/(\sqrt{3}\,\tau)$. □

EXAMPLE 8. (Derivative at the downcrossing.) For a stationary and continuously differentiable process x define a sequence of marked crossings $\{(t_n, x'(t_n))\}$, where t_n are the times for u-downcrossing of x, i.e. $H_n = x'(t_n)$. The density of a typical derivative at a downcrossing can be obtain from (50), i.e.

$$\begin{aligned} f_H(h) &= \frac{E[x'(0)^- \mid x(0)=u, x'(0)=h] f_{x(0),x'(0)}(u,h)}{E[x'(0)^- \mid x(0)=u] f_{x(0)}(u)}, \\ &= c \cdot |h| f_{x(0),x'(0)}(u,h), \qquad h<0, \end{aligned} \tag{51}$$

where c is a normalization constant. In the following we shall always denote by R a random variable with the density $c \cdot |h| f_{x(0),x'(0)}(u,h)$, $h<0$, with an interpretation as a typical derivative at the downcrossing of the level u by the process x. The reason for choosing the symbol "R" is that for Gaussian processes a typical derivative is minus a

Rayleigh variable. More precisely, it is an important property of stationary Gaussian processes that $x(0), x'(0)$ are independent Gaussian variables ($E[x'(0)] = 0$) and hence

$$f_R(r) = c|r|e^{-\frac{r^2}{2\sigma^2}} f_{x(0)}(u) = \frac{|r|}{\sigma^2} e^{-\frac{r^2}{2\sigma^2}}, \qquad r < 0, \tag{52}$$

where $\sigma^2 = V[x'(0)]$. □

EXAMPLE 9. (The height of a typical maximum.) Consider a twice continuously differentiable stationary process y. Define $x(t) = y'(t)$ and assume that the intensity of zero-downcrossings of $x(t)$ is finite. Using notation of Definition 7, $\{(t_n, y(t_n))\}$ is a marked sequence of local maxima of y. Consequently, with max $= H$, the density of the height of a typical maximum is given by

$$\begin{aligned} f_{\max}(z) &= \frac{E[x'(0)^- \mid x(0) = 0, y(0) = z] f_{x(0),y(0)}(0, z)}{E[x'(0)^- \mid x(0) = 0] f_{x(0)}(0)} \\ &= c \cdot \int_{-\infty}^{0} |r| f_{y''(0)y'(0),y(0)}(r, 0, z)\, dr. \end{aligned} \tag{53}$$

□

In the case of a typical derivative at a downcrossing we have been able to give an explicit formula for its density. However, the density of the height of a typical maximum is already a single integral. For more complicated marks $H(y)$ the joint density of $x(0), x'(0), H(y)$ can be difficult to obtain, even if the joint densities of $x(0), x'(0), y(\cdot)$ are known explicitly, e.g. x, y are jointly Gaussian, and has to be computed approximatively. One general method is the so called regression approximation which approximates the sample paths of the y process by a conditional expectation $y_n(\cdot) = E[y(\cdot)|Y_1, Y_2, \ldots, Y_n]$, where $Y_1, \ldots, Y_n$, are suitably chosen random variables. Then the density of $x(0), x'(0), H(y)$ is approximated by the density of $x(0), x'(0), H(y_n)$. The approximation of the intensity distribution of H is an $n+1$ dimensional integral. However for many types of marks H small values of n gives very good approximation of the density of H, see [16], for more detailed discussion.

4.3. Slepian model processes

In previous subsections a mark H was defined as any vector valued function defined on the process y. Here we shall allow a mark to be a function, e.g. $H_n = y(t_n + \cdot)$. The intensity distribution of H describes a typical variation of the process y around any downcrossing of the level u by x. The process y_u, which distribution is equal to the intensity distribution of marks $y(t_n + \cdot)$, is called a Slepian model process for y around downcrossing of the level u by x.

More precisely, assume that for any $k > 0$, the joint densities of $x(0), x'(0)$, $y(s_1)$, $\dots$, $y(s_k)$ can be computed, where s_i are fixed time points. Then one can define all finite dimensional conditional distributions of the process $y(\cdot)$ given $x(0) = u$, $x'(0) = r$. Let us denote the process with these distributions by $y_{u,r}(\cdot)$. Consider a k-dimensional mark $H(y) = (y(s_1), \dots, y(s_k))$. Then the intensity distribution of H is equal to the distribution of the vector $(y_{u,R}(s_1), y_{u,R}(s_2), \dots, y_{u,R}(s_n))$, where R is a "typical" derivative at the u-downcrossing of x, defined in Example 8, independent of the process $y_{u,r}(\cdot)$. The density of R is given by (51). Consequently the process $y_{u,R}(\cdot)$ has the intensity interpretation as a typical behavior of the process y after downcrossing of the level u by the process x. The process

$$y_u(s) = y_{u,R}(s) \tag{54}$$

is called a *Slepian model process* for y around the downcrossing of u by x.

In the following remarks we give some properties of Slepian models. For a detailed discussion of the properties and applications of Slepian models, see [16].

REMARK. Consider any marked crossing sequence $\{(t_n, H_n)\}$, where $H_n = H(y(t_n + \cdot))$. Using the concept of Slepian models the intensity distribution of a mark H is equal to the distribution of $H(y_u)$. □

REMARK. The distribution of the Slepian model process y_u is equal to the limit of a conditional distribution of y given that the process x passes a horizontal window at time 0 as the window width goes to zero. □

REMARK. Assume that x, y are jointly ergodic processes. For any sequence of marked crossings $\{(t_n, H_n)\}$, where $H_n = H(y(t_n + \cdot))$, the empirical distribution of H defined as

$$F_H(h) = \lim_{S\to\infty} \frac{\#\{t_n \in [0,S];\ H_n < h\}}{\#\{t_n \in [0,S]\}} = \frac{E[\#\{t_n \in [0,1];\ H_n < h\}]}{E[\#\{t_n \in [0,1]\}]} = F_H^{\text{int}}(h).$$

Hence the distribution of $H(y_u)$ is equal to the empirical distribution of H. □

4.4. Slepian model process for Gaussian processes

In the following theorem we give explicit formulas for Slepian model processes y_u under Gaussian assumption for the x, y processes.

THEOREM 7. *Let x, y be stationary, zeromean and jointly Gaussian processes. Denote by $r_y(s-t) = C[y(t), y(s)]$, $r_x(s-t) = C[x(t), x(s)]$ and $r_{xy}(s-t) = C[x(t), y(s)]$. For any fixed level u the Slepian model process y_u for y around downcrossing of level u by x is given by*

$$y_u(s) = ua(s) + Rb(s) + \Delta(s), \tag{55}$$

where R is a Rayleigh distributed random variable, with $f_R(r) = re^{-r^2/2}$, independent of a zeromean Gaussian process Δ with covariance function

$$C[\Delta(t), \Delta(s)] = r_y(s-t) - \frac{r_{xy}(t)r_{xy}(s)}{r_x(0)} - \frac{r'_{xy}(t)r'_{xy}(s)}{-r''_x(0)}. \tag{56}$$

(Note, $-r''_x(0) = V[x'(0)]$.) The deterministic functions $a(s)$, $b(s)$ are given by

$$a(s) = \frac{r_{xy}(s)}{r_x(0)}, \qquad b(s) = \frac{r'_{xy}(s)}{\sqrt{-r''_x(0)}}. \tag{57}$$

The theorem is a simple consequence of the following well known result for Gaussian vectors. The proof of the theorem is left to the reader.

LEMMA 8. *Let $(X, \mathbf{Y}) = (X, Y_1, \ldots, Y_n)$ be a Gaussian vector and denote the cross covariance vector $C[\mathbf{Y}, X] = (C[Y_1, X], \ldots, C[Y_n, X])$ and the mean $E[\mathbf{Y}] = (E[Y_1], \ldots, E[Y_n])$. Then*

$$E[\mathbf{Y} \mid X = x] = E[\mathbf{Y}] + \frac{C[\mathbf{Y}, X]}{V[X]}(x - E[X]),$$
$$C[\mathbf{Y}, \mathbf{Y} \mid \mathbf{X} = \mathbf{x}] = C[\mathbf{Y}, \mathbf{Y}] - \frac{C[\mathbf{Y}, X]\, C[\mathbf{Y}, X]^T}{V[X]}. \tag{58}$$

Furthermore the conditional distribution of $\mathbf{Y} \mid X$ is Gaussian with mean $E[\mathbf{Y} \mid X = x]$ and covariance matrix $C[\mathbf{Y}, \mathbf{Y} \mid X = x]$.

In the following examples we give Slepian models which we shall use in the next section.

EXAMPLE 10. (Slepian model for y around its downcrossings of level u.) Assume y is a zeromean continuously differentiable Gaussian process with covariance function $C[y(t), y(s)] = r(s-t)$. Since $x(t) = y(t)$, the covariances are $r_y(\tau) = r_x(\tau) = r_{xy}(\tau) = r(\tau)$ and the Slepian model y_u is given by

$$y_u(s) = ua(s) + Rb(s) + \Delta(s), \tag{59}$$

where R is a Rayleigh distributed random variable, independent of a zeromean Gaussian process Δ with covariance function

$$C[\Delta(t), \Delta(s)] = r(s-t) - \frac{r(t)r(s)}{r(0)} - \frac{r'(t)r'(s)}{-r''(0)}. \tag{60}$$

The functions $a(s)$, $b(s)$ are given by

$$a(s) = \frac{r(s)}{r(0)}, \qquad b(s) = \frac{r'(s)}{\sqrt{-r''(0)}}. \tag{61}$$

□

EXAMPLE 11. (Slepian model for y around its local maximum.) Assume y to be a zeromean twice continuously differentiable Gaussian process with covariance function $C[y(t), y(s)] = r(s-t)$. Since the times of local maxima are zero downcrossings of $y'(t)$, let $x(t) = y'(t)$, and hence $r_y(\tau) = r(\tau)$, $r_x(\tau) = -r''(\tau)$ and $r_{xy}(\tau) = -r'(\tau)$. Now the Slepian model for y around its local maximum is given by

$$y_0(s) = Rb(s) + \Delta(s), \tag{62}$$

where R is a Rayleigh distributed random variable independent of a zeromean Gaussian process Δ with covariance function

$$C[\Delta(t), \Delta(s)] = r(s-t) - \frac{r'(t)r'(s)}{-r''(0)} - \frac{r''(t)r''(s)}{r^{IV}(0)}. \tag{63}$$

The function $b(s) = -\frac{r''(s)}{\sqrt{r^{IV}(0)}}$. (Note, that $r^{IV}(0) = V[y''(0)]$.) Since $V[\Delta(0)] = r(0) - \frac{r''(0)^2}{r^{IV}(0)} \geq 0$ the height of a typical maximum of y, Max, say, is distributed as follows

$$\text{Max} = b(0)\, R + V[\Delta(0)]^{1/2}\, X,$$

where R, X are independent random variables, Rayleigh and Gaussian distributed, respectively, i.e.

$$f_R(r) = re^{-\frac{r^2}{2}}, \quad r > 0, \qquad f_X(x) = \frac{1}{\sqrt{2\pi}} e^{-x^2/2}.$$

For narrow band processes $V[\Delta(0)] = V[y(0)] - \frac{V[y'(0)]^2}{V[y''(0)]} \approx 0$ giving a well known result that a typical maximum of a zeromean stationary and narrow band Gaussian process is approximatively Rayleigh distributed. □

4.5. Additive wave functionals, mean and variance

Let $x(t)$, $0 \leq t \leq T$, be a stationary random process with finite number of cycles $\{(x,y)_{t_n}\}$. As before, if $f(x,y)$ is the damage caused by a cycle (x,y), then the total damage $D(t)$ is given by

$$D(t) = \sum_{t_n \leq t} f((x,y)_{t_n}).$$

A straightforward generalization of the total damage to any marked crossing sequence are *additive wave functionals* defined as follows.

If $\{(t_n, H_n)\}$ is a marked crossing sequence, then an additive wave functional $H(t)$ is given by

$$H(t) = \sum_{t_n \leq t} H_n.$$

In the following we shall give explicit formulas for mean value and the covariance function of $H(t)$. Proofs are based on frequent use of Fubini's theorem and the intensity measure of a mark H_n. We shall not give them here.

We begin with $E[H(t)]$,

$$E[H(t)] = T \cdot E[x'(0)^- H(y(\cdot)) \mid x(0) = u] f_{x(0)}(u) = T \cdot m_H. \tag{64}$$

In order to compute the covariance function of $H(t)$ we have to describe the joint typical behavior of two marks attached to two different time points. Define a joint intensity

$$\delta_H(t,s) = E[x'(t)^- x'(s)^- H(y(t+\cdot))H(y(s+\cdot)) \mid x(t) = u, x(s) = u] f_{x(t),x(s)}(u,u), \tag{65}$$

and let $\delta_H(t,t) = 0$. (Since x, y are jointly stationary processes, the function $\delta_H(t,s)$ depends only on the time difference $s-t$.) Let m_{H^2} be given by

$$m_{H^2} = E[x'(0)^- H(y(\cdot))^2 \mid x(0) = u] f_{x(0)}(u). \tag{66}$$

Consider the following mixed intensity

$$c_H(t,s) = \delta_H(t,s) - m_H^2. \tag{67}$$

Then the covariance of $H(t)$ is given by

$$C[H(s), H(t)] = \int_0^t \int_0^s c_H(s_1, s_2)\, ds_1\, ds_2 + \min(t,s) \cdot m_{H^2}. \tag{68}$$

Since the total damage $D(T)$ is an additive wave functional, then, using (68), the variance $V[D(T)] = C[D(T), D(T)]$ can be written in a closed form. Although in the case of rainflow count, the Formulas (64)–(68) are hard to evaluate exactly, they can be bases for approximations. We shall not further discuss this problem here.

Finally, one can show that for Gaussian processes, under some additional assumptions, $V[D(T)]^{-1/2} \cdot (D(T) - E[D(T)])$ is asymptotically a zeromean Gaussian variable with variance 1, see [17].

5. RAINFLOW COUNT IN GAUSSIAN LOADS

In this section we shall be concerned with computation of expected damage using rainflow count in the case of Gaussian loads or elongation processes. We begin with a well known narrow band (Rayleigh) approximation to the expected damage. Then we present Markov chain approach, introduced in Section 3, and finally discuss a general algorithm to compute the rainflow count intensity μ^{RFC}.

5.1. Narrow band approximation

The common hypothesis in the literature is that the expected damage due to stationary Gaussian loads can be conservatively estimated by approximating the load by a Gaussian narrow-band process, see [18, 19, 20]. In [11], we proved the hypothesis for rainflow, peak-valley and zero-crossings amplitude counts. We begin with a short review of the hypothesis.

Consider a Gaussian cosine load defined by

$$x(t) = R\cos(t+\phi), \tag{69}$$

where R, ϕ are independent standard Rayleigh, and $U[0, 2\pi]$ (uniform on the interval $[0, 2\pi]$) distributed variables, respectively.

Denote the inverse of the N_s, where N_s is the cycle life obtained from S-N data, by $g(s)$, i.e. $g(s) = N_s^{-1}$, and assume that $g(0) = 0$. Usually N_s is given in explicit form, e.g. $N_s = K \cdot s^{-m}, \quad K > 0, \ m \geq 1$, giving the $g(s)$ function

$$g(s) = \frac{1}{K} \cdot s^m. \tag{70}$$

For the simple cosine load (69), most of the counting methods give a cycle amplitude $S = 2R$. Consequently, the expectation of the damage $D(t)$, defined by (1), is given by

$$E[D(t)] = \frac{t}{2\pi} E[g(2R)].$$

(For the function $g(s)$ given by (70), the expected damage is proportional to moments of the Rayleigh variable R, e.g. $E[D(t)] = \frac{t}{2\pi}\frac{2^m}{K}E[R^m]$.)

Let $x(t)$ be a zeromean stationary twice continuously differentiable Gaussian load with spectral density function $S(\lambda)$, and denote by λ_0, λ_2, λ_4 the spectral moments,

$$\begin{aligned} \lambda_0 &= \textstyle\int_{-\infty}^{+\infty} S(\lambda)\, d\lambda = V[x(0)], \\ \lambda_2 &= \textstyle\int_{-\infty}^{+\infty} \lambda^2 S(\lambda)\, d\lambda = V[x'(0)], \\ \lambda_4 &= \textstyle\int_{-\infty}^{+\infty} \lambda^4 S(\lambda)\, d\lambda = V[x''(0)]. \end{aligned} \tag{71}$$

(Observe that for the simple cosine process, defined by (69), the spectral moments $\lambda_0 = \lambda_2 = \lambda_4 = 1$, since $E[R^2] = 2$.) Denote by $y(t)$ the normalized process $x(t)$, i.e.

$$y(t) = \frac{1}{\sqrt{\lambda_0}} x\left(\sqrt{\tfrac{\lambda_0}{\lambda_2}}t\right). \tag{72}$$

(Obviously, $V[y(0)] = V[y'(0)] = 1$.)

For $g(s)$ given by (70), the expected damage of x can be easily obtained from the expected damage due to the load y, i.e.

$$E_x[D(t)] = \sqrt{\tfrac{\lambda_2}{\lambda_0}}(\sqrt{\lambda_0}\,)^m \cdot E_y[D(t)].$$

The *narrow band approximation* of the expected damage $E_x[D(t)]$ is obtained by approximating in the last formula the y-process by the simple cosine process (69) leading to the following approximation of the expected damage $E_x[D(t)]$,

$$E_x[D(t)] \approx \frac{t}{2\pi}\frac{1}{K}\sqrt{\tfrac{\lambda_2}{\lambda_0}}(2\sqrt{\lambda_0}\,)^m E[R^m]. \tag{73}$$

A simple check shows that the narrow band approximation is simply the D_1^+, bound for expected damage, see Formula (20) in Section 2. Hence "$\approx$" in (73) can be replaced by "$\leq$" for rainflow, peak-valley, mean-crossing cycle and any crossing consistent count.

The narrow band bound (73) depends only on two spectral moments λ_0 and λ_2, however, the expected damage $E_x[D(t)]$ is a complex function of the spectral density of the process x and one can not construct any sharper bounds, which depend only on the spectral moments λ_0, λ_2, λ_4. This is illuminated by an example given in [11] where for any $\lambda_4 > 1$ a Gaussian process is constructed which expected damage is close to the upper bound (73). For completeness we give the narrow band bound in form of a theorem.

THEOREM 9. *If x is a zero mean stationary Gaussian load, the damage intensity $d = E[D(t)]/t$ due to rainflow count is bounded by*

$$d \leq \frac{1}{2\pi}\sqrt{\tfrac{\lambda_2}{\lambda_0}}E[g(2\sqrt{\lambda_0}\,R)],$$

where R is a standard Rayleigh variable with density $f_R(r) = re^{-0.5r^2}$.

5.2. Markov method

In Section 3 we have presented a Markov chain method to compute expected damage for rainflow count. The method approximates the sequence of local extremes in the load by a Markov chain. In previous works we have used Slepian model for a stationary Gaussian load around its local extremum, see Example 11, to compute the transition probabilities from maximum to the following minimum and then approximated the sequence of extremes by a Markov chain to compute the intensity distribution $F_S^{\text{int}}(s)$ of rainflow cycle amplitude S, say, see e.g. [21]). Then the expected damage due to a typical rainflow amplitude can be computed. (Observe that the sequence of local extremes in Gaussian loads are not Markov chains.) The method requires that the intensity of local extremes in the load is finite, e.g. $\lambda_4 < +\infty$, which is a too restrictive assumption. Here we shall present an alternative algorithm based on Formulas (29) and (30) in Lemma 4 of Section 3, which is both more general and usually gives more accurate approximations.

More precisely, consider the rainflow and peak-valley count intensities $\mu^{\mathrm{RFC}}(u,v)$, $\mu^{\mathrm{PV}}(u,v)$. Since for any $u \geq v$, $\mu^{\mathrm{PV}}(u,v) \leq \mu^{\mathrm{RFC}}(u,v)$ we have

$$\mu^{\mathrm{RFC}}(u,v) = \mu^{\mathrm{PV}}(u,v) + \kappa(u,v),$$

where κ is a nonnegative function of u, v.

Assume that the $\mu^{\mathrm{PV}}(u,v)$ density is known. Then by approximating the residual term κ we obtain an approximation to the rainflow count intensity and hence approximation of the expected damage. Since $\mu^{\mathrm{PV}}(u,v)$ is an intensity distribution of a typical maximum and the following minimum we can compute (from μ^{PV}) the transition probabilities from a typical maximum to the following minimum. (A simple algorithm is available in [9].) Consequently, we can use the Markov chain approximation method to estimate the function κ, see Theorem 6 in Section 3.

In the following two examples we have simulated sample paths of a Gaussian process and estimated the intensity $\mu^{\mathrm{PV}}(u,v)$, from the sample paths, by computing

$$\hat{\mu}^{\mathrm{PV}}(u,v) = \frac{N_T^{\mathrm{PV}}(u,v)}{T}. \tag{74}$$

Then we have used the Markov algorithm to approximate the κ-function by $\kappa_{\mathrm{Markov}}(u,v)$, say. Denote by $\hat{\mu}$ the Markov approximation of the intensity $\mu^{\mathrm{RFC}}(u,v)$, i.e.

$$\hat{\mu}(u,v) = \hat{\mu}^{\mathrm{PV}}(u,v) + \kappa_{\mathrm{Markov}}(u,v). \tag{75}$$

As before let

$$\hat{N}(u,v) = \frac{N_T^{\mathrm{RFC}}(u,v)}{T} \tag{76}$$

be an estimate of the intensity $\mu^{\mathrm{RFC}}(u,v)$, obtained from a sample path of x. The agreement between the approximated intensity $\mu^{\mathrm{RFC}}(u,v)$, $\hat{\mu}(u,v)$ and estimated from a simulated sample path of x is excellent.

EXAMPLE 12. (Gaussian load with two-peaked rectangular spectrum.) In the example we consider a zeromean Gaussian load with the following spectrum

$$S(\lambda) = \begin{cases} 1/6, & |\lambda| \in [3,6] \cup [9,12], \\ 0, & \text{otherwise.} \end{cases}$$

Figure 15 shows the excellent agreement between $\hat{\mu}(u,v)$, $\hat{N}(u,v)$. We checked the results for other simulated sample paths and obtained the same results. □

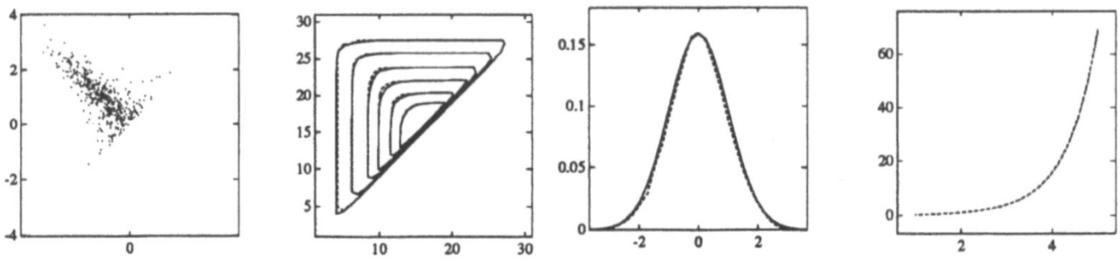

Figure 15: Gaussian load process with two-peaked spectrum; a) Rainflow count. b) Contour plots of $\hat{N}(u,v)$ (dashed line) and $\hat{\mu}(u,v)$ (solid line). c) Theoretical expected downcrossing intensity (solid line) and $\hat{\mu}(u,u)$ (dashed line). d) Estimated expected damage (dashed line) and computed damage using Markov approximation (dotted line).

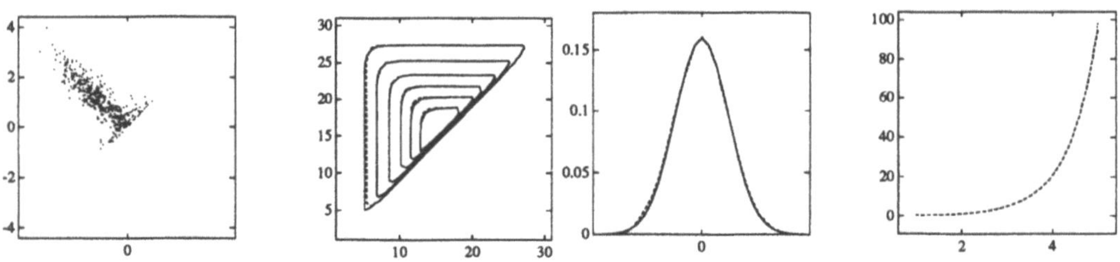

Figure 16: Gaussian load process with JONSWAP spectrum; a) Rainflow count. b) Contour plots of $\hat{N}(u,v)$ (dashed line) and $\hat{\mu}(u,v)$ (solid line). c) Theoretical expected downcrossing intensity (solid line) and $\hat{\mu}(u,u)$ (dashed line). d) Estimated expected damage (dashed line) and computed damage using Markov approximation (dotted line).

EXAMPLE 1, (contd.) (Gaussian load process with JONSWAP spectrum.) The results presented in Figure 16 show the applicability of the Markov method for the load. □

Although the examples show that the algorithm based on Markov approximation of κ function gives good results, the method requires knowledge of the intensity $\mu^{\mathrm{PV}}(u,v)$. We turn now to the problem how to compute the $\mu^{\mathrm{PV}}(u,v)$-intensity from the covariance function of a load.

Assume x is a zeromean stationary Gaussian process. Using formula (47) of Section 4, the density $\mu^{\mathrm{PV}}(u,v)$ is given by

$$\mu^{\mathrm{PV}}(u,v) = E[x'(0)^{-}1_{\{x'(s)<0 \text{ for all } 0<s<S(x)\}} \mid x(0)=u] f_{x(0)}(u), \tag{77}$$

where $S(x)$ is the first passage time of v by x, i.e. $S(x)=\inf\{s>0;\ x(s)=v\}$. Using the so called Durbin's formula for the first passage density, see [22] and [23], Formula

(77) can be rewritten as follows

$$\mu^{\mathrm{PV}}(u,v) = \int_0^{+\infty} E[x'(0)^- x'(t)^- 1_{\{x'(s) < 0 \text{ for all } 0 < s < t\}} \\ \mid x(0) = u, x(t) = v] f_{x(0),x(t)}(u,v)\, dt. \tag{78}$$

As before, let $\mu(u)$ be the u-downcrossing intensity of x, e.g.

$$\mu(u) = E[x'(0)^- \mid x(0) = u] f_{x(0)}(u) = \frac{1}{2\pi}\sqrt{\frac{\lambda_2}{\lambda_0}} e^{-\frac{u^2}{2\lambda_0}}. \tag{79}$$

In order to use the Slepian model approach to compute μ^{PV} we have to normalize the conditional expectation in (78) by $\mu(u)$, i.e.

$$\mu^{\mathrm{PV}}(u,v) = \mu(u) \int_0^{+\infty} \mu(u)^{-1} \cdot E[x'(0)^- x'(t)^- 1_{\{x'(s) < 0 \text{ for all } 0 < s < t\}} \\ \mid x(0) = u, x(t) = v] f_{x(0),x(t)}(u,v)\, dt. \tag{80}$$

Now, by the definition of intensity distributions and Slepian model processes we obtain that

$$\mu^{\mathrm{PV}}(u,v) = \mu(u) \int_0^{+\infty} E[x_u'(t)^- 1_{\{x_u'(s) < 0 \text{ for all } 0 < s < t\}} \mid x_u(t) = v] f_{x_u(t)}(v)\, dt, \tag{81}$$

where x_u is a Slepian model process for x around its downcrossing of level u, see Example 10 of Section 4.

Although the integration (81) is over the hole positive axis, for many Gaussian processes there is a finite cutoff point T, say, e.g. two mean zerocrossing distances, such that $1_{\{x_u'(s) < 0 \text{ for all } 0 < s < t\}} = 0$ for most of the sample paths of x and all $t > T$ and hence

$$\mu^{\mathrm{PV}}(u,v) \approx \mu(u) \int_0^{T} E[x_u'(t)^- 1_{\{x_u'(s) < 0 \text{ for all } 0 < s < t\}} \mid x_u(t) = v] f_{x_u(t)}(v)\, dt. \tag{82}$$

Finally in order to compute $E[x_u'(t)^- 1_{\{x_u'(s) < 0 \text{ for all } 0 < s < t\}} \mid x_u(t) = v] f_{x_u(t)}(v)$ we are using the regression method and the software (running on PC/AT) presented in [24].

Since the x process is time reversible and symmetrical we need only to compute $\mu^{\mathrm{PV}}(u,v)$ for $u \geq |v|$. For $v < -u$ we have $\mu^{\mathrm{PV}}(u,v) = \mu^{\mathrm{PV}}(-v,-u)$. The restriction of the domain of $\mu^{\mathrm{PV}}(u,v)$ is crucial for the speed of computations, since for most values of v the integral in (82) is equal to 1. Only for v close to $-u$ the integral is less than 1.

We propose to use the regression method to compute $\mu^{\mathrm{PV}}(u,v)$ and the Markov method to approximate the $\kappa(u,v)$-function, and then by summing them to obtain an approximation of $\mu^{\mathrm{RFC}}(u,v)$. The proposed method gives very good agreement with simulations for many Gaussian processes with different spectral densities. In the following example we illustrate the method.

EXAMPLE 13. (Gaussian low frequency white noise process.) In the example we consider a zeromean Gaussian load with the covariance function $r(\tau) = \sin(\sqrt{3}\tau)/(\sqrt{3}\tau)$, which has the spectrum

$$S(\lambda) = \begin{cases} 1/2\sqrt{3}, & \lambda \in [-\sqrt{3}, \sqrt{3}], \\ 0, & \text{otherwise.} \end{cases}$$

Figure 17a shows diagonals of $\hat{\mu}^{\mathrm{PV}}(u,v)$ and $\mu^{\mathrm{PV}}(u,v)$, which are the estimated downcrossing intensity of x and the theoretical one given by Formula (79), respectively. In Figure 17b, we compare contour lines of computed $\mu^{\mathrm{PV}}(u,v)$, using the regression method, with simulated $\hat{\mu}^{\mathrm{PV}}(u,v)$. (We checked the results for other simulated sample paths and obtained similar results.) The increasing variability of the estimate $\hat{\mu}^{\mathrm{PV}}(u,v)$ as u becomes high and v low, respectively, is caused by the fact that a smaller number of cycles are counted in $N_T^{\mathrm{PV}}(u,v)$ making estimates more noisy.

Figure 18 illustrates the approximation method of $\mu^{\mathrm{RFC}}(u,v)$. In Figure 18a,b, we show $\mu^{\mathrm{PV}}(u,v)$ and $\kappa_{\mathrm{Markov}}(u,v)$, respectively, while in Figure 18c the approximation $\hat{\mu}(u,v) = \mu^{\mathrm{PV}}(u,u) + \kappa_{\mathrm{Markov}}(u,v)$ is shown. (The contours in Figures 18a,b,c are drawn on different levels.) Finally, in Figure 18d, we show the computed damage intensity compared with the simulated. The approximative 95% confidance bands are also given. The bands cover the computed damage intensity. □

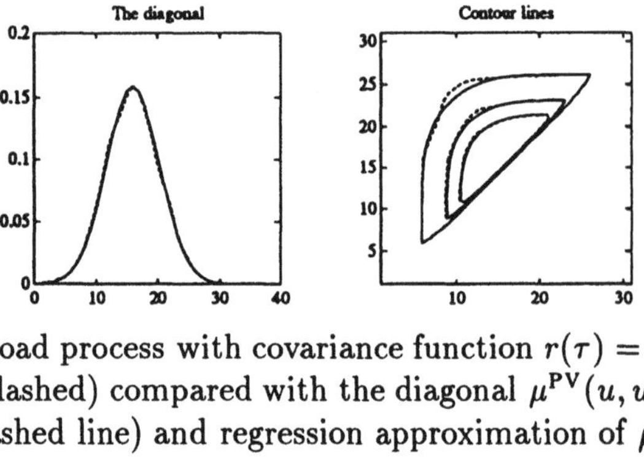

Figure 17: Gaussian load process with covariance function $r(\tau) = \sin(\sqrt{3}\,\tau)/(\sqrt{3}\,\tau)$; a) Diagonal $\hat{\mu}^{\mathrm{PV}}(u,u)$ (dashed) compared with the diagonal $\mu^{\mathrm{PV}}(u,u)$ (solid). b) Contour plots of $\hat{\mu}^{\mathrm{PV}}(u,v)$ (dashed line) and regression approximation of $\mu^{\mathrm{PV}}(u,v)$ (solid line).

5.3. Regression approximation of rainflow intensity $\mu^{\mathrm{RFC}}(u,v)$

As in the previous subsection assume that x is a zeromean stationary Gaussian process. The extension to non-stationary processes was discussed in [25]. The method is very similar to the stationary case, but the computations are much more difficult.

Consider the explicit Formula (46) of Section 4 for $\mu^{\mathrm{RFC}}(u,v)$, viz.

$$\begin{aligned} \mu^{\mathrm{RFC}}(u,v) &= E[x'(0)^{-} 1_{\{x \text{ crosses } v \text{ before } u\}} \mid x(0) = u] f_{x(0)}(u) \\ &= E[x'(0)^{-} 1_{\{x(s) < u \text{ for all } 0 < s < S(x)\}} \mid x(0) = u] f_{x(0)}(u) \end{aligned}$$

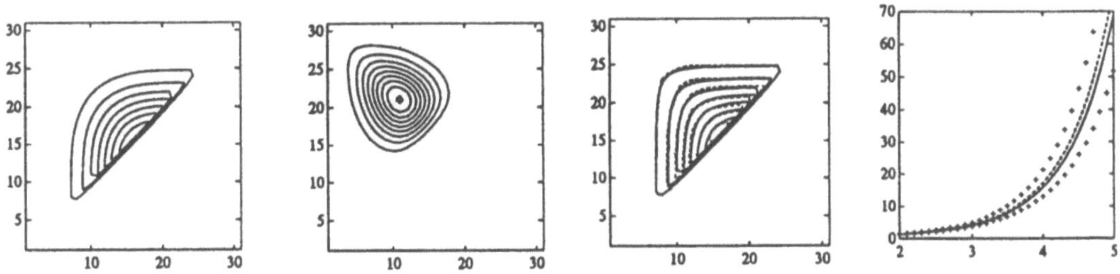

Figure 18: Gaussian load process with covariance function $r(\tau) = \sin(\sqrt{3}\,\tau)/(\sqrt{3}\,\tau)$; a) Regression approximation of $\mu^{PV}(u,v)$. b) Markov approximation of κ function, $\kappa_{\text{Markov}}(u,v)$. c) Contour plots of Markov approximation of rainflow intensity $\hat{\mu}(u,v) = \mu^{PV}(u,u) + \kappa_{\text{Markov}}(u,v)$ (solid line) and estimated $\hat{N}(u,v)$ (dashed line). d) Estimated expected damage (dashed line) with approximative 95% confidence bands and computed damage using Markov approximation (dotted line).

$$= \int_0^{+\infty} E[x'(0)^- x'(t)^- 1_{\{v < x(s) < u \text{ for all } 0 < s < t\}} \\ |x(0) = u, x(t) = v] f_{x(0),x(t)}(u,v)\,dt,$$

where $S(x)$ is the first passage time of the level v by x. Now proceeding as in the previous subsection we obtain

$$\mu^{RFC}(u,v) = \mu(u) \int_0^{+\infty} E[x_u'(t)^- 1_{\{v < x_u(s) < u \text{ for all } 0 < s < t\}} \mid x_u(t) = v] f_{x_u(t)}(v)\,dt,$$
$$\approx \mu(u) \int_0^{T} E[x_u'(t)^- 1_{\{v < x_u(s) < u \text{ for all } 0 < s < t\}} \mid x_u(t) = v] f_{x_u(t)}(v)\,dt,$$

where x_u is a Slepian model process for x around its downcrossing of the level u.

Observe, that the last formula and Formula (82) are very similar. (Only the condition $x_u'(s) < 0$ is replaced by $v < x_u(s) < u$.) Although the difference may look very marginal, the condition $x_u'(s) < 0$ for all $0 < s < t$ is much more restrictive than $v < x_u(s) < u$, for all $0 < s < t$, especially for high u and $v \approx -u$. Consequently the truncation constant T in the last formula has to be chosen much larger then in (82) making the computations very extensive, requiring additional approximations. The technical details can be found in [26, 25], where some numerical examples are given. Further develop is needed to reduce the complexity of computations.

EXAMPLE 13, (contd.) In Figure 19 we show the computed $\mu^{RFC}(u,v)$, using the computer programs presented in [26], compared with simulation. The agreement is excellent. □

References

[1] H. CRAMÉR AND M. R. LEADBETTER. *Stationary and Related Stochastic Procsses.* Wiley, New-York 1967.

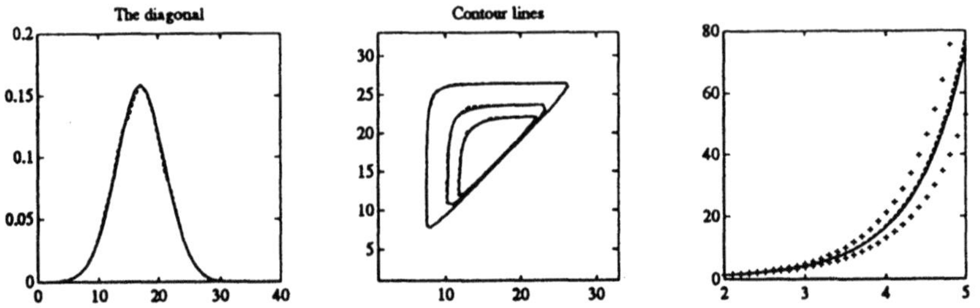

Figure 19: Gaussian load process with covariance function $r(\tau) = \sin(\sqrt{3}\,\tau)/(\sqrt{3}\,\tau)$; a) Diagonal $\hat{\mu}^{\text{RFC}}(u,u)$ (dashed) compared with the diagonal $\mu^{\text{RFC}}(u,u)$ (solid). b) Contour plots of $\hat{\mu}^{\text{RFC}}(u,v)$ (dashed line) and regression approximation of $\mu^{\text{RFC}}(u,v)$ (solid line). c) Estimated expected damage (dashed line) with approximative 95% confidence bands and computed damage using regression approximation $\hat{\mu}^{\text{RFC}}(u,u)$.

[2] V. BORMANN. Messungen von Fahrbahn – Uneberiheit paralleller Fahrspuren und Anvendung der Ergebnisse. *Veh. Syst. Dynamics*, **7** (1978), 65–81.

[3] A. NAESS. Prediction of extremes of morison-type loading – an example of a general method. *Ocean Engng.*, **10** (1983), 313–324.

[4] K SOBCZYK. *Stochastic Differential Equations with Applications to Physics and Engineering.* KluwerAcademic Publisher 1991.

[5] J. A. COLLINS. *Failure of Materials in Design.* Wiley-Interscience, New-York 1981.

[6] N. E. DOWLING. Fatigue prediction for complicated stress-strain histories. *J. Materials*, **7** (1972), 71–87.

[7] T. ENDO, K. MITSUNAGA, AND H. NAKAGAWA. Fatigue of metals subject to varying stress – prediction of fatigue lives. *Preliminary Proceedings of The Chugoku-Shikoku District Meeting*, (1967), 41–44. The Japan Soc. Mech. Engrs, Nov. 1967.

[8] S. D. DOWNING AND D. F. SOCIE. Simplified rainflow counting algorithms. *International Journal of Fatigue*, **4** (1982), 31–40.

[9] I. RYCHLIK. A new definition of the rainflow cycle counting method. *Int. J. Fatigue*, **9** (1987), 119–121.

[10] M. FRENDAHL AND I. RYCHLIK. Rainflow analysis – Markov method. Stat. Res. Report 1992:6. Dept. of Mathematical Statistics University of Lund. 1–60.

[11] I. RYCHLIK. Note on cycle counts in irregular loads. Stat. Res. Report 1992:1. Dept. of Mathematical Statistics University of Lund. 1–15.

[12] I. RYCHLIK. On the 'narrow-band' approximation for expected fatigue damage. *Probabilistic Engineering Mechanics*, **7** (1992). To appear.

[13] G. LINDGREN. Slepian models for χ^2-processes with dependent components with applications to envelope upcrossings. *J. Appl. Prob.*, **26** (1989), 36–49.

[14] M. KAC AND D. SLEPIAN. Large excursions of Gaussian processes. *Ann. Math. Statist.*, (1959), 304–310.

[15] D. SLEPIAN. On the zeros of gaussian noise. In M. Rosenblatt, editor, *Time Series Analysis* pages 104–115. Wiley, New York 1963.

[16] M. R. LEADBETTER, G. LINDGREN, AND H. ROOTZÉN. *Extremes and Related Properties of Random Sequences and Processes.* Springer-Verlag, New-York 1983.

[17] G. LINDGREN AND I. RYCHLIK. Slepian models and regression approximations in crossing and extreme value theory. *International Statistical Review*, **59** (1991), 195–225.

[18] V. PITEBARG AND I. RYCHLIK. Central limit theorem for wave-functionals of Gaussian processes. In preparation.

[19] P. H. WIRSCHING AND M. C. LIGHT. Fatigue under wide band random stresses. *J. Struct. Div., ASCE*, (1980), 1593-1607.

[20] P. H. WIRSCHING AND A. M. SHEHATA. Fatigue under wide band random stresses using the rainflow method. *J. Engng. Materials and Technology, Trans. ASME*, (), 205–211.

[21] G. JIAO AND T. MOAN. Probabilistic analysis of fatigue due to Gaussian load processes. *Probabilistic Engineering Mechanics*, (1990), 76–83.

[22] I. RYCHLIK. Rainflow cycle distribution for ergodic load processes. *SIAM J. Appl. Math*, **48** (1988), 662–679.

[23] J. DURBIN. The first-passage density of a continous Gaussian process to a general boundary. *J. Appl. Prob.*, **22** (1985), 99–122.

[24] I. RYCHLIK. A note on durbin's formula for the first passage density. *Statist. Prob. Letters*, **5** (1987), 425–428.

[25] G. LINDGREN AND I. RYCHLIK. CROSSREG – a technique for first passage and wave density analysis. *Probability in the Engineering and informational Sciences*, **7** (1993), 705–728.

[26] I. Rychlik. Rainflow cycles in random loads. In Y. Murakami, editor, *The rainflow method in fatigue* pages 21–30. Butterworth-Heinemann Ltd. 1992.

[27] I. Rychlik. Rainflow cycles in Gaussian loads. *Fatigue Fract. of Engng. Mater. and Struct.*, **15** (1992), 57–72.

PHYSICAL AND MICROMECHANICAL ASPECTS OF STOCHASTIC FATIGUE CRACK GROWTH

B. Michel, T. Winkler and L. Skurt
Centrum für Mikromechanik, Chemnitz, Germany

ABSTRACT

The authors try to "combine" physically-based knowledge about defects and their interaction mechanisms with the more phenomenologically oriented approach to fatigue crack growth. Thermodynamics and Statistical Physics become more and more important on the "mesoscopic" level of modelling. Dissipative terms occur in the expressions of a "generalized" energy release rate and lead to non-linear R-curve behaviour. Experimental methods are presented which enable to measure local deformation and fracture behaviour (laser technique, micro- moire method, X-ray diffraction and scanning tunnel microscopy) in direct connection with image processing. Coupling of experiments and simulation is essential. Stochastic FEM and first ideas of "fuzzy" FEM will be mentioned for the study of fatigue in the fields of microelectronics and microengineering.

1. INTRODUCTION

Empirical fracture relations have become very important tools in engineering applications but they may provide necessary information only within the region of measured and tested behaviour: extrapolation from these very specific regions using empirical equations "governing" the fatigue crack behaviour, however, is dangerous. Reliable predictions on crack growth and fatigue behaviour can be made only if they are based on well-accepted physical principles.

It is necessary, however, frankly to say that until now there is no unified general theory of fatigue which is able to describe the great variety of fatigue phenomena on a commonly accepted basis of fundamental principles. The problem is that a gap between the theoretical framework based on microstructural mechanisms and the "laws" and equations of phenomenological description of fatigue is widely observed.

Though it is well accepted among specialists that the defects are responsible for nearly all the crack and fatigue phenomena in different materials the variety of effects in this field leads to a lot of separate and until now incommensurable theories for very specific materials and applications.

The collective behaviour of defects (including the defect interaction phenomena) can be described to a certain extend on the basis of defect statistics assuming weak defect interaction and small deviations from thermodynamic equilibrium. But, of course, this hypothesis does not hold in all situations which then makes the problem very difficult to solve. If the approximation , however, is valid, then the framework of statistical mechanics as well as thermodynamics can be applied and some important effects of fatigue on a microstructural level may be understood. These ideas are far from being complete. Another microscopic approach is possible to start directly from the point of view of so-called rate theories. They are very close to the physical understanding of interatomic (lattice) effects of the defects, include some aspects of non-equilibrium thermodynamics depending on the level of the theoretical concept concerned and provide a good basis for advanced fracture and fatigue kinetic theories. The random walk concepts of crack growth and related theoretical concepts today can be considered as special branches of probabilistic fracture kinetic models which may be related with a stochastic approach to the fatigue and fracture phenomena. The rate theories have the advantage to be quite close to the real material behaviour also in the case of physical nonlinear constitutive equations (e.g. creep fatigue, relaxation etc.).

An outstanding parameter of all the theoretical concepts related to the microstructure is the temperature. Within defect statistics it is related via BOLTZMANN's constant k with one or another term of activation energy. On a more macroscopic scale (e.g. continuum modelling) the state variable "temperature" becomes a direct field quantity and last but not least it enters most of the phenomenological laws as a field quantity or a parameter. Advanced coupled field theories are based upon the mechanical equilibrium equations (LAME, NAVIER etc.) on the one hand and on the heat equation on the other, both being more or less strongly coupled through the time-dependent mechanical and thermal field quantities and through the term of heat dissipation. Fluctuations in these processes which are observed experimentally can only be described by means of stochastic processes.

Fatigue phenomena in real materials and under complex loading conditions require a more precise analysis of these fluctuations. The temperature field is an "interrelating field quantity" which appears both on the macroscopic (phenomenological) and on the microscopic (e.g. micromechanical) or "defect" level, but in different ways, the fluctuations beeing strongly influenced by this field quantity.

In the following chapter we start with a short survey of micromechanical behaviour of defects and their interactions becoming important for the principle understanding of stochastic fatigue and fracture phenomena. The formulae to be presented have to be understood only as examples for the variety of possible defect interaction phenomena described in a lot of monographs. We can only mention some basic ideas of this "micromechanics of defects and their interactions" within this lecture. But it is our strong opinion that a principle understanding of stochastic fatigue is impossible without taking into account the defects and the scatter bands of physical quantities influenced by activated defect mechanisms.

2. PHYSICS OF FATIGUE: PROBLEMS OF DEFECT MECHANISMS AND RELATED QUESTIONS

2.1. Micromechanics of Defects and their Interactions in Solids

The crystal (lattice) structure plays a major role in determining many properties of materials, in particular their mechanical behaviour. Consider a certain lattice structure with a characteristic linear dimension a (a can be a lattice paramater, the distance between the centres of nearest neighbours etc.). The total energy $E(a)$ shall have a minimum value at a_{min}.

The position a_{min} gives the equilibrium parameter. The curvature at a_{min} is related to the bulk modulus K by

$$K = -V(\partial p / \partial V) = V(\partial^2 E / \partial V^2) = V(\partial^2 E / \partial a^2)(\partial a / \partial V)^2 \tag{1}$$

where V is the volume of the sample and p the pressure.

The binding energy E_b is evaluated by

$$E_b = E(a \to \infty) - E(a_{min}) \tag{2}$$

2.2. Formation Energy, Formation Volume and Related Defect Quantities

The formation energy E_d of a defect is the energy difference between a defect and the perfect (undistorted) solid containing the same number of atoms. The formation volume V_d of the defect is defined as the difference in volume between the same two reference bodies as for the energy considerations.
From thermodynamics it follows

$$V_d = (\partial G_d / \partial p)_T \tag{3}$$

G_d - GIBBS free energy of the defected region, T - temperature and p - pressure.

Let N be the number of lattice sites. The number of defects in the equilibrium state is given by the relation

$$N_d \approx N \exp (S'_d / k) \exp \left(-\frac{H_d}{kT} \right) . \tag{4}$$

Here it is assumed that the defect concentration is small enough

$$N_d << N \tag{5}$$

with k - BOLTZMANN's constant. H_d is HELMHOLTZ' enthalpy.

The entropy term S'_d can be divided into two parts:

$$S'_d = S_g + S_d \tag{6}$$

$$S_g = k \ln (z_d) , \tag{7}$$

where z_d is the number of possible defect configurations associated with a particular site in the lattice.
S_g is a temperature-independent geometrical part of the entropy and S_d is the normal defect entropy term.

For harmonic lattice vibration (high temperature approximation) the entropy can be estimated by the formula

$$S_d = 3 N k \ln (\theta(0) / \theta_d(0)) . \tag{8}$$

$\theta_d(0)$ and $\theta(0)$ are the Debye temperature for the crystals with and without a vacancy, respectively.

2.3. Elastic Interactions Between Defects

If we consider a dilute substitutional solid solution with concentration c of the "solute" atoms the equilibrium concentration of the vacancies c_v in an alloy is approximated by LOMER's relation

$$c_v = c_v^0 (1 - z c) + c_v^0 z c \exp (\tilde{G} / kt) \tag{9}$$

Here $\tilde{G}$ is the GIBBS free energy of binding between the vacancy and the solute atom. An approximation often neglects the entropy term and replaces G by the binding energy E_b (see equ. (2)). c_v^0 is the normalization factor.
For special materials a lot of distinguished formulae was derived.

FLINN and MARADUDIN were able to calculate the self stress energy for impurities in fcc metals including the effect of elastic anisotropy:

$$U_s = 0.271 \mu (1 + \nu) (1 - \nu)^{-1} a_0^3 \varepsilon^2 \tag{10}$$

with
μ, ν - elastic constants
a_0 - lattice parameter and
ε - surplus of the radius of impurity atom over the initial atomic site at the place of the impurity.

Besides the self energy of defects also their elastic interaction energies become important.
ESHELBY obtained the following formula for elastic interaction energy for two defects in a cubic lattice

$$U_{int} = const\, \Delta V_1\, \Delta V_2 / d^3 \tag{11}$$

where ΔV_1 and ΔV_2 are the "dilatation strength" originating from the defects and d denotes the distance between the two defects.

In fatigue of metallic materials besides the point defects the behaviour of dislocations and voids shows a strong influence on the failure mechanisms. We cannot go into the details here but give some expressions for the defect interaction energetic terms.

If an isotropic body with elastic constants μ *and* K (μ - shear modulus, K - bulk modulus) contains a region of volume V whose elastic constants μ' and K' differ from μ *and* K but which fits into the body without difficulty, the region in question is an inhomogeneity. If no external stress (or internal stress, e.g. self stress excerted on the inhomogeneity by other defects) acts on the body, the entire solid is free of stresses. In the case of an external stress, however, (because of the differences in the elastic constants of inclusion and matrix), the defect interacts with the stress field. The latter might be the stress due to external loading in a fatigue test. The interaction of the stress with the material inhomogeneity is the "modulus effect". According to investigations by ESHELBY, the interaction energy between a small spherical volume defect and the stress field σ^A is

$$E_{i\,mod} = (-1/2)\, V_h \,(A\,(K' - K)\,(\sigma^A)^2 + B(\mu' - \mu)\,\Sigma\, '\sigma_{ij}^A\, '\sigma_{ij}^A) \qquad (12)$$

where

$$\sigma^A = \sigma_{xx}^A + \sigma_{yy}^A + \sigma_{zz}^A \qquad (13)$$

is the first invariant of the stress and

$$'\sigma_{ij}^A = \sigma_{ij}^A - \frac{1}{3}\sigma^A\,\delta_{ij} \qquad (14)$$

are the deviatoric stresses. V_h is the volume of the defect. A and B are material parameters (see e.g.[14]).

If we apply formula (12) for the case of an edge dislocation with absolute value b of the BURGERS vector the following expression is obtained for the interaction energy

$$E_i^{(V,d)} = (-\mu / 8\pi^2)\, b^2\, V_h\, (1-\nu)^{-2}\, r^{-2}\, \{[(1-\nu-2\nu^2)(K'-K)A$$
$$+(2\nu^2-2\nu-1)(\mu'-\mu)B](2/3)\sin^2\theta+(\mu'-\mu)B\} \qquad (15)$$

A and B are parameters which depend on the POISSON's ratio only.

From equation (15) the following expression for the interaction of a dislocation with an interstitial is obtained:

$$E_i^{(i,d)} = -D\mu b^2 V_h (1-0.2(1+6\nu-5\nu^2)\sin^2\theta)\, r^{-2} \qquad (16)$$

D is a parameter depending on the POISSON's ratio and r is the distance between the dislocation and the interstitial.

Because $E_i^{(i,d)}$ is negative for any orientation angle θ this means that the edge dislocation attracts the vacancy from every direction. This fact is one physical reason for the appearence of so-called "COTTRELL atmospheres", i.e. regions of higher defect concentrations near dislocations in some alloys. It is an interesting mechanism to be taken into account, if one tries to understand the physical reason for defect accumulation in fatigue tests from the microscopic point of view (see e.g. ref. [20].).

More extended investigations show that the defect concentration also changes with plastic deformation. The dislocation density in most cases is proportional to the plastic work. From this it follows that also the defect interaction should increase with plastic work. This must influence the degree of damage accumulation in fatigue. The widely used MINER hypothesis in fatigue tests, that is why, can only be a more or less rough approximation.

A free boundary surface, an interface or interface layer, and a crack flank may have a more or less strong influence on defect interaction. This can be seen having a closer look at the surface influence on the interaction energy expressions. It can be studied in the case of elastic interaction between two dilatation centres (or more generally: "volume defects" such as inclusions, precipitations etc.).

It is a well-known fact that in an unbounded isotropic elastic medium two dilatation centres do not interact mechanically (more exactly in the frame of linear theory of elasticity), i.e. their interaction energy is equal to zero. Near a flat boundary surface this statement does not hold. The whole interaction energy then must be due to the influence of the surface.

In the fatigue investigations surface-near regions have to be considered very carefully because fatige crack initiation often starts from the surface of the test specimen.

From micromechanics it has become known that the elastic interaction energy of a spherical defect of radius r_d with a second volume defect with volume V (the shape of the second defect need not be a sphere, but can be arbitrarily chosen) near an interface can be calculated analytically. It is given by the expression

$$E_{int}{}^{(d,v)} = (8\pi/27)\, G\, (\gamma-1)\, B(V)\, r_d^3\, f(V)\, \eta \tag{17}$$

Here $\gamma = G_i/G$ denotes the ratio of the shear moduli of the interface and the matrix, respectively. η is the local "quasiplastic" or "stress-free" strain of the defect which can be described in geometric terms of the defect [20]. $f(V)$ is a shape function which characterizes the volume defect and $B(\nu,\gamma)$ is a material-specific function:

$$B(\nu,\gamma) = (1+\nu)^2\, (1-\nu)^{-1}\, [1+\gamma(3-4\nu)]^{-1} \tag{18}$$

For spherical and spheroidal defects $f(V)$ can be given in a closed analytical form. For the special case of elastic interaction of two spherical defects the following expression is obtained from (17) and (18):

$$E_{int}^{(d_1,d_2)} = (32\pi/81)\, G\, (1+\nu)^2 (1-\nu)^{-1}(\gamma-1)[1+\gamma(3-4\nu)]^{-1}$$

$$\times r_1^{\,3} r_2^{\,3} (\eta)^2 (8-\alpha^2)(4+\alpha^2)^{-5/2} \xi^{-3} \tag{19}$$

or

$$E_{int}^{(d_1,d_2)} = const\, r_1^{\,3} r_2^{\,3} (8-\alpha)^2 (4+\alpha^2)^{-5/2} \xi^{-3} \tag{20}$$

where $\alpha = 2a(\xi_1+\xi_2)^{-1}$ is a geometric parameter, a is the distance between the two defects, ξ_1 and ξ_2 denote the distances of the defects from the interface (see e.g.[22]-[24]).

For a hard interface, i.e. $\gamma >> 1$, a repulsive force between the defects is the result for small distances a. For a soft layer we have $\gamma < 1$.This leads to an attractive force between the defects.

Near a free boundary surface of a solid ($\gamma = 0$) we have

$$E_{int}^{(d_1,d_2)} = -(32\pi/81)\, G\, (1+\nu)^2 (1-\nu)^{-1} r_1^{\,3} r_2^{\,3} \eta^2 (8-\alpha^2)(4+\alpha^2)^{-5/2} \xi^{-3} \tag{21}$$

The far field approximation ($\alpha >> 1$, i.e. $a >> (\xi_1+\xi_2)$) leads to the well-known result

$$E_{int}^{(d_1,d_2)} \sim a^{-3} \tag{22}$$

For $\alpha \to 1$ there yields

$$E_{int}^{(d_1,d_2)} \sim \xi^{-3} \tag{23}$$

This is the famous ξ^{-3}- law for the elastic defect interaction of two point defects near a free boundary surface.

It can be shown in addition that the following approximation is valid for two point defects which are immediately located at the surface

$$E_{int}^{(d_1,d_2)} \approx E_0/4 \tag{24}$$

The expression

$$E_0 = (2G/9)\, (1+\nu)\, (1-\nu)^{-1} V_D\, \eta^2 \tag{25}$$

is the CRUM-NABARRO energy for an elastic defect in an isotropic homogeneous unbounded solid ([22],[24]).

The energy in formula (24) is also a suitable approximation for the crack-flank interaction energy of point defects.

3. ENERGETIC APPROACH, THERMODYNAMICAL AND DEFECT STATISTICAL ASPECTS

3.1. Energy Balance in the Crack Tip Region

Fatigue crack growth can be described to a certain extend in terms of fracture mechanics quantities. For this reason the energetic situation around the cracks is studied.

The energy balance of a crack within a small thin disc of thickness ΔX_3, of surface A and volume ΔV ($=A\Delta X_3$) reads as

$$\Delta W^{\cdot V} + \Delta W^{\cdot A} = \Delta \dot{K} + \Delta \dot{U} + \Delta\Phi \tag{26}$$

$W^{\cdot V}$ and $W^{\cdot A}$ represent the local power supply per unit volume performed due to external sources. $\Delta K^{\cdot}$ and $\Delta U^{\cdot}$ denote the changes of kinetic and internal energies per unit time of the material related to the volume ΔV, respectively.

$\Delta\Phi$ corresponds to the energy flux (generalized energy release rate due to crack propagation per unit time) across the interface enveloping an internal process region A_0 around the fatigue crack tip.

Under certain circumstances a so-called effective energy release rate $\tilde{G}$ is obtained as

$$\tilde{G} = \tilde{J}_k v_k{}^c + I \tag{27}$$

(see e.g. [25],[26],[27]).

Here I is related also with some terms of dissipation which dominate the processes inside the process region in many materials and fracture mechanisms.

From eq. (27) it is possible to derive special relationships for a given crack growth mechanism.

WILL (see e.g. [27]) has shown that a good approximation for $\tilde{G}$ is given by the formula

$$\tilde{G} = \tilde{J} + e_0 \frac{dA_0}{da} \tag{28}$$

Here $\tilde{J}$ is the absolute value of a generalized $\tilde{\mathcal{J}}$-vector, e_0 is a material-specific energetic term which can be related to the definition of shape and size of the process region.

A_0 is the area of the process region (remember that our model assumes the investigated fracture region to be a thin disc, that is why A_0 may be defined in a simple manner).

It can be shown then that under certain circumstances the proportionality

$$e_0 \frac{dA_0}{da} \sim \frac{d}{da}(\tilde{J}^2) \tag{29}$$

is valid.

For stable crack propagation the term $e_0 dA_0/da$ can be expressed by

$$e_0 \frac{dA_0}{da} = \frac{\alpha}{2\pi} \tilde{J} \tilde{T}_J \tag{30}$$

where α is a normalization factor depending e.g. on the thickness effect. $\tilde{T}_J$ is the generalized tearing modulus.

Relation (30) is similar to the T_W-criterion developed by SAKA et al (28).

Using eq. (30) an instability criterion was derived by WILL:

$$\tilde{J} \tilde{T}_J \geq 0.5\, E \frac{C_R}{Y^2} \quad \text{for} \quad \begin{matrix} \text{unstable} \\ \text{stable} \end{matrix} \quad \text{crack growth} \tag{31}$$

C_R is a critical crack propagation parameter and Y denotes the yield stress of the material.

Relation (31) associated with stable crack growth implies a specific nonlinearity of J-resistance curves for a wide range of materials subjected to dissipative crack propagation. This will also lead to some important corrections in the PARIS-ERDOGAN equation, if a definite relation between K and J is valid.

Integration of relationship (31) leads to a nonlinear expression for $\tilde{J}$:

$$\tilde{J} = \left(\tilde{J}_c^{\,2} + C_R\,(a - a_c)\right)^{1/2}, \quad a \geq a_c \tag{32}$$

$\tilde{J}_c$ and a_c are material parameters, and a here denotes the crack length.

The generalized integrals such as $\tilde{J}$ are related to energetic quantities. This is the reason for their applicability also in the field of micromechanical (and microphysical) description of fracture and fatigue problems, as the energy terms both can be physically interpreted on the microscopic and the macroscopic (i.e. phenomenological) level of modeling.

3.2. Defect Statistics

Defect statistics is commonly related with the concept of activation energies. The typical activation factor

$$A = const \ \exp\left(-\frac{U_A}{kT}\right) \tag{33}$$

is very often called ARRHENIUS factor in phenomenological applications. It appears in most of the rate equations and represents a major conceptual point in several rate theories.

But here in this moment we do not consider equation (33) from the physical point of view, but assume that U_A is an activation energy for a special defect mechanism or a process connected with such a mechanism.

Then eq. (33) may be referred to as a "BOLTZMANN factor" coming out of a statistical approach.

From statistical thermodynamics we know that the probability for an energy level E_n to be occupied is expressed by the formula

$$W_n = \exp\left(-\frac{F-E_n}{kT}\right) \tag{34}$$

where F is HELMHOLTZ' free energy of the system.

The distribution which leads to (34) is called "canonical".

E_n may be dependent on the crack length, on position and orientation of the defects (e.g. microcrack orientation interacting with a main crack) and the geometrical quantities which appear in the formulae in chapter 2 of this article.

Eq. (33) is equivalent to

$$F = -kT \ln Z \tag{35}$$

the fundamental equation for Helmholtz free energy, Z beeing the "partition function":

$$Z = \sum \exp\left(-\frac{E_n}{kT}\right) \tag{36}$$

The sum in eq. (36) can be approximated in case of using continuum theory by the partition integral.

Now we are going to consider small stochastic changes in the internal energy E of the system. From "thermodynamical perturbation theory" we are able to get an approximation for the scatter of the free energy up to the second order term

$$F = F_0 + \Delta E - (\tfrac{1}{2}kT)\,\overline{(\Delta E - \Delta \overline{E})^2} \tag{37}$$

Equation (37) is important for an estimation of the variations of F due to the real scatter of energy levels occupied by the defects.

The equations (34)-(37) are based upon the idealized assumption for the defect interaction energy to be comparatively small with respect to the selfenergies of the defects. If we depart from the so-called "ideal defect gas" to the real one we have as a first step the relation

$$F = F_{id} \rightarrow F_{real} = F_{id} + \tilde{F} \tag{38}$$

$\tilde{F}$ can be estimated in the continuum field approximation as follows

$$\tilde{F} = -\frac{kTN}{2V}\iint \left(\exp\left(-E_{int}{}^{(1,2)}\right) - 1\right) dV_1\, dV_2 \tag{39}$$

with N - number of defects.

The integral term has to be computed numerically. This, however, is a very complicated procedure. The interaction term is dependent on all the geometrical parameters of the defects.

From eq. (39) a distribution function under equilibrium conditions for the defects follows which allows - in principle- to derive the relevant mechanical and physical quantities as "mean values".

Up to this moment a fixed number of defects N was prescribed. During (stochastic) fatigue ,however, defects may be generated as well as annihilated (e.g. generation and coalescence of microvoids, initiation and "healing" of vacancies etc.).

For a variable number of defects N in the given volume V we then have to apply the grand canonical potential Ω of the system (i.e. instead of the canonical partition function we start with the grand canonical partition function).

The relation between F and Ω is

$$\Omega = F - \mu_C N \tag{40}$$

with μ_C - chemical potential.

If we denote by E_{nN} the nth energy level which may be occupied by N defects it follows

$$\Omega = -kT \sum_N \ln Z^* \tag{41}$$

where

$$Z^* = \exp\left\{\frac{\mu_C}{kT} \sum_n \exp\left(-\frac{E_{nN}}{kT}\right)\right\} \tag{42}$$

In the crack tip near region voids, dislocations and point defects may be taken into account. The description of the relevant processes of formation and annihilation is a complicated procedure which requires knowledge from different branches of solid state physics up to the quantum mechanical defect level.

It should be possible also to develop a concept which includes the statistics of microcracks into such an approach.

Many open questions have remained until now to get quantitative results from the physics of defects involved in the fatigue processes. But there is no effective tool to incorporate the micromechanical aspects of defects into the fatigue damage analysis without application of defect statistics in one or another manner.

3.3. Elements of Defect Kinetics

In its very simple form, the crack velocity v can be expressed as

$$v = n\, a_0\, K \tag{43}$$

where the integer n indicates that the crack (microcrack) may occur at each step by one or more interatomic distances a_0 .

K is the rate constant as defined by rate theory and represents the number of steps per unit time. In brittle fracture the total energy of the atomic (lattice) system increases after each step by the amount ΔG_b of the surface energy (some additional term of "internal volumetric heat dissipation" being neglected here).

The rate to overcome the energy barrier (the frequency of steps is assumed to be unchanged) is given by

$$K_b = \frac{kT}{h} \exp\left(-\frac{\Delta G_b - W_b}{kT}\right) \tag{44}$$

with h - PLANCK's constant. W_b is the work which contributes to the rearrangement of the atomic configuration while the crack is growing from the energy valley to the top G_b of the real energy barrier. The work W_b is provided by the external load and by the change in the stored elastic energy (see e.g. [30]).

After KRAUSZ (see e.g. [29]) the elementary rate constant K is a fundamentally important quantity, because all time- and temperature-dependent crack growth processes are functions or functionals of K. W_b depends on the forces or "generalized forces" that describe crack growth (e.g. K_I, J, $\tilde{J}$, C^* etc.).

In a more advanced rate theory we have to consider a multi- barrier kinetics associated with several energy levels ΔG_{bi} and W_{bi} for the i-th step of elementary crack advance.

For n barriers in m parallel branches we have

$$v = \sum_{j=1}^{m} L_j P_j \tag{45}$$

with

$$P_j = \left[\sum_{i=1}^{n} L_i^{-1} \frac{h}{kt} \exp\left(\frac{\Delta G_{bi} - W_{bi}}{kt}\right)\right]_j^{-1} \tag{46}$$

L_j are local distances over which the crack moves after each step of activation (for details see [20]).

The next step is to improve the kinetic model itself taking into account the stochastic character of the elementary mechanism underlying the cracking.

Probabilistic fracture kinetics considers that crack growth is a thermally activated time- and temperature-dependent process controlled by the random thermal fluctuations of the atoms. To establish the mathematical model the MARKOV-chain analysis can be applied which leads to a differential equation system of the following kind :

$$\frac{dn_i}{dt} = K_{b_{i-1}} n_{i-1} - (K_{b_i} + K_{h_i}) n_i + K_{h_{i+1}} n_{i+1} \tag{47}$$

n_i is the number of cracks of size a_i, t is the time, and the subscripts i-1, i and $i+1$ indicate that the related quantities belong to the crack length a_{i-1}, a_i and a_{i+1}, respectively.

Because of the random character of the frequency of steps the crack sizes are also probabilistic quantities. In our example (see equation (47)) a defect activation mechanism is supposed, for simplicity reasons, which consists of two different processes.

K_b and K_h are the related elementary rate constants. K_b denotes the K-factor for the brittle fracture bond breaking mechanism and K_h characterizes the healing mechanism ("bond healing").

KRAUSZ et al. [20] applied this probabilistic fracture kinetics models to the phenomenon of corrosion fatigue. Corrosion fatigue results from simultaneous action of a fluctuation mechanical load and the degradation of the material in the crack-tip region. We assume that the corrosion process is controlled by rate constant K_c. The crack propagation steps should be controlled by the rate constants for bond breaking, $K_{b'}$ and for healing, $K_{h'}$ respectively. The distribution of the cracks between the principle states of corrosion or cracking is governed by a random mechanism. The fraction of cracks in the corrosion process state n_{corr} and in the crack growth process state n_{crack} can be given as

$$\begin{aligned} n_{corr} &= A \ \exp(-\lambda_2 t) + B \ \exp(-\lambda_3 t) + C \\ n_{crack} &= D \ \exp(-\lambda_2 t) - B \ \exp(-\lambda_3 t) + E \end{aligned} \qquad (48)$$

where A , B , C , D and E are functions given in terms of the rate constants and λ_2 and λ_3 are given by the relationships $\lambda_2 = K_c$, $\lambda_3 = K_c + K_b + K_h$.

3.4 Modelling of Fatigue Processes leading to Random Damage Structures

In most engineering problems of fatigue which are modelled on a macroscopic level the increase of damage can be observed by the growth of a single, dominating fatigue crack. The foregoing period of incubation and microcrack initiation has to be modelled at least on a mesoscopic level. In this period there is usually be found not only one single defect. There exist a field of small defects in the material which can be activated during the fatigue process. The modelling of such fatigue processes is a very complex problem. A special statistical approach useful for the investigation of such problems will be presented in this chapter. The approach is based on a statistical characterization of the random damage structure. The statistical characterization enables a comparison of a simulation model for the fatigue process with experimental results. The procedure will be explained by applying it on an example.

Thermal fatigue experiments with smooth specimens of austenitic steel give rise to an irregular pattern of small cracks on the specimen surface [16]. An example of an observed crack pattern is shown in the left part of Fig. 1. On the one hand, modern image processing systems with appropriate hardware and software can be used to analyse the geometric patterns. On the other hand, the application of more sophisticated fatigue models requires the identification of the pattern in terms of the model.

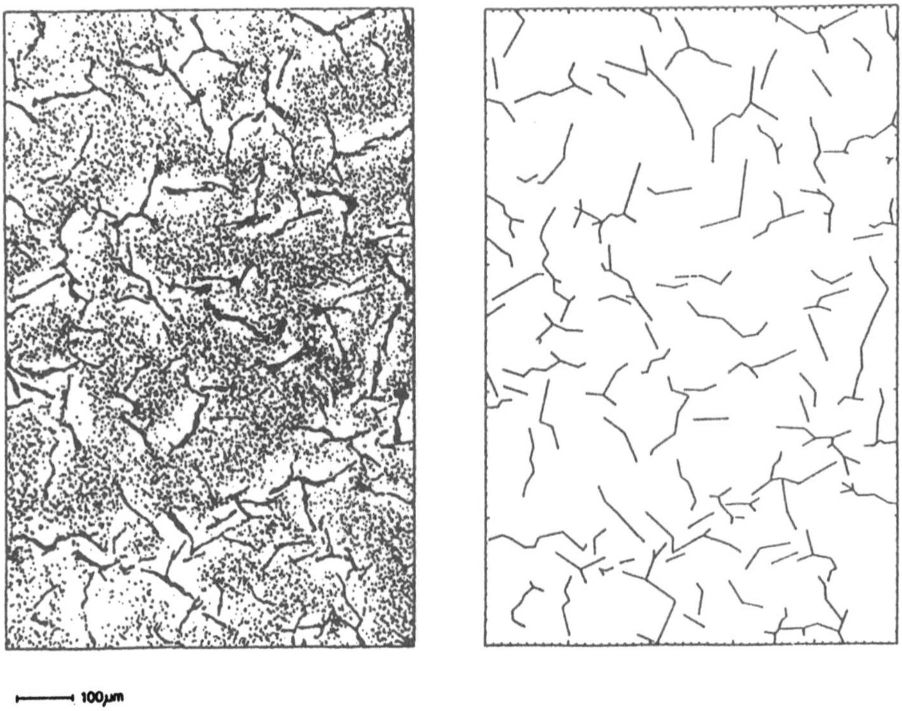

Fig. 1: Experimentally observed pattern due to thermal fatigue (left) and digitized pattern (right)

Therefore the pattern was digitized for statistical analysis. In the right part of Fig. 1, the digitized pattern corresponding to the observed one is presented.

The experiments show an evolution of the damage process in sucessive stages:

1. nucleation of a field of microcracks

2. microcrack growth and coalescence

3. growth and coalescence of cracks formed by the linked microcracks.

The crack pattern varies strongly from specimen to specimen and depends on the number of load cycles. Additionally, crack initiation, growth and coalescence seem to be dependent on local interaction because heavily and less heavily damaged regions can be observed.

A model for the evolution of such damage processes has to take into account these two effects: First, the random character due to the underlying material structure, and second, the field character which causes interaction effects.

The basic idea of the statistical approach is presented in Fig. 2. It can be summarized as follow:

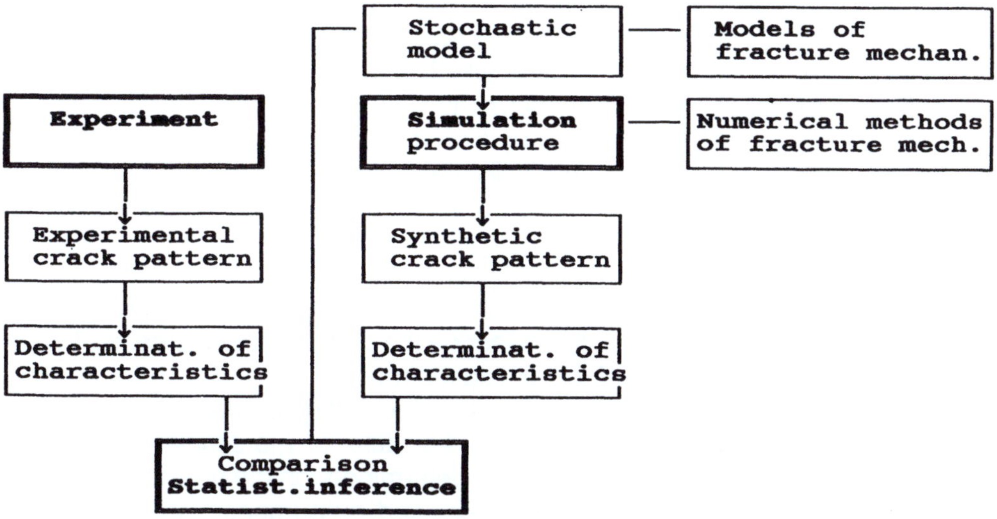

Fig. 2: Verification and calibration of a stochastic model by means of simulation

Realizations of the random crack patterns are obtained from the experiments as well as from simulation runs using a simulation model for the fatigue process. A statistical characterization of the patterns enables the comparison of experiment and simulation model and its verification.

In a first part of statistical analysis characteristic quantities of conventional descriptive statistics can be determined from the patterns. Such characteristics are, for instance, the number of cracks or the magnitude of the cracks. The mean value of the quantity as well as the corresponding frequency distribution can be determined and used for a comparison.
However, the field character of the patterns cannot be described in this way.

Models and methods of the Stochastic Geometry [17] supply with more sophisticated statistical quantities. They are suitable to characterize the correlation effects in a certain neighbourhood of a defect. Examples for modified models of stochastic geometry applicable to fatigue problems are:

1. Model of a random point process for the microcrack initiation sites. The points of microcrack initiation form a random point process on the surface or in the volume, respectively. Several types of random point processes including distinct levels of interaction are available to be fitted to experimental results.
2. Model of a random marked point process for the position, length and orientation of microcracks. The microcracks are modelled as straight lines with random length, random orientation and random midpoint. The midpoints form a random point process as mentioned above. The length and the orientation of the corresponding line segment are defined as so-called marks of the point [18]. Statistical methods are available to characterize correlations between points or marks in a certain neighbourhood.

3. Model of a random mosaic for possible paths of crack growth. It is assumed that there exist weak spots in the material which are favoured for microcrack initiation and following growth of microcracks, as well. This underlying structure in the material is modelled, for instance, by a random mosaic. A random plane mosaic forms a partition of the plane into cells, the cells are separated by straight lines, the edges. These edges form the potential crack initiation sites and possible paths of crack growth [16].

The statistical quantities of the crack patterns have to be related to physical quantities. Then they can be used to describe the trends in the increase of damage and to verify the assumptions of the simulation model.

It is important to notify, that the models of stochastic geometry do not serve only for the statistical analysis. They also form the base for the simulation model. The parameters of the simulation model can be determined, on the one hand, empirically from the experimental results, on the other hand by numerical considerations of fracture mechanics. In the example described above the failure probability of a potential microcrack initiation site depends on the local stress state.

3.5. Some Remarks Concerning the Short Fatigue Crack Problem

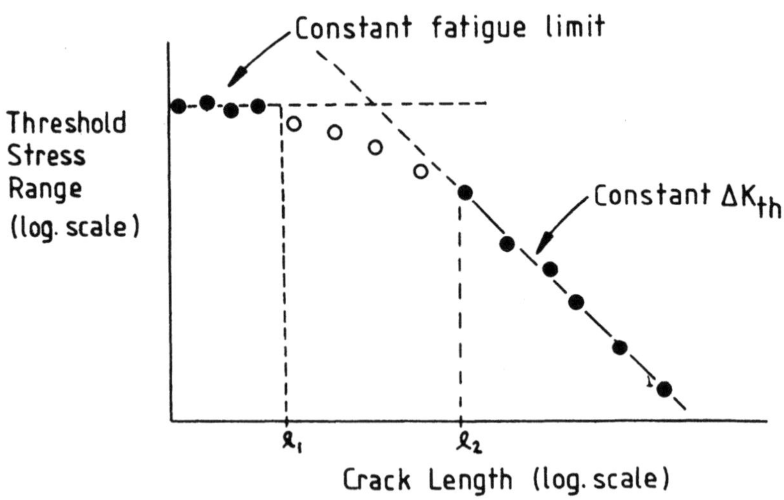

Fig. 3: Typical short crack fatigue; the open circles indicate the region of non-conservative behavior [44]

Since the early observations by LANGE [31] on the apparently anomalous growth of fatigue microcracks and the subsequent findings of PEARSON [32] that the fatigue cracks show the characteristic long crack growth behaviour only beyond

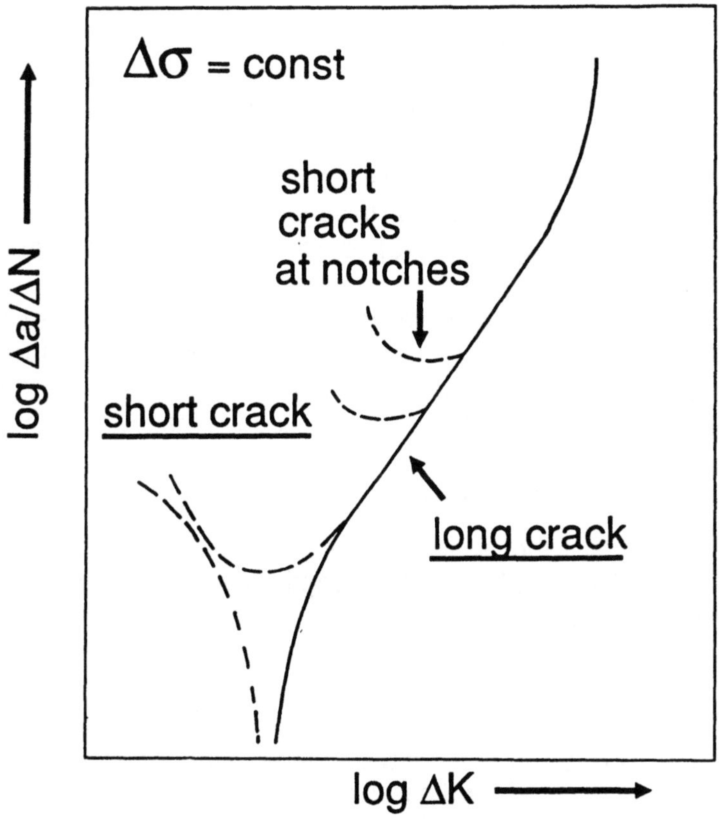

Fig. 4: Crack propagation in fatigue (schematic)

a certain crack length, the phenomenon of short crack behaviour has attracted considerable interest all over the world. The magnitude of research efforts and the increasing number of publications reveal not only academic interest but also technical relevance and engineering consequences of the short crack problem [33], [34].

Possibly the most important point to emerge is that, where short cracks are concerned, present design procedures whether based on S/N type data or on fatigue crack propagation data, are in danger of being non-conservative.The Fig. 3 shows the typical short crack fatigue results described in dependence on the crack length (TAYLOR [44]).In Fig. 4 the usual representation of a cyclic fatigue curve is given. The short crack behaviour can be observed for small values of ΔK .

The length l_2 (see Fig. 3) can be approximated by

$$l_2 = 10\,d \tag{49}$$

Here d is a "characteristic microstructural dimension" which is usually equal to the grain size but becomes carbide lath width or precipitate spacing in microstructures dominated by fine precipitates.

d is similar to the "effective grain size" l defined by YODER et al [45]. It was shown that a quite good correlation exists between 10 d and l_2 for a wide range of materials.

However, the microstructural argument is not the only one to apply. A number of other hypotheses have been advanced to account for short crack behaviour, these include the effect of crack closure [46], crack-tip stress distribution [47], crack deflection [48] and constraint [46].

These various models can be divided into two priciple types: those based directly or indirectly on microstructure, for which the parameter d is important, and those based on plasticity for which the appropriate parameter is the plastic zone size. The extent of the so-called reversed plastic zone r_p used in fatigue can be estimated as $0.04 \left(\frac{\Delta K}{Y_C}\right)^2$, where Y_C is the cyclic yield strength.

It can be shown that the plasticity based arguments can be reduced to the condition that, for LEFM to apply, the crack length must be greater than 10 r_p; hence a second estimate of l_2 is

$$l_2 \approx 10\,r_p \tag{50}$$

Equations (49) and (50) represent two independent conditions on the value of l_2. Thus, for any given material l_2 will be whichever is the larger r_p or d. For instance, if microstructural considerations lead to a prediction of, say, $l_2 = 0.05$mm, but a crack closure argument predicts $l_2 = 0.1$mm, then 0.1mm will be the relevant value (to be used for reliability analysis) and the microstructural effects can be discounted as far as l_2 is concerned. Since a conservative prediction is looked for in practical applications, a sufficient condition is all that is required.

MILLER has shown that the formula

$$\frac{da}{dN} = A\,(\Delta\gamma_p)^n\,(d-a) \tag{51}$$

with $\Delta\gamma_p$ - plastic strain measure
is a good approximation for short cracks which propagate in carbon steels. In the case investigated experimentally the parameters were found to be $A = 6$, $n = 2.24$.

Also in this case it could be shown that the parameter d which fits the experimentally da/dN slope in the best way comes quite near the crack length threshold.

A more detailed analysis of short fatigue crack problems is given in the monograph [19] ed. by MILLER and de los RIOS.

4. EXPERIMENTAL STUDY OF MICROMECHANICAL DEFORMATION BEHAVIOUR OF FATIGUE SPECIMENS AND COMPONENTS

4.1. Laser Experiments Combined with Electronic Image Processing

By means of laser metrology it is possible to measure the two- or three-dimensional displacement fields very accurately. There are several methods used practically. The holography (holographic interferometry) is applied to obtain the displacement vector $\underline{u}$ from the image patterns of a fracture specimen surface subjected to the laser light before and after deformation. For the interference patterns a set of linear equations of the following kind holds

$$N\lambda = (\underline{e}_0 - \underline{e}_l)\,\underline{u} \tag{52}$$

where

$\underline{u}$ - displacement vector

$\underline{e}_0$ - unit vector (direction of "viewing")

$\underline{e}_l$ - unit vector (direction of illumination by laser)

λ - wavelength of laser light

N - order of "fringe patterns" in the holographic experiment

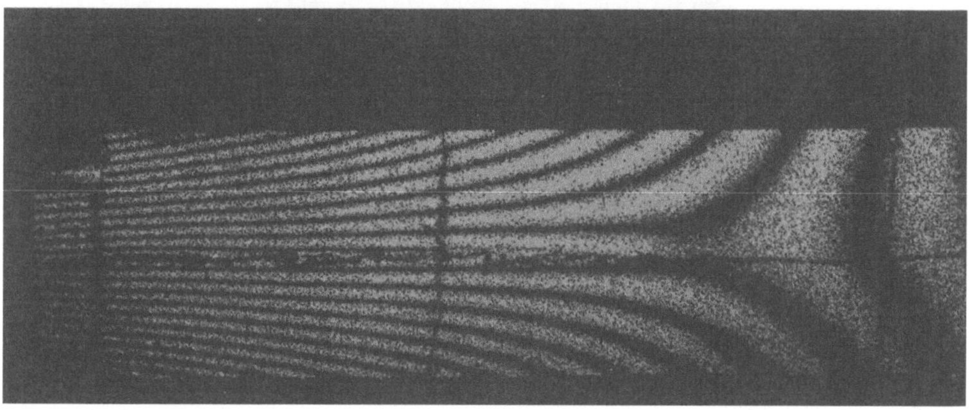

Fig. 5: Holographic interferogram of a double toursion specimen of ceramic material

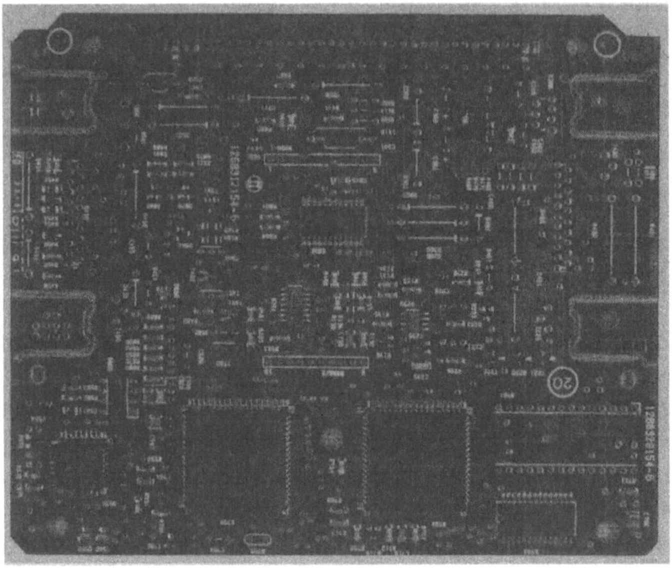

Fig. 6: Printed board containing a lot of ceramic capacitors prior to mechanical failure

Fig. 5 shows a holographic interferogram of a double torsion fracture test specimen of a ceramic material. The specimen is directly evaluated via image processing. The measuring data then were fed into the FEM computer code ASTOR

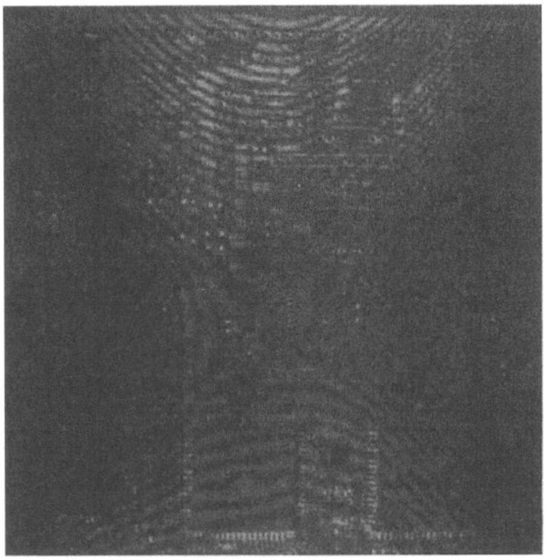

Fig. 7: Selected part of holographic interferogram of the SMD component seen in Fig. 6 in case of higher temperature

(this is a special FEM programme sytem which enables to combine experiments and Finite Element calculations). As a result we obtain all the stress and strain components of the specimen. For this it is necessary to know the constitutive law of the material. If we know e.g. the elastic-plastic stress-strain curve or a visco-plastic law, we also may obtain besides the stresses all the fracture quantities (e.g. K, J, C^*, ΔK, ΔJ etc.). In the case of rapid dynamic fracture problems a double pulse laser is used instead of the continous laser which is sufficient for the static or quasi-static deformation analysis. The fatigue testing for small electronic and microelectronic components can also be carried out with high precision by an advanced laser measurement technique combined with image processing units. Fig. 6 shows

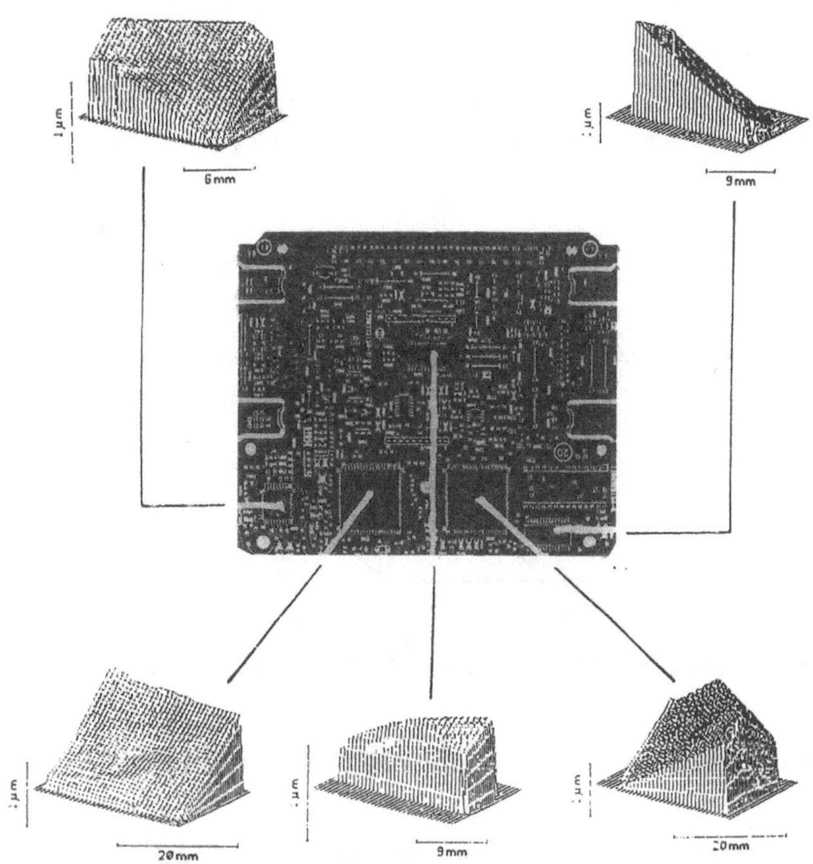

Fig. 8: Plots of the local displacements obtained from the holographic experiments using image processing. The displacement field is the "input-quantity" for the FEM analysis directly to the generalized fracture integrals discussed above in this paper

a printed circuit board prior to deformation. Fig. 7 shows a selected part of the holographic interferogram. Fig. 8 shows a plot of the local displacement field obtained from the laser experiment using image processing.

4.2. X-Ray Diffraction Method as a Suitable Means for Surface Near Stress Analysis

The use of the X-ray diffraction method to determine stresses (external stresses due to loading and unloading as well as internal or self-stresses) is a well-established procedure. The formula (53) below gives the principle insight in the connection between measured strain components as mean value over all reflecting crystallites covered by the X-ray (see e.g.[40]) and the geometrical quantities (angles) of the X- ray experiment.

$$e_{\varphi\psi} = \left(e_{11}\cos^2\varphi + e_{12}\sin 2\varphi + e_{22}\sin^2\varphi - e_{33}\right)\sin^2\psi + e_{33} + \left(e_{12}\cos\psi + e_{23}\sin\varphi\right)\sin 2\psi \qquad (53)$$

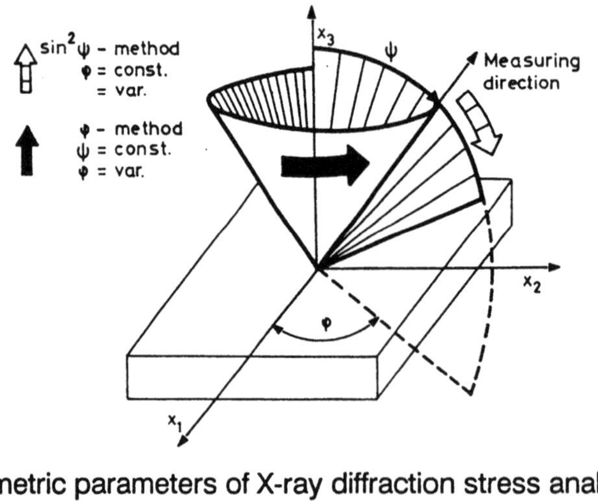

Fig. 9: Geometric parameters of X-ray diffraction stress analysis

The e_{ij} are the strain components. The explanation of the geometrical parameters is given in Fig. 9.

If the elastic constants of the material have been determined (the so-called "X-ray elastic constants") then the determination of stresses is reduced to X-ray experiments for different angles of diffraction. To get all the stress components a series of measurements in different φ - and ψ - directions are needed. Nevertheless, partial solutions are already possible by changing the φ - position only, the ψ - position being constant. This special method - called φ - rotation - has been qualified by KAEMPFE [50]. The variation of measurement position is experimentally realized by turning the sample around its surface normal in steps of $\Delta\varphi$. These measurements are easy to carry out and lead also by the usual ω-diffractometers to good results, if a focusing slit is used.

Fig. 10 gives the lattice spacing obtained by X-ray diffraction method for steel C15 after grinding, where the described method has been applied. The strains and stresses are given in Fig. 11, Fig. 12 shows the stresses measured by X-ray diffraction

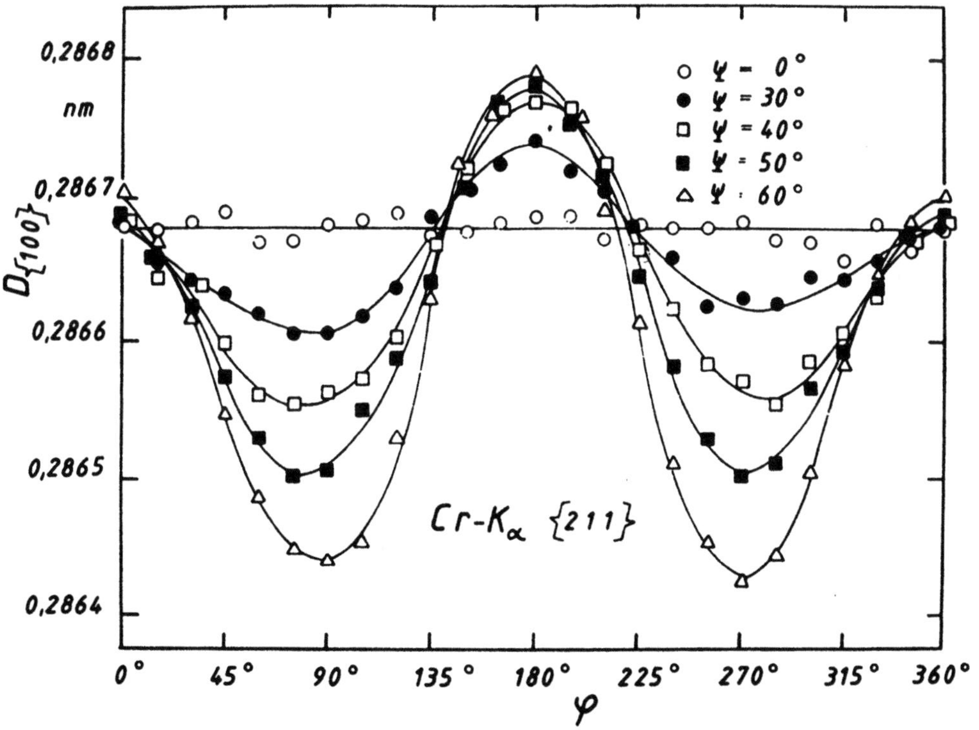

Fig. 10: Residual stresses in grinding steel C15 obtained by X-ray diffraction

in a crack tip region of a CT-specimen (along the *x*- direction of the ligament).

4.3. Scanning Micro Moire Experiments for the Evaluation of Local Deformation and Fracture Quantities

A very modern means, suited for extreme accurate measurements of local displacement components and deformation field quantities in the region smaller than 1mm^2 is the so-called Micro Moire Technique (MMT) developed by KUEHNERT (see e.g. [41]-[43]).

Cr - K_α {211} φ = variable, $\psi = 50^0 \ldots 60^0$,

penetration depth τ = 2.8 μm

$$\overline{\varepsilon}_{ij} = \begin{bmatrix} 3.39 & -0.52 & -1.56 \\ -0.52 & -10.70 & -0.05 \\ -1.56 & -0.05 & 2.34 \end{bmatrix} +/- \begin{bmatrix} 0.18 & 0.07 & 0.05 \\ 0.07 & 0.18 & 0.05 \\ 0.05 & 0.05 & 0.34 \end{bmatrix} \quad (10^{-4})$$

$$\overline{\sigma}_{ij} = \begin{bmatrix} 5 & -9 & -27 \\ -9 & -239 & -1 \\ 27 & 1 & -13 \end{bmatrix} +/- \begin{bmatrix} 11 & 1 & 1 \\ 1 & 11 & 1 \\ 1 & 1 & 8 \end{bmatrix} \quad (\text{MPa})$$

Fig. 11: Stress and strain derived from the experimental results given in Fig. 10

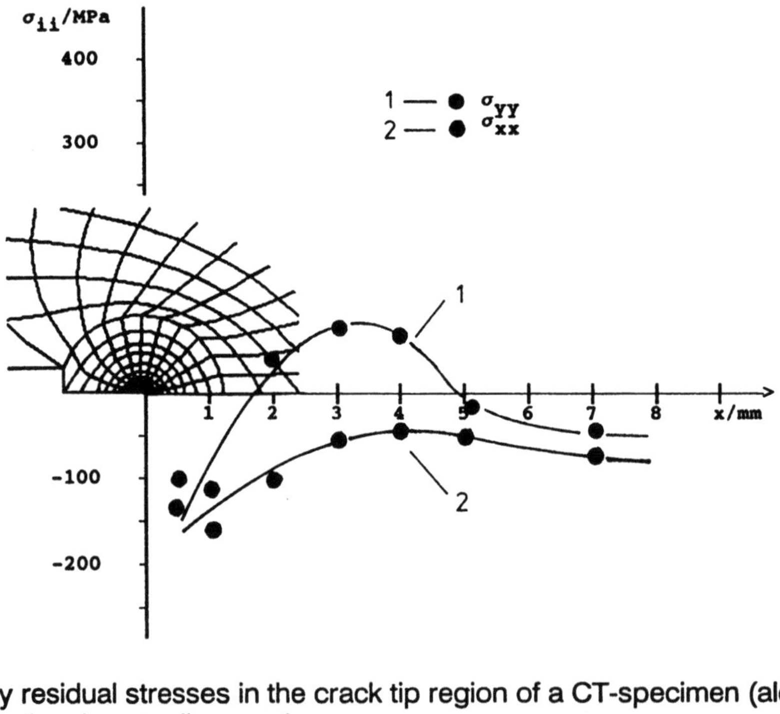

Fig. 12: X-ray residual stresses in the crack tip region of a CT-specimen (along the x-direction at a ligament)

This method represents a combination of the classical moire technique and the principle of scanning electron microscopy (SEM). To utilize the resulting fringe patterns in the experimental procedure a pattern sequential technique was applied that works very effectively in automatic image processing. In the experiment a periodic grid is necessary to be produced at the specimen surface. If both the periodically modulated topographic contrast as object structure (active grating) and the periodically modulated primary electron intensity in the SEM as reference structure (reference grating) are represented as Fourier series, then the generated moire fringe patterns can be described in a simple analytical manner. The theoretical background of this method is developed in [43],[42].

Fig. 13 shows an experimental moire pattern of a CT specimen near the crack tip produced in a scanning electron microscope.

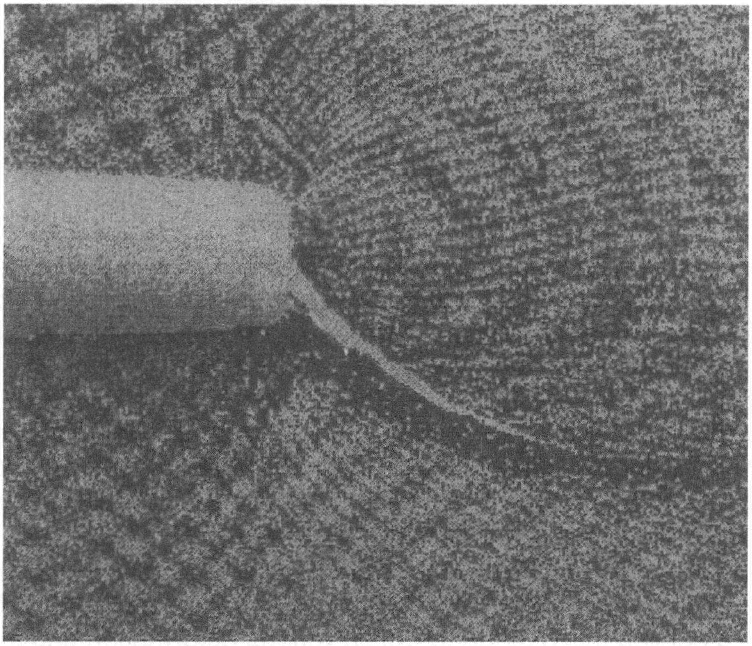

Fig. 13: Micro moire patterns around fatigure crack tip produced in a scanning electron microscope

Fig. 14 shows a quasi-periodic moire structure which was obtained in a scanning tunnel microscope, being a late high- tech application of the micro moire method. The area considered is much smaller than that of the specimen in the scanning electron microscope. This could provide very good possibilities for quantitative fracture and deformation analysis in the sub μm range (necessary for application of fracture mechanics in the fields of microelectronics and microtechnologies, e.g. microengineering, see e.g. [25]).

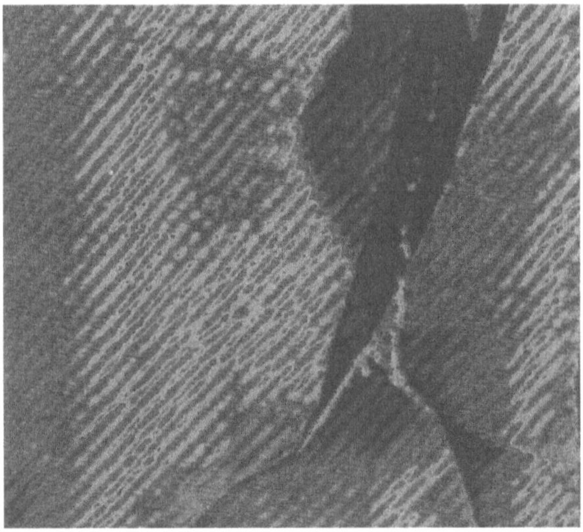

Fig. 14: Micro moire interference pattern in scanning tunnel microscope

To avoid some difficulties of fringe peak determination like the determination of the fringe sign in the MMT or the fringe ordering every increment of loading to be analyzed is recorded as 2nd-order pattern, i.e. a superposition of fringe pattern is observed containing both plane fringe order components. Thus, the surface displacement at all nodal positions of the interference mesh can directly be determined from the node point displacements without any additional manipulations. So there is no necessity for fringe ordering and the requirements of pattern recognition and processing can be reduced to the recording of the interference nodal trajectories in the image plane for the whole deformation process.

Applying this approach an accuracy of recording displacement components can be achieved that is better than 0.2 μm within an analyzed area of about 1 mm^2. The authors have applied this method to obtain various deformation and fracture quantities for a lot of materials. The experimental results have been combined with a detailed FE analysis in many cases. So it was possible to use the micro moire data as input data directly for the FE calculations.

On the other hand it is also possible to compute the fringe patterns for a given material and a given specimen synthetically by FEM , i.e. simulating the experiment. From this it follows that the model can be improved successively comparing the experimental results and the theoretical input data of the model.

Fig.15 shows the fringe patterns obtained from FE calculations of a CT-specimen to be observed by MMT in the scanning electron microscope. It could be shown that the FE calculations are very close to the experimental results.

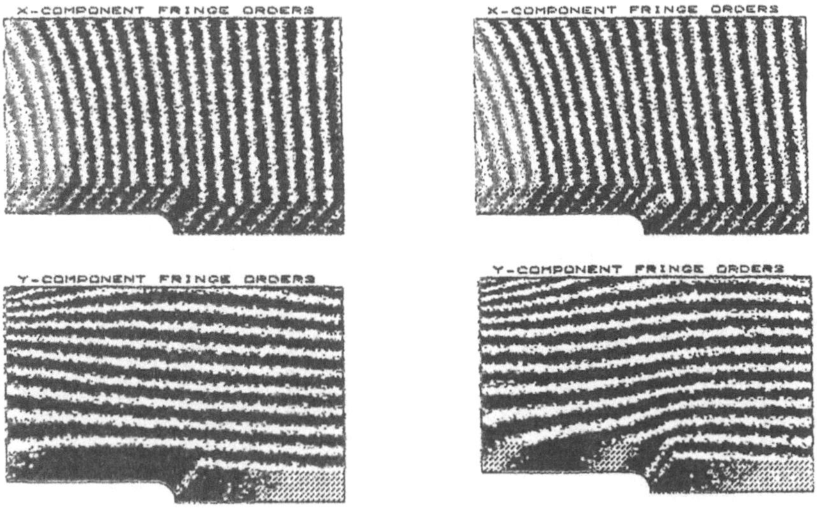

Fig. 15: Synthetic micro moire fringe orders for a CT-specimen loaded within a scanning electron microscope

Fig.s 16 and 17 show the stresses and some fracture quantities obtained by the micro moire method. A Perzyna-like visco-plastic constitutive material behaviour was assumed which proved to be a good approximation for the real material behaviour in the experiment considered.

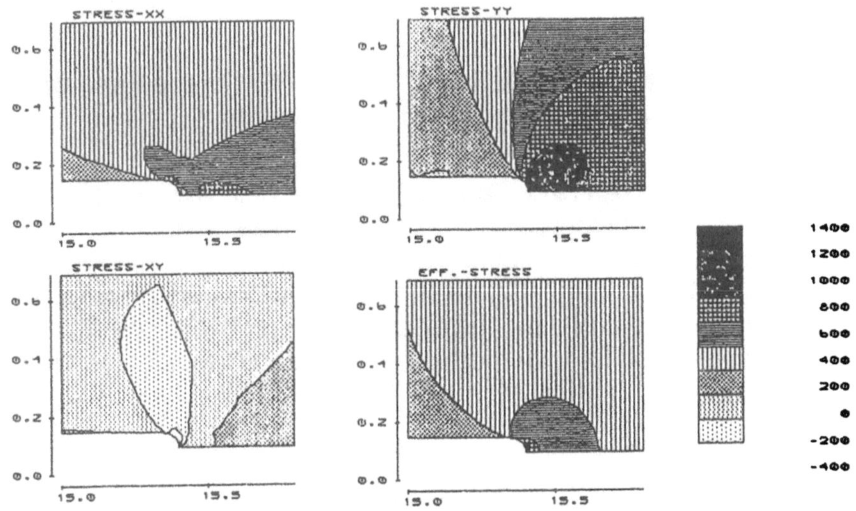

Fig. 16: Stress derived from the fringe patterns of Fig. 15 by means of image processing

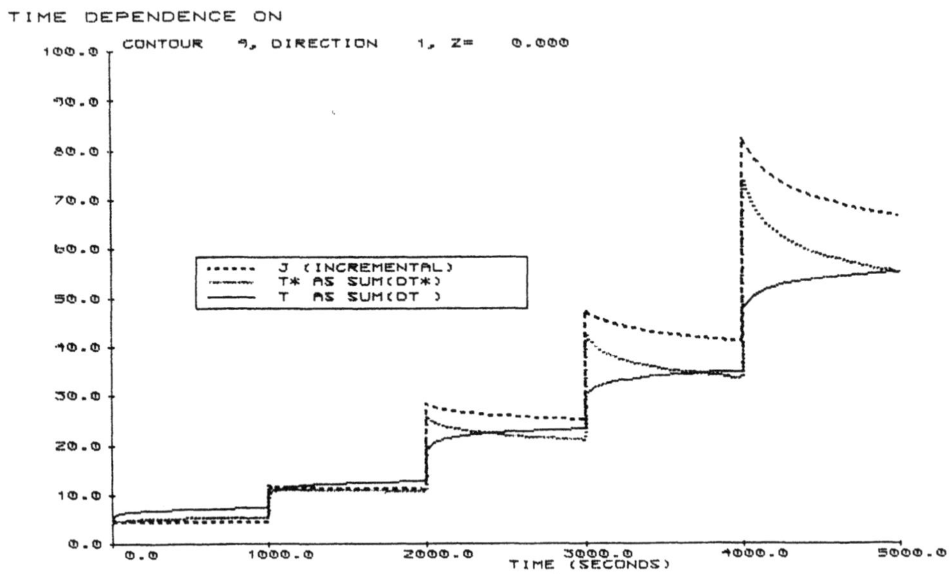

Fig. 17: Path-independent integrals *J, T* and T^* for different (time-dependent) load steps

5. APPLICATIONS OF FATIGUE MECHANICS IN THE FIELD OF MATERIAL ENGINEERING

5.1. Application of Stochastic FEM and "Fuzzy" FEM in the Field of Fracture Micromechanics of Microelectronic Components

We refer to the printed circuit board (PCB) the deformation field of which was measured by means of holographic interferometry (see chapter 4.1). On the board there are mounted several small ceramic capacitors. The technology is called SMT (Surface Mounted Technology). Fig. 18 shows a metallographic investigation of such a capacitor which is soldered on the board. The result is a complicated material transition region between three or more different materials (solder, ceramic capacitor and polymeric board).

A small crack is visible which is a dangerous failure reason of the PCB in the fatigue test.

To study the deformation and fracture behaviour of the PCB all the three methods which we have described in the previous lecture were applied and led to very interesting results [45].

Fig. 18: Metallographic picture of SMD component

We do not want to go deeper into the experimental details here but want to make some short remarks about a theoretical method which is related to a stochastic approach. A simple FE model of the PCB (plane model) is shown in Fig. 19 which is suited for an investigation of the crack near region in the gap between solder and board.

Before going to apply FEM to PCB let us give a short survey about the stochastic FE aproach.

The finite element method has found a wide application for problems in fracture mechanics. The random scatter of materials can be taken into account in FE calculations. In practical aplications not all input variables are deterministic. For example the crack length can be uncertain. The same is valid for the Young's modulus of certain structural components. In the same way the loads can be stochastically distributed. In these cases the normal finite element computer codes cannot directly be applied. To get probability evaluations regarding the displacements, strains, stresses, stress intensity factors etc. , and related fatigue and fracture quantities a large number of computations with various realizations of stochastic input variables such as the coordinates of nodal points, material properties or loading parameters is necessary. An advanced approach to this problem is a stochastic FEM for stochastic material and structural behaviour.

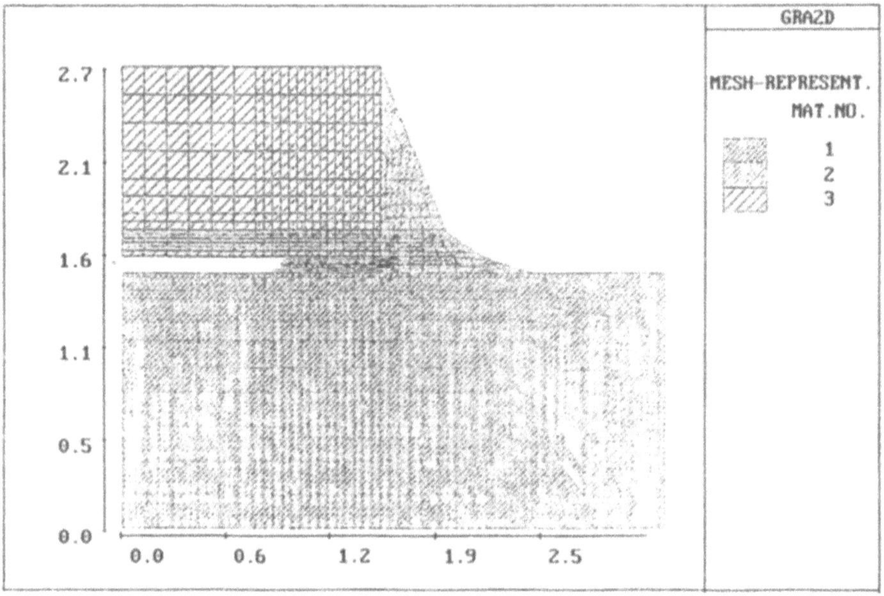

Fig.19: Micromechanical finite element model of a microcapacitor connected with the printed board

The stochastic FEM can be applied to problems of LEFM and non-linear fracture mechanics as well. We here outline one possible approach only (for details see e.g. [51] and [52]).

If we have, for example, some coordinates x_i which are assumed to be uncertain, then these stochastic coordinates $\underline{x_i}$ are represented by the formula

$$\underline{x_i} = x_i \, (1+\tilde{\underline{x_i}}) \tag{54}$$

where $\tilde{\underline{x_i}}$ are small stochastically distributed displacement corrections with a small value of variance ($D^2(\tilde{\underline{x_i}})<1$). The global stiffness matrix of the system can be constructed as a series of uncertain values $\underline{x_i}$ by means of equ. (54):

$$\underline{K} = K^0 + K^i\tilde{\underline{x_i}} + K^{ij}\tilde{\underline{x_i}}\,\tilde{\underline{x_j}} + K^{ijk}\tilde{\underline{x_i}}\,\tilde{\underline{x_j}}\,\tilde{\underline{x_k}} + \ldots \tag{55}$$

The Matrix K^0 corresponds to the stiffness matrix of the deterministic problem, when all $\underline{x_i}$ are equal to zero. This matrix is obtained from the finite element discretization with consideration of the stochastic values. It must be solved as a system of stochastic equations

$$\underline{K}\,\underline{u} = f \ ,$$

where $\underline{u}$ is the stochastic displacement vector and f takes into consideration special kind of boundary conditions.

The stochastic displacements are also described as a series of the uncertain values $\underline{x}_i$:

$$\underline{u} = u^0 + u^j\tilde{\underline{x}}_j + u^{ij}\tilde{\underline{x}}_i\tilde{\underline{x}}_j + u^{ijk}\tilde{\underline{x}}_i\tilde{\underline{x}}_j\tilde{\underline{x}}_k + \ldots \tag{56}$$

The coefficients $\left(u^j\right)$ of this series are computed by the help of perturbation theory. The vector u^0 is the solution of the corresponding deterministic problem. Its derivatives and related field quantities as strains, stresses and the fracture quantities are also represented in terms of the stochastic quantities.

The mean value, the variance and the moments of higher order can be computed using the stochastic series above. If the density functions of the stochastically distributed uncertain displacements are known, then the density and distribution functions of the results can be determined.

In order to test the method at first there were computed some standard fracture specimens. It was shown that the computing time of the stochastic procedure is in favour of the stochastic approach too, because a lot of single deterministic calculations necessary consumes much more time than one stochastic FEM simulation leading to the same result.

As an example for the stochastic FEM Fig. 20 shows a comparison between stochastic and deterministic calculations of the *J*-integral of a ceramic specimen (see e.g. [25]).

In Fig. 21 the FEM result for the printed circuit board (PCB) mentioned already above is shown. It was the task of this stochastic FE calculation to study the influence of a scatter in Young's moduli within different highly stressed regions of the PCB to get some more detailed information on the confidence range as the moduli are not very well known in practical applications. The uncertain values in this example are the Young's moduli of the capacitor, the solder and the board, respectively. A variation coefficient of 10 % was prescribed. Fig. 21 shows the contours of the stress component σ_{yy} . In addition the contours of so-called confidence regions are presented. The stresses depend on the random Young's moduli and are with the probability 0.9 in the range $\left(\sigma_{yy}-\Delta,\ \sigma_{yy}+\Delta\right)$. For the Young's moduli a normal distribution was assumed. From Fig. 21 we can draw the conclusion that a high sensitivity is observed within the highly stressed regions near the "gap" and at the outer surface of the solder.

Another approach to take into account the uncertainty of the elastic material parameters which might become important in microelectronic and microengineering is a fuzzy approach. If we assume the Young's moduli to be "fuzzy numbers" (see e.g. [53]) we also get interesting results for the same PCB which may be a good help for the evaluation of the fatigue and damage behaviour of materials and components. For details of this method we refer to the research report [45], which also includes other examples of technical application.

Fig. 22 shows the results obtained by SKURT [45] for the above mentioned PCB taking into account the Young's modulus *E* of the board to be a "fuzzy" number. The computer code ASTOR was applied again which contains a special software tool for "fuzzy"-FEM applications.

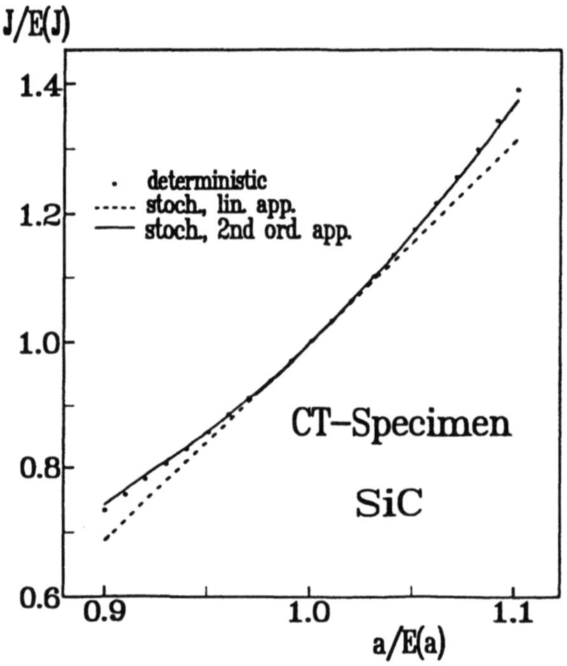

Fig. 20: Comparison between stochastic and deterministic calculation of *J*-Integral

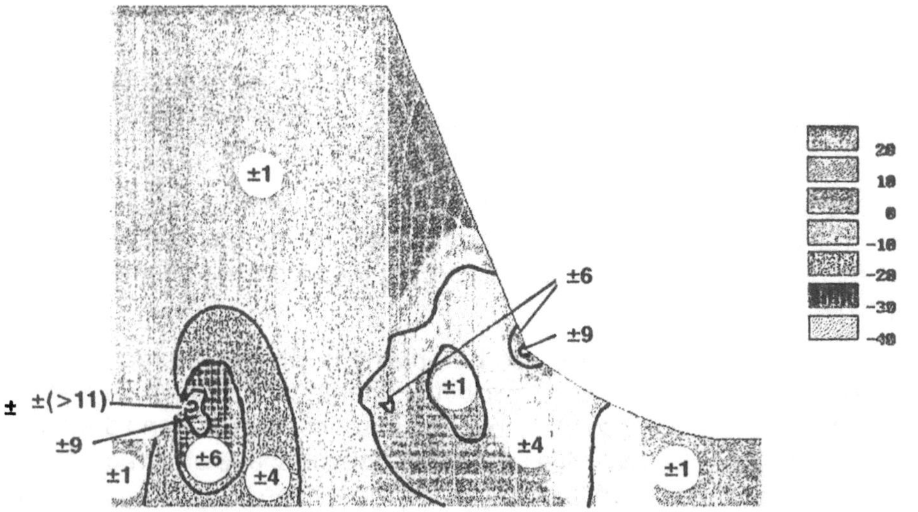

Fig. 21: Confidence range of stress component σ_{yy} at level 0.9

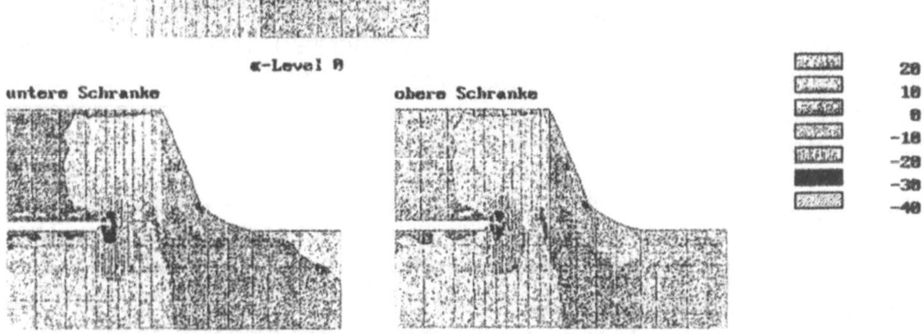

Fig. 22: Stress isolines obtained by "fuzzy FEM" calculations for a printed board

5.2. Characterization and Simulation of Random Crack Patterns Caused by Thermal Fatigue

The investigation of the thermal fatigue problem already mentioned in Chapter 3.4. will be discussed here in more detail (see also [36]-[38] for similar approaches). In a first part, the statistical analysis of the pattern is described (see [16]).

The statistical analysis has to serve two purposes: First to find quantities which characterize a crack pattern, including its random nature, for comparison of several patterns, and second to find quantities characterizing the increase in damage with increasing number of load cycles.

Two different classes of characteristics can be obtained: "pattern characteristics" measured in "number per unit area" and "crack characteristics" measured in "mean number per crack". The first class contains e.g. the number of cracks and the total (cumulative) length of all cracks. Quantities characterizing the shape of a crack, such as the number of branching points of a crack, or the magnitude of a crack, such as the total length of one crack or a suitable defined effective crack length, belong to the second class of characteristics.

It is important to note, that the quantities determined from a pattern have to be assessed in common and not separately. Global features of the damage process, such as the increase in the total amount of damage, can be assessed by the use of pattern characteristics. For the assessment of local features, the use of crack characteristics gives hints about underlying trends which control the initiation and propagation of single cracks within the pattern. Also, interaction effects may be analysed without knowing the details of crack propagation.

In the following the simulation of the damage process is described (see [39]). The simulation is performed in two steps:

First, a random geometrical model of "weak spots" on the surface of the material, representing possible sites of crack initiation, is generated. This is done with the help of the so-called Poisson-Voronoi-mosaic (see e.g. [17]). Application of a Poisson-Voronoi-mosaic to random patterns leads to meaningful results even in cases where the random mosaic is not strictly a Poisson-Voronoi-mosaic.

In a second step, microcrack initiation and growth is modelled as failure of the edges of the polygons of the mosaic according to some stochastic mechanism. Then, an initial damage configuration and succesive damage steps are simulated in order to evaluate the damage evolution. In a first level of approximation, the failure probability of an edge is assumed to depend from the status of the adjacent edges [39].

The next step is to define suitable local loading parameters that allow a fracture mechanics modelling of edge failure (KULLIG [40]). This will lead finally to the development of a physically based model of damage evolution.

5.3. Application of JT_J Concept for Dissipative Fatigue Crack Growth Mechanism

Starting with the energy balance of a fatigue crack surrounded by a dissipative process region WILL and MICHEL [54] developed a model which enables to explain the nonlinearities of the R-curves of many materials, e.g. ceramics and composites such as metal matrix fibre composites.

From the JT_J -concept, which was already introduced in Chapter 3.1 of this article, taking into account the material behaviour of PSZ-ceramics WILL obtained the following relationship:

$$J\frac{dJ}{da} = 0.5\,C_R \exp\left(-\frac{\Delta a}{a^*}\right) \tag{57}$$

Here C_R and a^* are parameters.

The integration of eq. (57) under certain conditions regarding the material and the crack leads to

$$a - a_0 = a^* \ln\left[(J_c - J_0)\,/\,(J_c - J)\right] \tag{58}$$

with

$$J_c^{\,2} = J_0^{\,2} + C_R\, a^* \tag{59}$$

Here, J_0 is the threshold of J and J_c is the critical engineering value of final failure in the nonlinear R-curve.

Energy dissipation attributed to microcracking below J_c justifies the application of a J-based analysis for the case of a ceramic fatigue crack too. WILL showed the formula

$$da\,/\,dN = a^* \ln\left\{[1 - c_2\,(\Delta J/J_c)^2]\,/\,(1 - c_3\,(\Delta J/J_c)^2)\right\} \tag{60}$$

where

$$J_0 \leq J_{min} < J < J_{max} \leq J_c \tag{61}$$

N ist the number of cycles in the fatigue test. The parameters c_2 and c_3 are defined by the ratio R of minimum to maximum stress intensity factors:

$$c_2 = [R / (1-R)]^{2(1+n)} \tag{62}$$

$$c_3 = [1 / (1-R)]^{2(1+n)} \tag{63}$$

with $R = (K_{min} / K_{max})$.

n can be described in terms of an effective cyclic hardening exponent. It was shown by WILL and HELBIG [55] that it is possible to combine the energy dissipated crack fatigue model of this kind with a stochastic approach for fibre composite materials using a Weibull-type fibre strength distribution. This will lead to damage models based on a good micromechanical basis of simulation.

References

1. Blauel, J.G. and K.-H. Schwalbe: Defect Assessment in Components - Fundamentals and Applications, Mechanical Engineering Publications Limited, London 1991,1140pp.
2. John, S.J. and C.E. Turner: Alternative Representations for Scaled R Curves in Titanium Alloy, in [1], 229-318.
3. Pineau, A and P. Joly: Local Versus Global Approaches to Elastic-Plastic Fracture Mechanics, in [1], 381-414.
4. Wallin, K.: Statistical Modelling of Fracture in the Ductile-to-Brittle Transition Region, in [1], 415-445.
5. Brückner-Foit, A.; D. Munz and B. Trolldenier: Micromechanical Implications of the Weakest Link Model for the Ductile-Brittle Transition Region, in [1], 477-488.
6. Kharin, V.S.: Nucleation and Growth of Microcracks: An Improved Dislocational Model and Implications for Ductile/ Brittle Behaviour Analysis, in [1], 489-500.
7. Gibson, G.P.; M. Capel and S.G. Druce: Effect of Heat Treatment on the FractureToughness Transition Properties of an A508 Class 3 Steel, in [1], 587-611.
8. Machida, S.; T. Miyata; Y. Yoshinari and Y. Suzuki: A Statistical Study of the Effect of Local Brittle Zone (LBZ) on the Fracture Toughness (CTOD) of Weldments, in: [1], 633-658.
9. Sobczyk, K. and B.F. Spencer: Random Fatigue: From Data to Theory, Academic Press Inc., Boston, San Diego 1992, 266 pp.
10. Bogdanoff, J.L. and F. Kozin: Probabilistic Models of Cumulative Damage, John Wiley and Sons Publ., New York 1985, 341pp.

11. Grimvall, G.: Thermophysical Properties of Materials, North Holland, Amsterdam 1986, 348 pp.
12. Munz,D. and T. Fett: Mechanisches Verhalten keramischer Werkstoffe, Springer Verlag Berlin, Heidelberg, New York 1989, 244 pp.
13. Schwalbe, K.H.: Bruchmechanik metallischer Werkstoffe, Carl Hanser Verlag München, Wien 1980, 847 pp.
14. Kovacs, I. and L. Zsoldos: Dislocations and Plastic Deformation, Akademiai Kiado, Budapest 1973, 342 pp.
15. Karlsson, B. and J. Wasen: Experimental Characterization and Geometrical Modelling of Fatigue Fracture Profiles,7th. European Conf. on Fracture (ECF7), Budapest, Oct. 1988, Proc. p. 3382.
16. Winkler, T.; A. Brueckner-Foit and H. Riesch-Oppermann: Statistical Characterization of Random Crack Pattern Caused by Thermal Fatigue. Fatigue Fract. Engng. Mat. Struct., 1992 (in print)
17. Stoyan, D.; W.S. Kandall and F. Mecke: Stochastic Geometry and its Applications. Akademie Berlin 1989, 345 pp.
18. Winkler, T.; B. Michel and H. Voigt: Simulation of Random Crack Initiation, Growth, and Coalescence. Materials Science Forum 62-64(1990),685-686
19. Miller, K.J. and E.R. de los Rios (ed.): The Behaviour of Short Fatigue Cracks, Mech. Eng. Publ. Ltd., London 1986, 560 pp.
20. Krausz, A.S. and K. Krausz: Fracture Kinetics of Crack Growth, Kluwer Academic Publishers, Dordrecht, Boston, London 1988, 181 pp.
21. Eshelby, J.D.: Progress in Solid Mechanics, Chapt. 3, North Holland Publ. Co., Amsterdam 1961, 231 pp.
22. Michel, B.: Elastic Interaction of Point Defects with a Precipitation Near a Solid Surface or an Interface, Crystal Research and Technology 14(1979)12,1483-1489.
23. Michel, B.: A Note on the Interface Interaction of Point Defects and Volume Defects, Physica Status Solidi (b) 89 (1978)K179-181.
24. Michel, B.: Über die Wechselbeziehungen zwischen Kontinuumsmechanik und Festkörperphysik, Fortschritte der Physik 29(1982)233-310.
25. Michel, B.; T. Winkler and J.-P. Sommer: Theoretical Concepts and Experiments in Fracture Mechanics - Macroscopic and Microscopic Aspects of Modelling, Plenary lecture held at 1st European Conference on Solid Mechanics, Munich, Sept 1991, in: European Journal of Mechanics, A/Solides, 11(1992), 115-133.
26. Michel, B.; P. Will ; R. Kuehnert and L. Skurt: Generalized and Nonlinear Fracture Concepts for Advanced Materials, Proc. 6th. Int. Conf. on Mechanical Behaviour of Materials (ICM 6), Kyoto,Japan, August 1991, vol. 4, Pergamon Press New York, Oxford 1991, 231-236.

27. Will, P.; B. Michel and U. Zerbst: JT_J -controlled Crack Growth Modification of *J-R* Testing and Failure Assessment Diagrams, Engineering Fracture Mechanics 28(1987)2,197-201.
28. Saka, M.; T. Shoji; H. Takehashi and H. Abe: A Criterion Based on Crack- Tip Energy Dissipation in Plane Strain Crack Growth Under Large Scale Yielding, ASTM STP 803 (1983)1130-1158.
29. Krausz, A.S. and H, Eyring: Deformation Kinetics, Wiley- Interscience 1975.
30. Fuller, E.R. and R.M. Thomson: Lattice Theories of Fracture, in: Fracture Mechanics of Ceramics (ed. R.C. Brand and D.F.H. Hasselmann), Plenum Press 1986, 333-340.
31. De Lange, R.G.: Plastic Replica Methods Applied to a Study of Fatigue Crack Propagation, Trans. AIME 230(1964) 644-648.
32. Pearson, S.: Investigation of Fatigue Cracks in Commercial Al Alloys and Subsequent Propagation of Very Short Fatigue Cracks, Engineering Fracture Mechanics 7(1975)235-247.
33. Schijve, J.: The Practical and Theoretical Significance of Small Cracks. An Evaluation, Fatigue 84, EMAS Publ., Warley 1984,751.
34. Jeal, R.H.: The Specification of Gas Turbine Disc Forgings, Metals Materials 1(1985)528-533.
35. Yoder, G.R.; L.A. Cooley and T.W. Crooker: A Critical Analysis of Grain Size and Yield-Strength Dependence of Near-Threshold Fatigue-Crack Growth in Steels, NRL Memo. Rep. No. 4576, Washington D.C., 1981.
36. Kitagawa, H.; Y. Nakasone, and S. Miyashita: Measurement of Fatigue Damage by Randomly Distributed Small Cracks. In: Fatigue Mechanisms: Advances in Quantitative Measurement of Physical Damage. ASTM STP 811 (ed. J. Lankford et al), 233-263. ASTM Philadelphia 1983
37. Hoshide, T. and D.F. Socie: Crack Nucleation and Growth Modelled in Biaxial Fatigue. Engng. Fract. Mech. 29(1988) 287-299
38. Suh, C.M.; J.J Lee and Y.G. Kang: Fatigue Microcracks in Type 304 Stainless Steel at Elevated Temperature. Fatigue Fract. Engng. Mater. Struct. 13(1990)487-496
39. Riesch-Oppermann, H.; A. Brückner-Foit, D. Munz and T. Winkler: Crack Initiation, Growth and Interaction in Thermal Cyclic Loading - A Statistical Approach. In: Proc. Int. Conf. Behaviour of Defects at High Temperatures, Sheffield 1992, ESIS-Publ. (to be published)
40. Kullig, E.; H. Riesch-Oppermann, T. Winkler and A. Brückner-Foit: Lifetime Prediction for Thermal Fatigue: Development of Stochastic Model. In: Low Cycle Fatigue and Elasto-Plastic Behaviour of Materials - 3, DVM-Verlag, Berlin 1992.
41. Kuehnert, R. and B. Michel: Second Order Moire Analysis, Physica Status Solidi 89(1988)K163-165.

42. Michel, B. and R. Kuehnert: Quantitative Deformationsanalyse im Mikrobereich, Proc. 2nd Symposium Materialforschung des Bundesmin. für Forschung und Technologie, Dresden, August 1991, 20 pp.
43. Kuehnert, R.: Thesis, in: Series Fracture Mechanics, Micromechanics, Coupled Fields (FMC), vol. 30, Institute of Mechanics Chemnitz 1987, 100 pp.
44. Taylor, D.: The Limitations of Fracture Mechanics, in: EGF-Publication No. 1 (ed. K.J. Miller and E.R. de los Rios), London 1986, 479.
45. B. Michel (ed.): Forschungsbericht Mechanische Zuverlässigkeit in der Mikrotechnik, erstellt im Auftrag der Fraunhofer-Gesellschaft München, Centrum für Mikromechanik Chemnitz, Juli 1992.
46. Suresh, S. and R.D. Ritchie: The Propagation of Short Fatigue Cracks, Report No. UCB/RP/83/1014, University of Berkeley, California, 1983.
47. llen, R.J. and J.C. Sinclair: The Behaviour of Short Cracks, Fatigue Fract. Engng. Mater. Struct. 5(1982),343
48. Suresh, S.: Crack Deflection: Implications for the Growth of Long and Short Fatigue Cracks, Met.Trans. 14A(1983)2375.
49. Miller, K.J.; H.J. Mohamed and E.R. de los Rios: Fatigue Damage Accumulation Above and Below the Fatigue Limit, in [19] 491.
50. Kaempfe, B. and B. Michel: A New Approach to X-Ray Diffraction Analysis of Stress States in Surface Layers, Acta Technica Hungarica 99(1986)3/4,313.
51. Skurt, L.: Anwendung einer Stochastischen Finite Elemente Methode in der Bruchmechanik, Series Fracture Mechanics, Micromechanics, Coupled Fields (FMC), No. 19, Chemnitz, 1986,45-54.
52. Skurt, L. and B. Michel: Eine stochastische Finite Elemente Methode, angewendet in der Bruchmechanik, Series Fracture Mechanics, Micromechanics, Coupled Fields (FMC), No. 23, Chemnitz 1986,17-28.
53. Bandemer, H. and S. Gottwald: Einführung in Fuzzy-Methoden, Akademie-Verlag, Berlin 1989, 22.
54. Will, P. and B. Michel: A Model for JT_J-controlled Fatigue Crack Growth, Intern. J. Fatigue 11(1989)2,126-128.
55. Will,P; S. Helbig and B. Michel: Mikromechanisches stochastisches Versagensmodell uniaxial faserverstärkter Verbundwerkstoffe, Materialwissenschaft und Werkstofftechnik, Darmstadt (to be published).

RANDOM INITIAL DEFECTS AND FATIGUE LIFE PREDICTION

V.V. Bolotin
Russian Academy of Sciences, Moscow, Russia

A systematic study is presented of influence of initial and newborn defects on the fatigue lifetime and reliability measures of structural components. A survey of models of fatigue crack initiation based on various accumulation concepts is given. The influence of the random material microstructure and specimen-to-specimen variability and mechanical properties on the fatigue crack growth and fatigue life is studied with the use of numerical simulation techniques. Methods of assessment of structural risk in the presence of sets of cracks and crack like defect are discussed with a special emphasis on the extrapolation of results in the area of small probabilities.

1. SURVEY OF STOCHASTIC FACTORS IN FATIGUE

All the phenomena related to fatigue, i.e. crack initiation, growth, kinking and branching, as well as the final failure considered as a loss of integrity, instability, or both, are subjected in a various degree to the influence of stochastic factors. These factors may be classified in several ways. For example, one may divide them in three groups: the randomness of material properties; random defects and imperfections of structural components; random loads, actions and environmental conditions.

Random factors, as a rule, are described with probabilistic models based on the statistical analysis of experimental data. In addition, so called "uncertainties" also may be included into probabilistic models using more or less appropriate apriori distributions and applying the Bayesian approaches when an additional information becomes available. In their turn, stochastic factors connected with material properties can be divided in two subgroups. The first subgroup includes the randomness inherent to the microstructure of materials, i.e. to nonhomogeneities, flaws and imperfections in the scale of grains, fibers, etc. This kind of randomness is inevitable due to the nature of a material and methods of its manufacturing. The second subgroup is associated with batch-to-batch scattering of material properties. Mechanical properties of commercial materials vary in a rather large range even in the case when they are strongly specified and controlled. A part of scatter enters under the influence of instabilities and imperfections of the manufacturing process. As it is well known to experimentators, there is a significant scatter of mechanical properties even if all specimens are made from the same sheet and carefully controlled before testing. Moreover, on the design stage of a structure, an engineer can not be sure that the specified properties will be realized in the future. This uncertainty can be described with probabilistic models, and parameters of the models, in principle, can be estimated from experimental data.

The scales of randomness of material properties vary within a wide range. The scale of crystal lattice, molecular structure of polymers, and dislocations is of the order of $10^{-10}, \dots, 10^{-6}$ m. On that level the randomness is a subject matter of solid state physics. Saying on microstructure, we mean the scales beginning from 10^{-6} m and more up to larger sizes that are nevertheless small compared with characteristic sizes of a structural component. It may be the order of 10^{-3} m, end even higher, as in the case of particles in large concrete structures. Essentially that all these elements of microstructures may be described in the framework of micromechanics understood as a branch of the continuum mechanics. Not only properties distributed in the volume of a body, but also properties of the surface, e.g. it small scale roughness can be attributed to this group of random factors.

Random fields with various properties are used to describe the randomness of microstructure on this level. The fields may be continuous as for amorphous polymers, piece-continuous as for multi-phase materials or materials with multiple microcracking, etc. The batch-to-batch or the specimen-to-specimen scatter is easily to introduce in such models just assuming that parameters characterizing the random fields are, in their turn, random too.

The next group of randomness is composed of various defects an imperfections on the scale of structural components. Among them are deviations from the measure in design, initial curvatures and bucklings, bad fittings born in assembling process, eccentricities, etc. Residual stresses born at the fabrication, assembling and service stages, ought to be attributed to this type of random factors too, although they are rather related to the third group of random factors. Some of these factors are described with simple probabilistic models; other require consideration of the random functions of coordinates. Small initial deviations of the middle surface of plates and shells from the ideal shape present a typical example of that kind of imperfections.

The most important type of random defects is composed of cracks and crack-like defects which sizes, opposite to microcracks from the first group, are not very small compared with the sizes of structural components. One of the classical examples are cracks and cavities in welded joints (initial, born in fabrication, as well as appeared on the transportation, assembling or service stages). Opposite to the sets of microcracks which influence on the bulk material properties may be assessed with the use of smoothing (e.g., with a self-consistent filed approach), macrocracks must be treated as single, separated formations in a body."The third group of random factors strongly influencing on the fatigue crack initiation, growth and fatigue life, is composed of loads and actions, including the environmental effects. Cyclic and slowly varying sustained loads are usually random functions of time (random processes) or random functions of time and space coordinates (random time depending fields). The aerodynamic noise pressure on the surface of an aircraft is an example of the latter kind of random fields. This comment concerns to environmental actions, too: temperature, active media, irradiation. Thermal actions influence both material properties and stress-strain fields. Active media result in corrosion, embrittlement and other kind of material's deteriation. Together with pure mechanical actions, they provoke corrosion fatigue and stress corrosion cracking. Since the environmental conditions vary randomly in time and, generally, in space, the models of fatigue damage and fracture with account of environment require the application of models which include stochastic processes and random fields.

Generally, random factors from all three groups enter together into fatigue damage and fracture phenomena. The effect of most of them, taken alone, has been discussed by many authors. Intuition and the common sense help to predict their joint effect in a more or less reliable way, but only qualitatively. As an example, not very happy attempts to describe the corrosion fatigue in the frame of the superposition principle are to be mentioned. In fact, some of random factors interact in a nontrivial way, and their synergistic effects are to be studied to make the fatigue life prediction more convincing.

2. INITIATION OF MACROSCOPIC FATIGUE CRACKS

2.1. Definition of macroscopic cracks

Fatigue failure, generally, consists of the two phases. The first one includes the microdamage accumulation, cracking and deteriation on the microlevel. This phase terminates when a macroscopic crack is born as a result of coalescence of microcracks, microvoids and other microdefects. The second stage is in fact the macroscopic fatigue crack propagation up to the final failure.

The border between micro- and macrocracks is rather vague. From the conceptional viewpoint microcracks are flaws that are to be treated in the framework of micromechanics. From the practical viewpoint, it is expedient to put the border with account of detectability of cracks with the common non-destructive inspection techniques. Then the border depends on the current state of techniques. It depends also on the type of structures that prescribes this or that thoroughness of inspection as well as on the scale of the material's microstructure. For example, the defectable sizes of cracks in aircraft engineering are of the order 1 mm, and those in large concrete structures may be larger, say, of the order 10 mm. In addition, the distinction between micro- and macrocracks is complicated with the problem of so-called "short cracks" that fill the domain between the cracks related surely to micro- and macrocracks.

Later on, we assume that for materials and structural components the initial size a_* of macroscopic cracks is stated, and we call the corresponding crack a nucleus of a macroscopic crack. In the context of micromechanics, we will consider as a nucleus a set a certain number n_* of closely situated microflaws which interaction almost surely results into the macrocrack's propagation. All cracks with sizes more than a_*, or consisting more than of n_* neighboring microflaws, will be attributed to macroscopic cracks and treated correspondingly in the framework of the theory of fatigue cracks growth.

2.2. Phenomenological models of macrocracks initiation

On the earlier stages of the studies of fatigue phenomena, standard testes were performed widely of specially prepared specimens in cyclic loading with a given stress or strain extremal values kept constant during all the test process. The test results, being presented in the form of fatigue curves (Wöhler's or S-N curves) were used in engineering design against fatigue in more complicated conditions including non-stationary cyclic loading. The typical example is the Palmgren-Miner's summation rule

$$\sum_{k=1}^{N} \frac{n(s_k)}{N(s_k)} = 1. \tag{1}$$

Here $N(s_k)$ is the cycle number up to the failure at a given characteristic stress level s_k, and $n(s_k)$ is the cycle number on the same loading level during the designed service life of a structure. Generally, s_k might mean a set of parameters characterizing the cyclic loading, say, the stress range $\Delta s = s_{\max} - s_{\min}$, and the stress ratio $R = s_{\min}/s_{\max}$. In some cases, e.g. for the low-cycle fatigue, stress parameters are to be replaced with analogous strain ones.

A number of generalizations of the Miner's rule was suggested during the last decades: nonlinear summation rules, the rules including not only fatigue but corrosion, creep, aging, etc. The most proper way of generalization is to introduce the proper damage measures and the proper kinetic equations governing the damage accumulation. Usually, the scalar measures D is introduced relation to a specimen, or a structural component, or a certain critical domain in a structure primarily responsible for the structural reliability (later on the term "body" is used). For a nondamaged body $D = 0$, and for a completely damaged $D = 1$. The kinetic equation with respect to the damage measures may be taken in the form

$$\frac{dD}{dn} = f(s(n), D(n)). \tag{2}$$

Here n is the running cycle number treated as a continuous argument, and the right-hand side is a non-negative function that depends on a cycle parameter (or a set of such parameters) treated as a continuous or peace-continuous function of n. Under the known conditions [1] Equation (2) results into Equation (1), more precisely, into

$$\int_0^N \frac{dn(s)}{N(s)} = 1 \tag{3}$$

where $N(s)$ is to be taken from the standard S-N curves.

Generally, the question remains open, what is to understood under the term "complete damage". On the pre-fracture-mechanics stage, when the engineering safe-life analysis has been based mostly on the standard fatigue tests of specially fabricated specimens, failure of a specimen has been usually associated with the observation of the first macroscopic crack. Actually, for initially "flawsless" specimens the initiation phases takes 80, ..., 90% of the total fatigue life, and with account of a large statistical scatter of the latter, duration of the second phase has been used to be considered insignificant compared with the first phase. In any case, in the discussion below we treat $N(s)$ in Equations (1)–(3) as the cycle number up to the macrocrack initiation.

Probabilistic models based on Equations (1)–(3) were studied by many authors. The survey of different approachs to the randomization of these equations could be found in a number of books [1–6]. Two ways of randomization are to be summarized.

Firstly, it is the randomization of material properties, and namely, substituting of $N(s)$ in Equations (1)–(3) with an appropriate random function. It is illustrated in Figure 1a. Here a set of fatigue curves is shown representing the stochastic scatter of

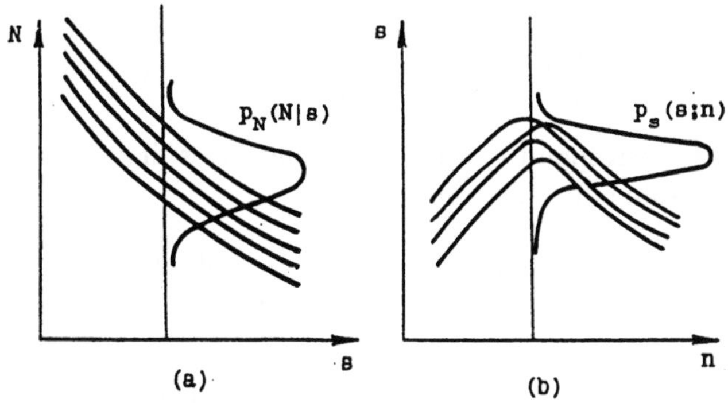

Figure 1: Randomization of the standard fatigue life analysis: a) stochastic S-N curves and the probability density function $p_N(N \mid S)$; b) stochastic loading process $S(t)$ and the probability density function $p_S(s,n)$.

fatigue test results. Analytically, this scatter is described with the cumulative probability function $F_N(N \mid s)$ of the critical cycle number N under the characteristic stress is assumed constant up to the macrocrack initiation. A typical assumption is Weibull distribution given with the probability distribution function

$$F_N(N \mid s) = 1 - \exp\Big(-\frac{M}{M_0}\Big(\frac{s - s_0}{s_c}\Big)^{\alpha}\Big(\frac{N - N_0}{N_c}\Big)^{\beta}\Big). \tag{4}$$

Here N_c and s_c are characteristic cycle and stress parameters; N_0 and s_0 are the corresponding threshold values; α and β are empirical exponent factors. The measure M of the specimen (or that of the structural component) enter into Equation (4), and the corresponding calibration constant M_0 (usually taken equal to unity), too. If cracks initiate from the surface only, M is to be understood as the area of the surface, For cables, pipings, etc. the role of M may be given to the length of a structural component, etc.

Fractiles of the distribution, i.e. the root of the equation $F_N(N \mid s) = \gamma$ are

$$N = N_0 + N_C\Big(\frac{M_0}{M}\Big)^{1/\beta}\Big(\frac{S_c}{S - S_0}\Big)^m \ln\Big(\frac{1}{1-\gamma}\Big) \tag{5}$$

with the power index

$$m = \alpha/\beta \tag{6}$$

similar to the exponent of the standard S-N curves plotted in log-log scale. The mean fatigue life is

$$E[N(s)] = N_0 + N_C\Big(\frac{M_0}{M}\Big)^{1/\beta}\Big(\frac{S_c}{S - S_0}\Big)^m \Gamma\Big(1 + \frac{1}{\beta}\Big) \tag{7}$$

where $\Gamma(\cdot)$ is the gamma-function.

The standard deviation of the fatigue life is

$$\sigma_N = N_C\Big(\frac{M_0}{M}\Big)^{1/\beta}\Big(\frac{S_c}{S - S_0}\Big)^m \sqrt{\Gamma\Big(1 + \frac{2}{\beta}\Big) - \Gamma^2\Big(1 + \frac{1}{\beta}\Big)} \tag{8}$$

Equations (4)–(8) correspond to a homogeneous stress filed in the domain. A more general form of the probability distribution is

$$F_N(N \mid s) = 1 - \exp\Big(-\Big(\frac{N - N_0}{N_c}\Big)^{\beta}\int_M \Big(\frac{sg(x) - s_0}{s_c}\Big)^{\alpha}\frac{dM}{M_0}\Big) \tag{9}$$

with the stress field characterized with the function $sg(x)$. In [1] other suggestions can be found, e.g. a fatigue life distribution for the case when both surface and volume flaws are responsible for the macroscopic fatigue crack initiation.

If a structure consists of components with different mechanical properties and, generally, differently loaded, obvious generalizations of Equations (9) are to be used such as

$$f_N(N \mid s) = 1 - \exp\Big(-\sum_i \int_{M_i}\Big(\frac{N_i - N_{0i}}{N_{ci}}\Big)^{\beta_i}\Big(\frac{sg_i(x) - s_{0i}}{s_{ci}}\Big)^{\alpha_i}\frac{dM}{M_0}\Big). \tag{10}$$

Here M_i is the measure of the ith component or a part of a component, and index i at other variables entering in Equation (10) related to the ith component or its part, too. The manifolds M_1, M_2, ... may be one-, two- and three-dimensional. For example, beams, columns, etc. is usually treated as one-dimensional at least the two sources of fatigue crack initiation: surface flaws and flaws distributed in the volume of a body.

Note that Weibull distribution and its generalizations are not the only family of distributions used for randomizing phenomenological models of fatigue. For example, the log-normal distribution is used widely. There is a powerful argument in favor of the Weibull distributions: it is one of the asymptotic distributions of extremes [7,8], and, say, log-normal distribution is not. Therefore, many more sophisticated stochastic models if fatigue return us in a natural way to Weibull distribution or its generalizations [9,10].

The second way of randomization takes into account the random nature of in-service loading or real structures. This is illustrated in Figure 1b where samples of loading process are depicted as well as the probability density function $p_s(s;n)$ of the characteristic stress during the loading history. The simplest way to randomize Equation (3) is to replace it with the following

$$\int_0^N \int_0^\infty \frac{p_s(s;n)\,dn(s)\,ds}{N(s)} = 1. \tag{11}$$

Under additional assumptions on the analytical form of the functions $p_s(s;n)$ and $N(s)$, the left-hand side of Equation (11) can be presented in explicit formulas [1,2,4]. There is a number of approaches how to count cycle, schematize their structure, identify characteristic stress and perform the statistical analysis of cycles. If a loading is a narrow-band random process, the cycle counting and the statistical treatment are rather simple: one may to count maxima and minima, their differences, i.e. ranges, etc. Under some additional assumption, say, considering the loading as Gaussian process and the standard fatigue life following the Weibull distribution, the results can be presented in an analytical or at least a rather transparent form. In most general case the rain-flow or the related complete-cycle counting approaches are used. A number of computer codes have been developed to facilitate the practical use of such probabilistic models.

We do not go into further details referring to textbooks and manuals for engineers. We do not discuss also models of fatigue based on Markov processes, such as Markov chains, etc., since these models seem "too phenomenological". Instead we turn to more sophisticated probabilistic models.

2.3. Continuous damage fields

The next step to generalize the Miner's rule and related damage accumulation models is to introduce a measure characterizing local damage in a body, but not damage of the body as a whole as the measure D does. Introduce in the volume of the body V a continuous or piececontinuous function $\omega(x,t)$ of the reference vector x and time t (time as an independent variable in many aspects is preferable compared with the cycle number). Function $\omega(x,t)$ may be scalar, taking the values from the segment $[0,1]$, or tensorial. In particular, tensors of second rank were suggested to take into account the level of microcracking and the microcracks orientation [11]. Later on, we assume that the norm $||\omega||$ of the function $\omega(x,t)$ exists such at $||\omega|| = 0$ there is no microdamage in the considered point of the body at the considered time moment, and $||\omega|| = 1$ means complete microdamage in the corresponding point. Continualization of microdamage means that the volume of the body V may be divided into the parts ΔV sufficiently large compared with the microstructure scale, but sufficiently small to treat the macroscopic stress-strain field within ΔV approximately homogeneous.

The local damage measure is a hereditary functional $\Omega\{\cdot\}$ of the history of loading in considered point. Therefore,

$$\omega(x,t) = \Omega\{s(x,\tau);\ \tau \text{ passes from } 0 \text{ to } t\} \tag{12}$$

where the field $s(x,t)$, generally, includes not only the local stress tensor but temperature and other actions effecting the fatigue life. The special form of Equation (12) analogous to Equation (2) is

$$\frac{\partial \omega(x,t)}{\partial t} = f(s(x,t), \omega(x,t)). \tag{13}$$

Equations (12) and (13) are a special case of more general constitutive equations for deformable solids with microdamage. A natural way to generalize Equation (12) is to consider jointly the microdamage and deformation processes. The set of governing equations takes the form

$$\begin{aligned} \epsilon(x,t) &= E\big[s(x,\tau), \omega(x,\tau)\big], \\ \omega(x,t) &= \Omega\big[s(x,\tau), \epsilon(x,\tau)\big], \end{aligned} \qquad \tau \text{ passes from } 0 \text{ to } t, \tag{14}$$

where $\epsilon(x,t)$ is the strain tensor. Equations (14), generally, contains two hereditary functionals, and one of them describes inelastic phenomena including those depending on the history of microdamage accumulation.

A theory of fatigue macrocrack initiation based on the concept of microdamage treated as a continuous field was suggested in [12]. The crucial point of this theory is the interpretation of the damage measure ω in terms of formation of nuclei of macroscopic cracks. The number of cracks in the area M is a random variable taking the values $k = 0, 1, 2, \ldots$. On the other side, the measure ω under a given, deterministic loading $s(t)$ is a deterministic time function. Let the value $||\omega|| = 1$ where $||\cdot||$ is a certain norm of microdamage, corresponds to a damaged specimen of a standard size M_0. Experimentators usually consider a specimen as damage when they defect the first macroscopic crack. Therefore, if $||\omega|| = 1$ in the whole area with the measure M_0, we may expect that in average one macroscopic crack appears in this area. Moreover, we suggest that the mathematical expectation of the number of macrocracks initiated in the area M_0 on the segment $[0, t]$ is a function of the norm of the measure $\omega(t)$, which values, generally, are extended over the interval $(1, \infty)$. Denoting the pointwise macrocracks initiation process $k(t)$ and its expectation $\mu(t) = E[k(t)]$, we postulate that

$$\mu(t) = f(||\omega(t)||). \tag{15}$$

The function f is differentiable and satisfies the conditions $f' > 0$, $f(0) = 0$ and $f(1) = 1$. A typical example is

$$f(\omega) = \omega^\beta \tag{16}$$

where ω is a scalar measure, and $\beta > 1$.

Equation (15) corresponds to a homogeneous distribution of microdamage in the area with the measure M_0. The natural generalization of Equation (15) upon nonhomogeneous fields is

$$\mu(t) = \int_M f(||\omega(x,t)||)\frac{dM}{M_0} \tag{17}$$

and, if the region M consists of subregions M_1, M_2, ... of generally, different dimensions,

$$\mu(t) = \sum_i \int_{M_i} f_i(||\omega_i(x,t)||)\frac{dM_i}{M_0}. \tag{18}$$

As a rule, in the further discussion we will refer to Equation (15) with the scalar measure ω such as $||\omega|| = \omega$. Generalizations based on Equations (17), (18) and tensorial measures ω are obvious.

To link the probability of formation of macrocracks with the expected value of their number, a Poisson model was used in [12]. In fact, the formation of nuclei of macroscopic cracks is s rare event, their characteristic size a_* is small compared with the sizes of the body, and the interaction between neighboring nuclei may be considered as negligible. The the probability of formation of k macrocracks in the area M_0 is determined as

$$Q_k = \frac{\mu^k}{k!}e^{-\mu}, \qquad k = 0, 1, \ldots. \tag{19}$$

Substitution of Equation (15) in Equation (19) at $||\omega(x,t)|| = \omega(t)$ yields into

$$Q_k(t) = \frac{f^k(\omega(t))}{k!}e^{-f(\omega(t))}, \qquad k = 0, 1, \ldots. \tag{20}$$

Probability of the event that at time instant t at least macrocrack initiates is

$$Q_k(t) = 1 - e^{-f(\omega(t))}. \tag{21}$$

The right-hand side of Equation (21) coincides with the distribution function $F_*(t_*)$ of times t_* up to the appearance of the first macrocrack.

In a more general case when the damage field $\omega(x,t)$ is not homogeneous in the domain M but varies there sufficiently slowly, we obtain with the use of Equation (17) that

$$F_*(t_*) = 1 - \exp\Big(-\int_M f(\omega(t_*))\frac{dM}{M_0}\Big). \tag{22}$$

Here the notation of the left hand side is changed putting $Q(t_*) \equiv F_*(t_*)$.

The suggested model is illustrated in Figure 2a showing the damage measure $\omega(t)$ and the expected number of macrocracks $\mu(t)$ varying in time. A realization of the step wise process $k(t)$ is presented there also. It is assumed that Equation (15) is $\mu = \omega^2$. Figure 2b shows the growth of the sizes of macroscopic cracks from the initial value a_* up to the critical value a_{**}.

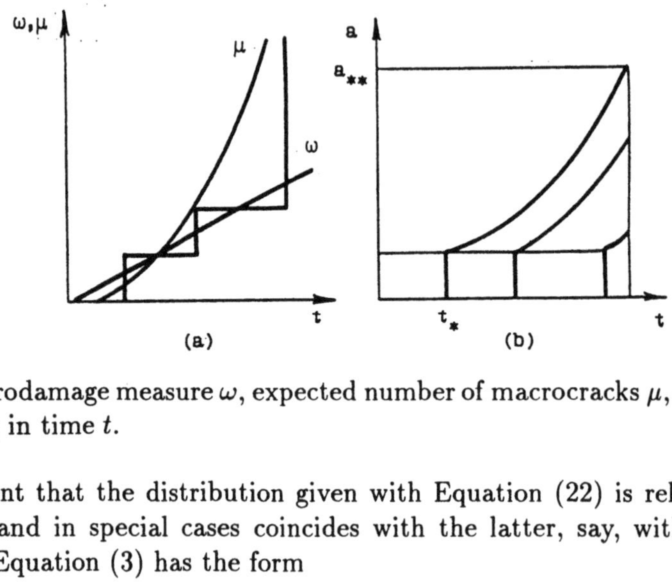

Figure 2: Microdamage measure ω, expected number of macrocracks μ, and macrocrack size a varying in time t.

It is evident that the distribution given with Equation (22) is related to Weibull distribution, and in special cases coincides with the latter, say, with Equation (4). Actually, let Equation (3) has the form

$$\frac{\partial \omega}{\partial t} = \begin{cases} \frac{1}{t_c}\left(\frac{s-s_0}{s_c}\right), & s > s_0, t > t_0, \\ 0, & s \leq s_0 \text{ or } t \leq t_0, \end{cases} \tag{23}$$

where $t_c > 0$ is a time constant (e.g. the duration of a single cycle), and $t_0 \geq 0$ is a certain threshold time necessary to initiate the damage accumulation process. Other parameters are of the same meaning as in Equation (4). At $s = \text{const.} > s_0$, $\omega(0) = 0$, Equation (23) results into

$$\omega(t) = \Big(\frac{s - s_0}{s_c}\Big)^m\Big(\frac{t - t_0}{t_c}\Big). \tag{24}$$

Substituting Equations (16) and (24) into Equation (21) we obtain

$$Q(t) = 1 - \exp\Big(-\Big(\frac{s - s_0}{s_c}\Big)^{\alpha}\Big(\frac{t - t_0}{t_c}\Big)^{\beta}\Big), \tag{25}$$

where similar to Equation (6) notation is used $\alpha = m\beta$. If instead of time t we use the cycle number N and the associated material constants N_c and N_0, Equation (25) transfers into Equation (4). The only difference is that the multiplier M/M_0 is absent. It is obvious that replacing $\omega^{\beta}(t)$ with $(M/M_0)\omega^{\beta}(t)$ or using Equation (22), we construct Equation (4) completely.

Generally, Equation (15 is not a unique consistent relationship between the number of macroscopic cracks nucleated due to microdamage and the microdamage measure. Moreover, it the microdamage $\omega(x,t)$ is a random function of space and/or time (for example, if loading is stochastic), Equation (15) becomes invalid since $\mu(t)$ is by definition a deterministic function. At least three ways to overcome the obstacle may be listed under which the basic relationship (15) remains valid for deterministic functions $\omega(t)$. Firstly, $\omega(t)$ can be replaced with its expected value. Secondly, the averaging procedure can be applied to the right hand side of Equation (15). In the third, most consistent way, one can interpret the probabilities $Q_k(t)$, $Q(t)$ and $F_*(t)$ in Equations (19)-(22) as conditional. For example, in the latter approach instead of Equation (21) we obtain

$$Q(t) = 1 - \int_0^{\infty} e^{-f(\omega)} dF_{\omega}(\omega, t) \tag{26}$$

where $F_{\omega}(\omega, t)$ is the cumulative distribution function of the measure ω at the instant t. A more general equation for the probability $Q(t)$ is as follows:

$$Q(t) = P[\sup \|\omega(x,t)\| \geq \omega_*;\ x \in M]. \tag{27}$$

Here both randomness of the microdamage field, and its nonhomogeneity in the domain M are taken into account. Note that special forms of governing equations such as Equations (16), (23), etc. are given here as examples only to demonstrate possibilities to obtain the final equations similar to commonly used in the statistical treatment of fatigue data.

2.4. Microstructural approach to macrocrack initiation

In the preceding analysis, macrocrack initiation has been treated in the framework of continuum mechanics. Now we are going the show that analogous results can be obtained considering the damage accumulation and macrocrack formation on the microstructural level.

Assume that a body with a standard measure M_0 consists of a large number N_0 of microstructural elements (later on – of elements). Let in the beginning the nominal stress be distributed in the body uniformly. Introduce the damage measure ϕ as the ratio of the number of fractured elements to their total number, i.e.

$$\phi = n/N_0. \tag{28}$$

Note that the measure ϕ, generally, does not coincides with the measure ω in Equation (16) etc. To achieve the correspondence, ϕ is to be properly normalized, i.e. $\omega = \phi/\phi_{**}$ where ϕ_{**} is the critical magnitude of ϕ.

The further discussion is done under the assumption that damage density is sufficiently small, so that it would be possible to neglect the interaction of fractured elements. At the same time, the number of fractured elements n is supposed to be large, For example, a standard specimen of carbon steel contains $10^6, \ldots, 10^7$ grains. Then even at $\phi = 10^{-2}$ the number $n + \phi N_0$ is very large. Thus, we assume that

$$N_0 \gg 1, \quad \phi N_0 \gg 1, \quad \phi \ll 1. \tag{29}$$

Considering the damage accumulation measure as a function of time t, and denoting the lifetime of elements with τ_k, Equation (28) results into

$$\phi(t) = \frac{1}{N_0} \sum_{k=1}^{N_0} H(t - \tau_k) \tag{30}$$

where $H(\cdot)$ is Heaviside function. Due to the assumption of mutual independence of fractured elements, and with account of Equation (29) we use the Bernoulli scheme and Moivre-Laplace theorem. The asymptotic probability distribution function for the damage measure at the time t takes the form

$$F_\phi(\phi, t) \approx \Phi\left(\frac{(\phi - F_\tau(t))\sqrt{N_0}}{\sqrt{F_\tau(t)(1 - F_\tau(t))}}\right). \tag{31}$$

Here $\Phi(z)$ is the normalized Gauss distribution function, i.e.

$$\Phi(z) = \frac{1}{\sqrt{2\pi}} \int_{-\infty}^{z} e^{-u^2/2}\, du. \tag{32}$$

The probability distribution function $F_\tau(\tau)$ of times up to fracture of arbitrary chosen elements enter into Equation (31), and the step wise function given with Equation (30) is replaced with an appropriately smoothed function. It follows from Equations (30) and (31) that the variance of ϕ is very small. Ignoring the scatter of ϕ, we come to he semi-deterministic approximation

$$\phi(t) \approx F_\tau(t) \tag{33}$$

which will be used in the later discussion.

Returning to the definition of the nucleus of a macroscopic crack given in section 2.1, we estimate the probability that in domain M at least one nucleus (a set of n_* neighboring fractured elements) will be born up to the time t. Generally, it is a rather complicated problem since various combinations of fractured elements and the stress concentration around the elements are to be taken into account. An approximate, simplified approach was discussed in [13,14] in the connection with fracture of fiber composite materials. The time t_* up to the macrocrack initiation has been evaluated based on the following scheme Initiation occurs when in the considered domain at least one fractured element appears with $n_* - 1$ fractured neighbors. Assuming the all these events take place at the time t_* the probability distribution function $F_*(t_*)$ of times up to initiation may be determined as follows

$$F_*(t_*) = 1 - \left(1 - \phi(t_*)^{n_*-1}\right)^{\phi(t_*)N_0}.$$

The asymptotic equivalent of this equation is

$$F_*(t_*) = 1 - \exp\left(-N_0\phi^{n_*}(t_*)\right). \tag{34}$$

Generalizing the equation upon nonhomogeneous filed and using Equation (33), we obtain the relationship

$$F_*(t_*) = 1 - \left(1 - \int_M F_\tau^{n_*}(t_*)\frac{dM}{M_0}\right). \tag{35}$$

Evidently, Equations (33) and (35) are analogous to Equations (21) and (22). This analogy is a natural consequence that both models connect the macrocrack initiation with the rupture of weakest elements of microstructure.

To use Equations (34) and (35), the distribution function $F_\tau(\tau)$ of the lifetime of elements is to be known. That requires a special model of the damage accumulation in a single randomly chosen element. If the corresponding damage measure $\chi(t)$ is a scalar, equations similar to Equation (23) may be used. For example, assume the equation

$$\frac{d\chi}{dt} = \begin{cases} \frac{1}{t_c}\left(\frac{s-s_0}{s_c}\right)^m, & s > s_0, t > t_0, \\ 0, & s \le s_0 \text{ or } t \le t_0 \end{cases} \tag{36}$$

where, opposite to Equation (23), parameters m, t_c and s_0 relate to an element of microstructure, say, to a grain of a fiber segment with the attached matrix [3,13]. Let the characteristic stress applied to the element s = const. The, repeating the calculations resulting in Equation (25), we obtain

$$F_\tau(t) = 1 - \exp\left(-\left(\frac{s-s_0}{s_c}\right)^\alpha\left(\frac{t-t_0}{t_c}\right)^\beta\right). \tag{37}$$

The right hand side of Equation (37) according to Equation (33), is the damage measure $\phi(t)$ attained at the instant t. Substitution of Equation (37) into Equation (35) at $\phi \ll 1$ yields

$$F_*(t_*) = 1 - \exp\left(- \int_M \left(\frac{s - s_0}{s_c}\right)^{n_*\alpha} \left(\frac{t - t_0}{t_c}\right)^{n_*\beta} \frac{dM}{M_0}\right). \tag{38}$$

The difference between Equation (25) and (38) is in the entering power indices. But it should be noted that the parameters entering in Equations (25) and (38) in fact relate to different levels of consideration and, therefore, they could take quite different magnitudes. By the way, for surface initial cracks in granular alloys the power index n_* takes magnitude $n_* = 3$ or maybe 4, and for internal cracks n_* is much larger, $n_* = 4, \ldots, 6$ and even more. For example, if the packing of grains is cubic centered, the number of neighbors $n_* - 1 = 8$, and that results into $n_* = 9$. Combining results of direct observations of the fracture process with the statistical treatment of fatigue tests and using relationships similar to Equation (38), one can estimate parameters entering into Equation (36).

The model presented here is useful for the prediction of properties of composite materials. The properties of components (fiber, particles, interface) are usually more or less known, but a composite very often appears as a new born material which properties must be predicted beforehand. Some examples can be found in [3,14].

3. FATIGUE CRACKS GROWTH IN RANDOM MEDIA

3.1. Introductory remarks

Although stochastic problems of fatigue cracks growth are to be discussed in the lectures by other authors and, therefore, do not enter into the main body of presented lecture, we discuss this subject, not going into the literature survey and details of calculations. In fact, the fatigue cracks initiation and their further propagation form a united process since the final conditions for the former stage play the part of the initial conditions for the latter one. The fatigue life of a structure consists of two parts, and durations of these parts, generally, may be of the same order. Therefore, to predict structural safety, integrity and reliability both stages are to be taken into consideration.

Generally, account of all kinds of randomless influencing on the fatigue crack growth can be provided by randomization of common (deterministic) equations of crack growth. The account of the batch-to-batch and specimen-to-specimen scatter of material properties requires a slight modification of deterministic equations by means of replacement of some material parameters with their stochastic analogues – random variables. But to take into consideration the random variability of properties within a specimen or a structural component, these parameters are to be replaced with random functions of coordinates measured along the cracks trajectories, and, maybe, of time too.

A typical equation for an one-parameter fatigue crack with the size a has the form

$$\frac{da}{dN} = f(s_{\max}, s_{\min}, a). \tag{39}$$

Here $s_{\max}$ and $s_{\min}$ are extremal magnitudes of the characteristic stress within each cycle usually replaced with the stress range $\Delta s = s_{\max} - s_{\min}$ and the stress ratio $R = s_{\min}/s_{\max}$. Moreover, the right hand side of Equation (39) is usually replaced with a function of the stress intensity factor $K = Ys\sqrt{a\pi}$, more exactly its range $\Delta K = Y\Delta s\sqrt{a\pi}$. Here Y is calibration factor, usually of the order of unity. The most popular is the Paris-Erdogan equation

$$\frac{da}{dN} = \text{const.}(\Delta K)^m \tag{40}$$

with power index m taking, as a rule, the magnitude two to four. Most proposals to randomize Equation (40) and related equations include replacement of the constant in Equation (40) either with a random variable or a random function of a, N or both. Other models are based on the well known probabilistic models suitable for irreversible processes. A large number of Markov process models are discussed in [5]. An almost comprehensive survey of literature can be found in [6].

3.2. Equations of fatigue cracks growth

In the further study we use the equations of fatigue cracks growth developed by author [16] and discussed in detail in [17]. These equations are in fact theoretical ones. They are based on the synthesis of micro- and macromechanics of fractured and a cracked body under loading is treated there as a mechanical system with unilateral constraints. According to the principles of virtual work for systems with unilateral constraints, a system under quasistatic loading is in equilibrium if the work produced on all small admissible displacements is non-positive:

$$\delta A \leq 0. \tag{41}$$

Later all generalized coordinates are divided into the two groups: those at all fixed crack parameters (Lagrangian coordinates), and those describing sizes and shapes of cracks. The latter generalized coordinates are named in [16] the Griffith's coordinates, and the further analysis of states of a system is performed with respect to Griffith's coordinates only. Denoting the Griffith's coordinates $a_1, \ldots, a_m$, rewrite Equation (41) in the form

$$\delta_G A = \sum_{j=1}^{m} (G_j - \Gamma_j)\delta a_j. \tag{42}$$

Here index at $\delta_G A$ denotes that only Griffith's coordinates are subjected to variations. The right hand side of Equation (42) contains two kinds of generalized forces: the active, driving forces $G_1, \ldots, G_m$, and the corresponding passive, resistance forces $\Gamma_1, \ldots, \Gamma_m$. In the simplest case of the standard Griffith-Irwin problem, the force G is in fact the energy release rate, and the force Γ is the corresponding critical value of that rate – fracture toughness characteristics denoted commonly G_c.

States of cracked bodies under loading were classified in [16] on the two levels: with respect to equilibrium, and with respect to stability. Depending on the sign of $\delta_G A$, subequilibrium, equilibrium and non-equilibrium states are to be distinguished. Subequilibrium states are stable, non-equilibrium unstable, and analysis of the Griffith's variations of $\delta_G A$ is required to find out, is an equilibrium state stable, unstable or neutral. In terms of generalized forces that classification takes the form of relationship between G_j and Γ_j and their first derivatives with respect to Griffith's coordinates [3,7]. To include fatigue crack growth into this set of concepts, it is sufficiently to take into account microdamage accumulation near the crack tips (and, maybe, in the far field, too). It means that all generalized forces and especially resistance forces are to be calculated with account of microdamage. The fatigue crack growth can be represented as an alternating transition of the system from one sub-equilibrium state to the next one going each time through an equilibrium state. If jumps are small enough, the process may be replaced with an continuous one. In that case equations of fatigue cracks growth with respect to m_1 generalized coordinates take the form

$$G_j = \Gamma_j, \qquad j = 1, \ldots, m_1, \tag{43}$$

under the condition of stability

$$\sum_{j=1}^{m_1} \sum_{k=1}^{m_1} \frac{\partial (G_j - \Gamma_j)}{\partial a_k} \delta a_j \delta a_k < 0, \tag{44}$$

at all $\delta a_j > 0$, $j = 1, \ldots, m_1$. Under some additional assumptions, Equation (43) can be reduced approximately to a set of differential equations with respect to $a_1, \ldots, a_{m_1}$, i.e. to thoses analogous of the well-known equations of fatigue crack growth, say, Paris-Erdogan equation (40). In the case of one-parameter crack, Equation (44) is equivalent to the condition of stability stated in the terms of "tearing moduli" of nonlinear fracture mechanics.

Consider a fatigue crack in the classical Griffith-Irwin problem (Figure 3). Assume the equation of microdamage accumulation in the form similar to Equation (23):

$$\frac{\partial \omega}{\partial N} = \begin{cases} 0, & \Delta\sigma < \Delta\sigma_{\mathrm{th}}, \\ \left(\frac{\Delta\sigma - \Delta\sigma_{\mathrm{th}}}{\sigma_f}\right)^m, & \Delta\sigma \geq \Delta\sigma_{\mathrm{th}}. \end{cases} \tag{45}$$

Here $\Delta\sigma$ is the range of the opening stress $\sigma(x, 0, t)$ during a cycle at the prolongation of the crack $|x| > a$, $y = 0$, σ_f is resistance stress against microdamage, $\Delta\sigma_{\mathrm{th}}$ is threshold resistance stress, and m is a positive number relative to power exponent entering into

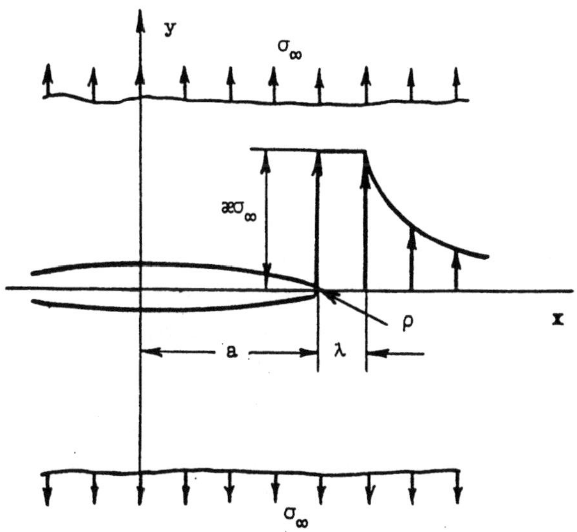

Figure 3: Fatigue crack growth in Griffith's problem.

S-N curves of Paris-Erdogan equation. Generally, these material parameters depend on temperature and stress ratio $R = \sigma_{\min}/\sigma_{\max}$.

The opening stress distribution at $|x| \geq a$, $y = 0$, is assumed $\sigma = \kappa\sigma_\infty$ at $a \leq |x| \leq a + \lambda$ and $\sigma + \sigma_\infty$ at $|x| > a + \lambda$. Here σ_∞ is applied stress and λ is length of the tip zone. The stress concentration factor is $\kappa = 1 - 2\sqrt{a/\rho}$ where ρ is interpreted as a characteristic radius of curvature on the crack tip. We may consider ρ as a measure of the stress concentration at the tip, or, if we want, as a characteristic length of an interior processing zone. Since $\rho \ll a$, so $\kappa \approx 2\sqrt{a/\rho}$. Instead of Equation (45) we obtain

$$\frac{\partial \omega}{\partial N} = \begin{cases} 0, & \Delta K < \Delta K_{\text{th}}, \\ \left(\frac{\Delta K - \Delta K_{\text{th}}}{K_f}\right)^m, & \Delta K \geq \Delta K_{\text{th}}. \end{cases} \tag{46}$$

Here ΔK is the range of the stress intensity factor during a cycle; K_f and K_{th} are material constants (at the given stress ratio, temperature and environmental conditions). Later on, we call K_f fatigue toughness (similar to fracture toughness K_{Ic}).

Neglecting the effect of microdamage on the bulk material properties, the energy release rate is given by Irwin formula, i.e. $G = K^2(1-\nu^2)/E$ where E is Young modulus and ν is Poisson ratio. The specific fracture work per unit crack area (without doubling of the upper and lower faces of the crack) let $\gamma = \gamma_0(1-\phi^\alpha)$. Here $\phi(N) = \omega(a(N), N)$, γ_0 is the specific fracture work for undamage materials, and $\alpha > 0$. Then Equation (43) at $m_1 = 1$ and Equation (46) results into

$$\frac{da}{dN} = \begin{cases} 0, & \Delta K < \Delta K_{\mathrm{th}}, \\ \dfrac{\lambda\Big((\Delta K - \Delta K_{\mathrm{th}})/K_f\Big)^m}{\Big(1 - (K_{\mathrm{max}}/K_{Ic})^2\Big)^{1/\alpha} - \omega_f(N)}, & \Delta K \geq \Delta K_{\mathrm{th}}, \end{cases} \tag{47}$$

where $K_{Ic} = \gamma_0 E[1 - \nu^2)$, and $\omega_f(N)$ is the measure of damage accumulated in the far field, before the crack tip and the related tip zone attain the material point with the coordinate $x = a(N)$. Equation (47) represents a sensible analytical model including all the stages of the crack growth diagramme. At $\Delta K_{\mathrm{th}} = 0$, $K^2_{\mathrm{max}} \ll K^2_{Ic}$, $\omega_f(N) \approx 0$ Equation (47) reduces to Equation (40), and the constant in the latter equation is λk_f^{-m}. Both lefthand branches of experimental fatigue crack growth diagrammes (at $\Delta K \approx \Delta K_{\mathrm{th}}$) and right hand branches (at $K_{\mathrm{max}} \approx K_{1c}$) are included into Equation (47). Other examples of equations of crack growth based on the synthesis of macro- and micromechanics of fracture can be found in [3,7].

3.3. Randomization of equations versus experimental data

For brevity of discussion, consider Equation (40) in notations of Equation (47),

$$\frac{da}{dN} = \lambda\Big(\frac{\Delta K}{K_f}\Big)^m. \tag{48}$$

Its right hand side depends on the applied stress range $\Delta\sigma_\infty$. Generally, $\Delta\sigma_\infty$ is a random function of the cycle number N and the crack size a (which is a random function of N with the distribution to be found). In addition, Equation (48) contains the material's parameters λ, K_f and m. The first of these parameters, characteristic length λ may be assumed deterministic with a proper choice of the other two parameters. since there is no enough information on the exponent m, it may be taken deterministic, Then we have to deal with the only random material's parameter K_f.

There are two sources of randomness of K_f. One is the variability of K_f along the crack's trajectory, and therefore K_f is a random function of the tip coordinate, i.e. of a. The second source has been called in Section 1. the batch-to-batch and/or specimen-to-specimen scatter. Hence, if we introduce a random function $K_f(x)$ of the running coordinate $x = a(N)$ it ought to be described with conditional distributions. For example, if $K_f(x)$ is taken as a homogeneous normal function (and that is an acceptable assumption if fluctuations of $K_f(x)$ are sufficiently small), the mean and the variance of this function ought to be considered as random as well, in the general case, parameters entering into its power spectral function. If a Markov model is used for $K_f(x)$, transition probabilities and initial probability distribution should be taken depending on random parameters that describe the scatter of properties between specimens or structural components.

Let reconsider from this viewpoint the relationship between experimental data and mathematical modeling of fatigue crack growth. The most popular are data by Virkler et al. [18]. There were obtained for central crack tension specimens of aluminum alloy cut from the same sheet and tested with the same initial crack size. Looking at the experimental diagrammes where the crack size a is plotted against the cycle number N (see Figure 4a where these diagrammes are shown schematically), one can not miss a striking point: the curves corresponding to various specimens intersect very rarely. It means that the scatter of crack growth is born not only from the point-to-point randomness of mechanical properties, but also, and not in a lesser degree, from the specimen-to-specimen scatter. Another explanation is that such a behavior of curves is caused by the scatter of initial conditions, i.e. by the resistance against crack growth at the beginning of loading. But in fact it is the same as specimen-to-specimen scattering.

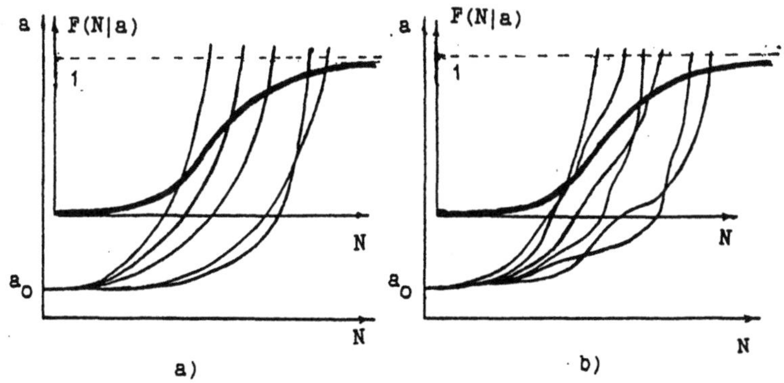

Figure 4: Comparison of experimental (a) and numerically simulated (b) fatigue crack growth curves.

This feature of the experimental fatigue crack growth curves is obvious. It is strange that, to author's knowledge, it has not been emphasized when comparison was made of experimental data and mathematical models. For example, in book [5] where a Markov type model was proposed for damage accumulation processes, both Virkler's and the corresponding numerically simulated curves are presented. The latters are shown schematically on Figure 4b. There is an evident difference in the behavior of sample functions in experiment and in simulation: opposite to experimental samples, simulated ones intersect violently. It means that the simulated random function $a(N)$ is well mixed, meanwhile the experimental results display a strong dependence of initial conditions. Those results, treated in the terms of mean and variance, are rather close since the model's parameters were estimated from the Virkler's data. The single point cumulative distribution function $F_N(N \mid a)$ also fits well (compare Figure 4ab). But in fact sample functions of quite different processes are presented in the diagrammes.

It is a trivial point to state that neither a single point distribution, or any set of single point moments can not provide the description of a continuous random function of a continuous independent variable in a unique way. Hence the problem arises of discriminating at least two different kinds of randomness of material properties. One may suppose that this problem is of no significance for engineering practice. But this point is easy to argue since engineers ought to understand clearly which properties are inherent to a certain material and can not be improved without profound chances of the manufacturing, and which properties are born, say, from the instability of manufacturing processes or from the poor quality control or both.

3.4. Numerical examples

Just to illustrate the above statements, consider a fatigue crack growth in an opening mode with account of the two sources of randomness. Let the properties of each specimen form a random field (a random function of x if a crack propagates in the x, z plane), and parameter of this field are random values. Taking the equation of fatigue crack growth in the form of Equation (47), use some special assumptions to reduce the number of random variable to minimum. In particular, let the loading process $\sigma_\infty(N)$ is deterministic and stationary, i.e. $\Delta\sigma_\infty = \text{const.}$ and $R = \text{const.}$ up to the final failure, and the material parameters λ and m are deterministic values. In addition, we express the remaining material parameters through a single random variable, say, K_f. For example, $\Delta K_{\text{th}} = 0.1 K_f$, $K_{Ic} = K_f$. Such assumptions are reasonable since all the parameters, K_f, ΔK_{th} and K_{Ic} characterize material's resistance against crack propagation and, therefore, are to correlate rather strongly. At last, although the far field damage $\omega_f(N)$ is easily to include into the numerical simulation using Equation (45), we put for simplification $\omega_f(N) = 0$.

The question is how to describe the conditional random function $K_f(x \mid I_0, I_1, \ldots)$ where I_0, I_1, ... are random parameters with joint probability density function $p(I_o, I_1, \ldots)$. There is a lot of ways to answer this question fitting experimental data quite well. For instance, $K_f(x)$ may be both a continuous of a discrete function. Although the microstructure of most of structural materials is discontinuous (from the viewpoint of continuum mechanics), a cracks, as a rule, propagates rather slowly. The crack front acts here as a kind of averager, smoothing irregularities of the microstructure. On the other hand, the computerized version of Equation (47) is a discrete one. For example, if we assume a continuous Markov model for $K_f(x)$, after discretization it becomes a kind of Markov model or semi-Markov discrete (chain) model. The well known situations in fatigue crack growth appear on the stage when elasto-plastic behavior is to be taken into account. The jump-like propagation of cracks is also covered with the mechanics of fatigue fracture [17]. To simulate this process, we must return to relationships between the generalized forces. The transition from one state of equilibrium to another is governed by equations similar to Equations (43) and (44) with a variable sign between the left- and right-hand sides. It is easily to simulate numerically this discrete process, but it is not all a Markov process.

As an example assume the random function $K_f(x)$ in the form

$$K_f(x) = I_0 + I_1 u(x). \tag{49}$$

Here I_0 and I_1 are random variables, and $u(x)$ is a normalized specified random function of the coordinate x measured along the crack path. Values I_0 and I_1 as well as values entering into the power spectral density $S_u(k)$ of the function $u(x)$ characterize the specimen-to-specimen scatter, meanwhile the within-a-specimen randomness is accounted with $u(x)$. If the latter is non-negative, then I_0 means the lower bound of the fatigue toughness parameter.

Let $u(x)$ is a nonlinear transform of normalized stationary ergodic Gauss process with the power spectral density

$$S_u(k) = \frac{2k_1 k_0}{\pi} \frac{1}{(k^2 - k_0^2) + 4k_1^2 k^2}. \tag{50}$$

Here k_0 is an analogue of frequency – the wave number connected with the characteristic length λ_0 of nonhomogeneties as $k_0 = 2\pi/\lambda_0$, and λ_1 describes the space correlation of nonhomogeneties. When k_0 and k_1 vary stochastically within a sample of specimens then they should be added to parameters I_0 and I_1. We consider for simplification k_0 and k_1 fixed. To obtain a further simplification, one may assume that either one of parameters I_0 and I_1 is deterministic or they are connected deterministically, say, I_1 is proportional to I_0. Then the only parameter I_0 remains characterizing the specimen-to-specimen scatter. When this random variable is two-side bounded, i.e. $I_- \leq I_0 \leq I_+$. Then β-distribution is natural to use for $v = I/(I_+ - I_-)$:

$$p(v) = \frac{\Gamma(\mu + \nu) v^{\mu - 1}(1 - v)^{\nu - 1}}{\Gamma(\mu)\Gamma(\nu)}. \tag{51}$$

Here $\mu > 0$ and $\nu > 0$.

In addition to Equations (47)-(51), an initial distribution of fatigue toughness is to be known, i.e. the probability density function of $K_f(a_0)$ where a_0 is the initial crack size. For example, one can assume that $K_f(a_0)$ follows the same distribution as $K_f(x)$ at $|x| > a_0$. It means that material properties are the same both near the tip of the initial crack and along the crack path, in the bulk of the material. In fact, it is not true even for laboratory specimens with initial cracks specially implanted before testing. In service, it is much more complicated situation due to the various origin of cracks, environmental actions, etc. The randomness of initial conditions that does not coincide with the randomness of material properties along the crack tip may be interpreted as the specimen-to-specimen scatter and treated correspondingly.

Some numerical results for a central crack with half length a in a plate of width b are presented in Figures 5–14. They correspond to the following set of data: $\Delta\sigma_\infty = 70$

MPa, $a_0 = 1$ mm, $b = 500$ mm, $\lambda = \lambda_0 = \lambda_1 = 1$ mm, $\Delta K_{th} = 0.05K_f$, $K_{Ic} = K_f$, $m = 4$, $\alpha = 1$, $I_1 = 0.1I_0$, $I_- = 50$ MPa $\sqrt{\mathrm{m}}$, $I_+ = 60$ MPa $\sqrt{\mathrm{m}}$, $\mu = \nu = 2$. The function $u(x)$ is taken in the form $u(x) = \sqrt{n_1^2(x) + n_2^2(x)}$ where $n_1(x)$ and $n_2(x)$ are two independent Gauss processes with zero mean and unity variance. Hence, $u(x)$ is a Rayleigh process. The single point probability density function $p(K_f)$ of the fatigue thoughness K_f is shown in Figure 5. There a sample density function obtained from the statistical treatment of resluts of the numerical simulation is plotted, too.

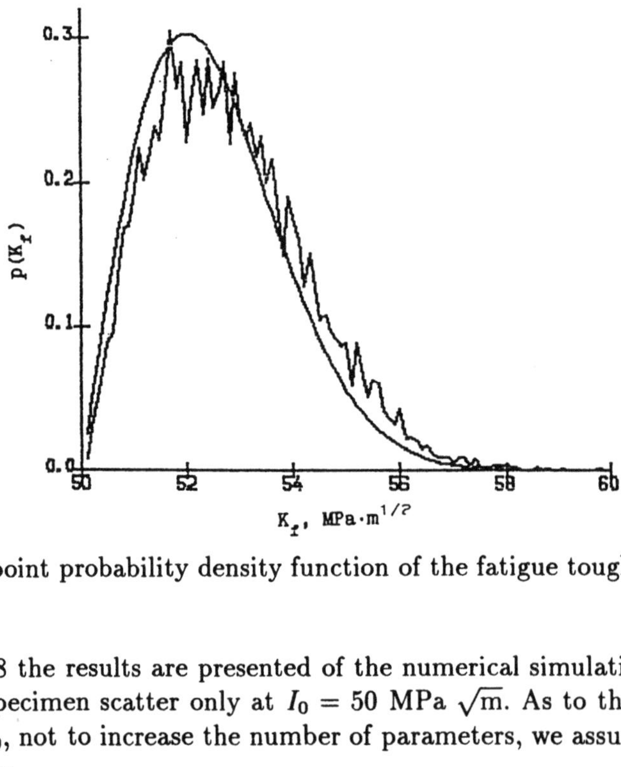

Figure 5: Single-point probability density function of the fatigue toughness and one of its samples.

In Figures 6–8 the results are presented of the numerical simulation with account of the within-a-specimen scatter only at $I_0 = 50$ MPa $\sqrt{\mathrm{m}}$. As to the distribution of $K_f(x)$ at $|x| = a_0$, not to increase the number of parameters, we assume that it is the same as at $|x| > a_0$.

A sample function $K_f(x)$ is shown in Figure 6. Fluctuations of $K_f(x)$ along the path of the crack result in variability of the crack growth rate, and, therefore, in variability of the crack size plotted against the cycle number N (Figure 7.). One of the sample functions is chosen randomly, and the pair of intersecting functions happens to be the only one among 15 trials. It is essential that the divergence of sample functions begins on the earlier stage of the crack growth. It signifies that the scatter of conditions near the initial crack tip plays a significant part in the within-a-specimen variability of fatigue crack growth. At last, Figure 8 shows the crack growth rate diagramme, where the rate da/dN is considered as a function of the range ΔK. As usually, the results are plotted in log-log scale which cover the scatter of results. By the way, all the three

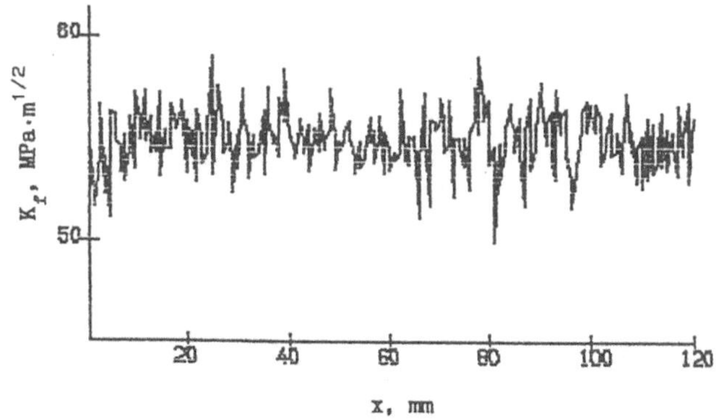

Figure 6: A sample function of fatigue toughness along the crack path due to the within-a-specimen scatter only.

stages of fatigue crack growth are represented in Figure 8. The scatter increases in approaching to the final failure.

Figures 9–11 present the same data obtained in the assumption that only the specimen-to-specimen scatter is present. It means that K_f is associated with the random value $v(I_+ - I_-)$ where v is distributed in accordance with Equation (51). The character of sample functions differs from that in Figure 6–8 significantly. In fact, crack growth sample functions do not intersect and the relationship between da/dN and ΔK plotted in log-log scale is more compact than in Figure 8. It ought to be noted that in Figure 10. (opposite to the corresponding Figure 7) the "best", the "worst", and "an average" sample functions $a(N)$ are plotted among 15 trials.

The interaction between the two kinds of randomness is illustrated in Figures 12–14 where both the within-a-specimen and specimen-to-specimen scatters are accounted. As in Figure 10, the two marginal sample functions are shown in Figure 13 as well as a pair of intersecting ones. The scatter of crack growth rate diagrammes is larger in Figure 14 than that in Figure 11, and is closer to Figure 8. By the way, all the three diagrammes are composed of several randomly chosen sample functions.

In the whole, the pictures presented in Figure 7, 8, 13 and 14 are vary alike to the published experimental data. Having in possession the assumed set of free parameters, it is possible to get their estimates that provide the necessary agreement with experiment not only in single point distributions and lower-order moments, but in more detailed statistics. We have not stated before ourselves such a task; it is enough that it looks easily attainable.

Assumptions concerning the probabilistic nature of the random function $u(x)$ and the random variable v may be varied in a wide scale. For example, Markov models can be used for $u(x)$. In that case the initial probabilistic density function $p_0(u)$ for $|x| + a_0$ and the transition probability function $p(u_k, x_k \mid u_{k-1}, x_{k-1})$ for all $|x_k| > a_0$

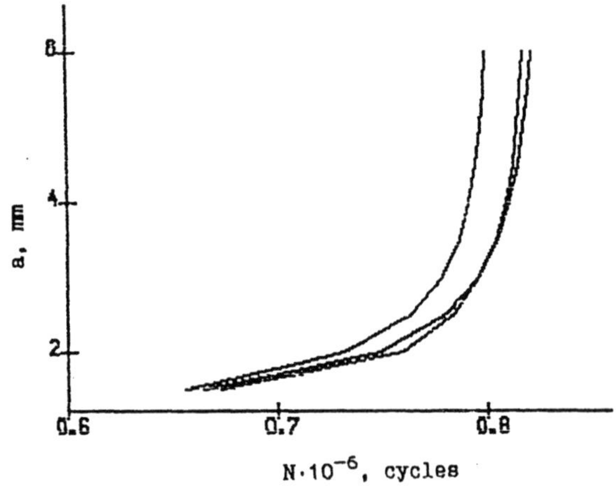

Figure 7: Sample functions of fatigue crack growth (the same conditions as in Figure 6.

are to be given (or estimated from experimental data). In numerical simulation, the discretization procedures result into models close to discrete Markov processes.

Let $K_f(x)$ follows Equation (49), but the random function $u(x)$ is a homogeneous Markovian with the transition probability density

$$p(u_k, x_k \mid u_{k-1}, x_{k-1}) = \frac{\Gamma(\mu_1 + \nu_1) u_k^{\mu_1 - 1}(1 - u_k)^{\nu_1 - 1}}{\Gamma(\mu_1)\Gamma(\nu_1)}. \tag{52}$$

Compared, say, with Equation (51) which also describes the β-distribution, the power indices $m_1 > 0$ and $m_2 > 0$ depend on the relationship between the consequent magnitudes u_k and u_{k-1} of the random variable u as well as on the relationship between the coordinates x_k and x_{k-1}. That dependence should satisfy some conditions arising from, generally speaking, semi-intuitive considerations. In particular, if in the preceding point x_{k-1} the value u_{k-1} has been close to the median of the unconditional distribution, one may await that the probability of transition to a distant value u_k is comparatively small. When u_{k-1} is near to the lower of the upper boundary of the interval $(0,1)$, the probability of moving to the central part of the distribution should be higher. Let the corresponding stationary distribution is symmetric, i.e. $\mu_1 = \nu_1 = \mu_0$, and $x_k - x_{k-1} =$ const. Then expressions

$$\begin{aligned} \mu_1 &= \mu_0(1 + C|u_{k-1} - 0.5|^n \operatorname{sign}(u_{k-1} - 0.5)), \\ \nu_1 &= \mu_0(1 - C|u_{k-1} - 0.5|^n \operatorname{sign}(u_{k-1} - 0.5)) \end{aligned} \tag{53}$$

with constant parameters $C > 0$, $n > 0$ satisfy the stated above (maybe rather vague) assumptions.

For numerical simulations divide the interval $(0,1)$ into equal parts, say, 0.1, 0.2, ..., 0.9. Each discrete value corresponds to a state of the system, i.e. to a certain

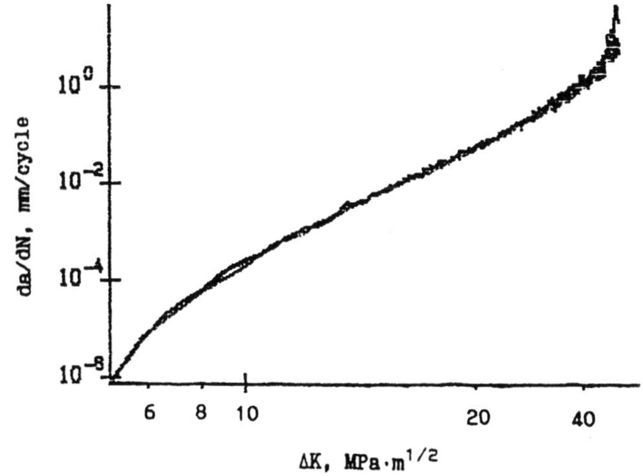

Figure 8: Sample functions of fatigue crack growth rate diagrammes (the same conditions as in Figure 6.

magnitude of the fatigue toughness $K_f(x)$ and other related parameters. Instead of the probability density given with Equation (52), we come to the transition probability matrix $\mathbf{P}$. For example, putting in Equation (53) $m_0 = 4$, $n = 1$, $C = 1$, the matrix $\mathbf{P}$ takes the form

$$\mathbf{P} = \begin{bmatrix} .25 & .28 & .21 & .13 & .07 & .03 & .01 & .00 & .00 \\ .14 & .24 & .24 & .18 & .11 & .05 & .02 & .00 & .00 \\ .07 & .17 & .23 & .22 & .16 & .10 & .04 & .01 & .00 \\ .03 & .11 & .18 & .22 & .20 & .15 & .08 & .03 & .00 \\ .01 & .06 & .13 & .19 & .22 & .19 & .13 & .06 & .01 \\ .00 & .03 & .08 & .15 & .20 & .22 & .18 & .11 & .03 \\ .00 & .01 & .04 & .10 & .16 & .22 & .23 & .17 & .07 \\ .00 & .00 & .02 & .05 & .11 & .18 & .24 & .24 & .14 \\ .00 & .00 & .01 & .03 & .07 & .13 & .21 & .28 & .25 \end{bmatrix}$$

where only first two decimal figures of the transition probabilities are given. Certainly, this matrix looks quite different from standard textbook matrices such as birth-and-death Markov chain matrix.

In Figure 16–18 some numerical results are presented for the case $I_- = 50$ MPa $\sqrt{\text{m}}$, I_+60 MPa $\sqrt{\text{m}}$. Other numerical data are the same as in the former example. Figures 16–18 show the stronger variability of $K_f(x)$ along the crack trajectory resulting, as it can be seen from Figures 17 and 18, in a larger scatter of crack growth. It should be stressed, however, that the numerical data used in the considered examples are not deliberately chosen to obtain an agreement between all the models. But the general conclusions made with respect to the former examples remain valid for this example,

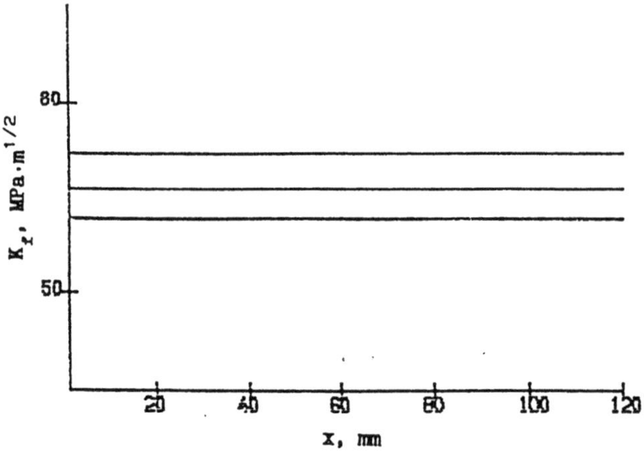

Figure 9: Sample functions of fatigue toughness due to the specimen-to-specimen scatter only.

too.

4. RELIABILITY PREDICTION IN THE PRESENCE OF SETS OF FATIGUE CRACKS

4.1. Sets of fatigue cracks

Reliability of structures under cyclic or/and long-acting loads depends essentially on the cracks' birth an growing during the lifetime of a structure. Speaking on cracks, we mean macroscopic cracks. Moreover, since we include into the analysis the crack's detection procedures, we consider a crack as macroscopic if its characteristic size a is no less than the lower detection threshold a_*. It makes the order of magnitude $a_* \approx 1$ mm and even more if conventional inspection procedures are used.

Two types of cracks are to be at least distinguished: initial crack born on the manufacturing, storage, transportation and constructions stages, and new-born cracks initiated on the in-service stage. A further differentiation is required if cracks detection during the lifetime of a structure is used. One must distinguish non-detected and detected cracks and among the latters – admitted and non-admitted ones. If some of structural components ought to be replaced during the lifetime, a new set of initial cracks enters, etc.

Initial cracks form a random sets whose properties are given by certain apriori distributions. Non-detected cracks remain to be the random item with aposteriori distributions that depend on the detection reliability. Detected cracks in principle can be identified and followed during the lifetime. But if the future loading is given in the probabilistic form, predicted sizes of detected cracks become the random values.

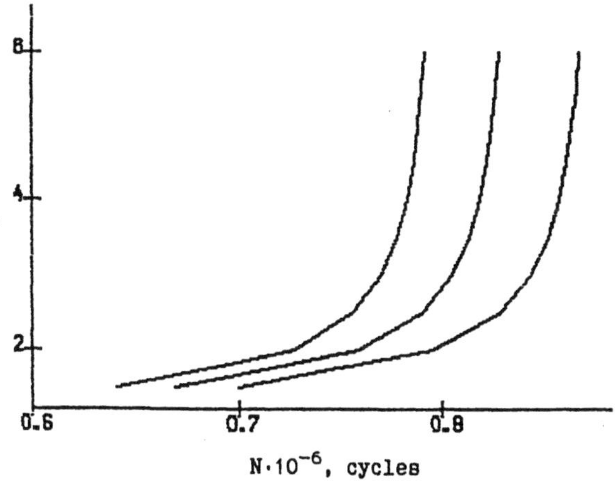

Figure 10: Sample functions of fatigue crack growth (the same conditions as in Figure 9).

Therefore, we may assume that this family of cracks form a random set, too.

Macrocracks in an operating structure are rather seldom items. As a rule, these cracks may be considered as non-interacting. Moreover, the Poisson model is applicable for the spatial distribution of cracks. The idea to use the Poisson distribution for fracture and fatigue problems generally, is natural and therefore not new. In particular, in [9] this idea was developed in a rather comprehensive way with applications to initial, new-born, non-detected, etc. cracks to build up most simple models for the assessment of reliability and lifetime of structures.

The notion of Poisson set of items located in a certain domain M_0 is well known. In terms of cracks they are as follows: all cracks in M_0 are non-interacting; the number of cracks placed in non-interacting subdomains ΔM of M_0 are independent; the probability of locating of a crack in a sufficiently small subdomain ΔM depends only on its measure with the error $\approx (\Delta M)^2$ and it is proportional to ΔM. Here, as in Sections 2.2 and 2.3, the same notations are used for domains and their measures. Probability that exactly k cracks are situated in the domain M_0 is given with an equation similar to Equation (19) where μ is the mathematical expectation of the number of cracks in M_0 at the considered time instant t. We have assumed that the nominal stress-strain state and the cracks distribution are homogeneous in M_0. Nonhomogeneous distributions can be considered as it has been done with respect the macrocracks initiation (see Section 2.2).

One of the properties of Poisson sets is their invariance with respect to a rather wide class of transformations such as screening of elements with respect to a certain property A, and mixing of independent sets. For example, let the property A is distributed homogeneously and independently among the elements of a Poisson set with probability

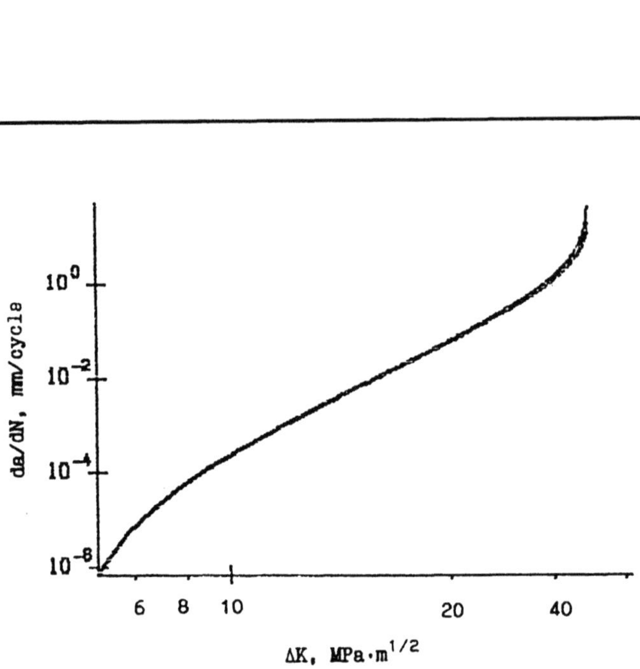

Figure 11: Sample functions of fatigue crack growth rate diagrammes (the same conditions as in Figure 9).

$P[A]$. Then the probability of locating k elements with the property A in the domain M_0 is

$$Q_k(A;t) = e^{-\mu(t)P[A]}\frac{\mu^k(t)P^k[A]}{k!}. \tag{54}$$

Similarly, if m Poisson sets are located in M_0 with expected values $\mu_1(t), \ldots, \mu_m(t)$, and probabilities of the property A are equal to $P_1[A], \ldots P_2[A]$, Equation (19) remains valid after replacing $\mu(t)P[A]$ with $\mu_1(t)P_1[A] + \cdots + \mu_1(t)P_m[A]$.

Assume that all the cracks families do not interact mutually and remain to be Poissonian after all procedures including inspection, decision making, repair, replacement and operation. For example, we assume that no change of instruments or operators is maintained during the cracks detection procedure after nonsatisfactory results of the current inspection become to be known. Restrictions of such a kind make the field of applicability of the proposed analytical model rather narrow. It is partially compensated by transparency of final equations.

4.2. Reliability with respect to fatigue fracture

The reliability and structural safety requirements put deterministic restrictions on the sizes.of cracks. Probability that these restrictions would not be violated at the time $[t_0, t]$ is the reliability function $R(t)$. Let a structure is divided into I manifolds M_1, $\ldots$, M_I, and J types of cracks are present in each manifold. Then

$$R(t) = P\Big[\max_{M_i} a_{ij}(t) < a_{ij}^*; \quad \begin{matrix} i = 1,2,\ldots,I, \\ j = 1,2,\ldots,J. \end{matrix}\Big] \tag{55}$$

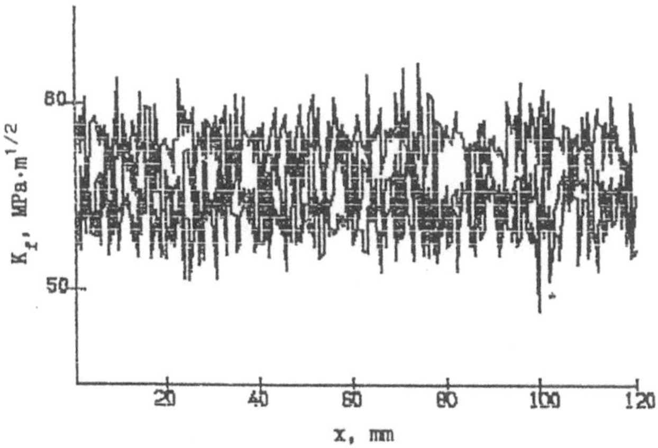

Figure 12: Sample functions of fatigue toughness due to the joint effect of the within-a-specimen and specimen-specimen scatters.

where $a_{ij}(t)$ are characteristic crack sizes and a^*_{ij} are corresponding limiting, specified sizes. Assume the expectations $\mu_{ij}(a_{ij};t)$ are known of the number of cracks from each family which size at the time t exceeds a_{ij}. Using Equation (54) and taking into account the fundamental property of Poisson sets, Equation (55) results into

$$R(t() = \exp\Big(-\sum_{i=1}^{I}\sum_{j=1}^{J}\int_{M_i}\mu_{ij}(a^*_{ij};t)\frac{dM_i}{M_{oi}}\Big). \tag{56}$$

To obtain Equation (56) each area M_i was divided into small parts ΔM_i, Equation (54) applied to each part, the independence of events used, and the transition from summation to intergration upon the areas M_i performed. If the total probability distribution functions $F_{ij}(a_{ij};t)$ are known, then

$$\mu_{ij}(a_{ij};t) = \mu_{ij}(t)\Big(1 - F_{ij}(a_{ij},t)\Big). \tag{57}$$

The difference $1 - F_{ij}(a_{ij},t)$ being put in Equation (56) play the same role as the probability $P[A]$ in Equation (54).

Situation becomes much more complicated if the limiting sizes a^*_{ij} depend on the load level at the considered time moment. Then instead of Equations (55) we must assume

$$R(t) = P\Big[\max_{M_i} a_{ij}(x,\tau) < a^*_{ij}(x,\tau);\ x \in M_i,\ \ \tau \in [t_0,t],\ \ \begin{smallmatrix} i=1,2,\ldots,I, \\ j=1,2,\ldots,J. \end{smallmatrix}\Big]. \tag{58}$$

It means that to asses the reliability function $R(t)$ we have to consider a problem of the first excursion of a multidimensional process out of an admissible region, generally, in

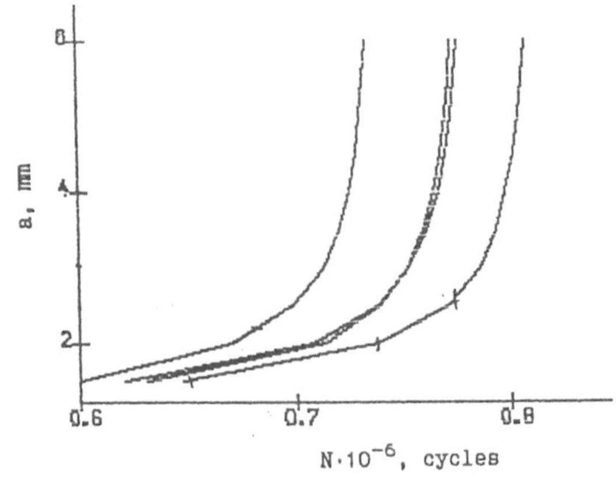

Figure 13: Sample functions of fatigue crack growth (the same conditions as in Figure 12).

the space of $I \times J$ dimensions. Simplifications are possible if we assume that excursions form a time-spatial Poisson process [2]. Then similar to Equation (56)

$$R(t() = \exp\Big(- \sum_{i=1}^{I} \sum_{j=1}^{J} \int_{t_0}^{t} \int_{M_i} \lambda_{ij}(x, \tau) \frac{dM_i \, d\tau}{M_{oi}} \Big) \tag{59}$$

where $\lambda_{ij}(x, \tau)$ are intensities of the steams of events equal to the expected values of violations of inequalities $a_{ij}^*(x, \tau) - a_{ij}(x, \tau) > 0$ per space and time unit.

4.3. Reliability of inspection procedures

The model of crack detection (most compatible with the discussed approach) is as follows. Let both detected and missed cracks form Poisson sets. It means that inspection procedures are supposed do not effect on properties of sets which under other conditions are Poissonian. For example, probability of non-detection in the region M_0 exactly k cracks which size is larger than a may be presented in the form

$$Q_k(a) = e^{-\mu_{nd}(a)} \frac{\mu_{nd}^k(a)}{k!}. \tag{60}$$

Here $\mu_{nd}(a)$ is the expected number of missed cracks, and the time dependence is omitted. The next step is to find $\mu_{nd}(a)$.

Consider the probability $P[\, D \mid a\,]$ to detect a crack of a given size a under the condition that this crack is situated in the field of an inspection device. The probability $P[\, D \mid a\,]$ can be assessed by statistical treatment of testing samples of specimens with planted defect of known sizes. Probability $P[\, D \mid a\,]$ characterizes properties of the

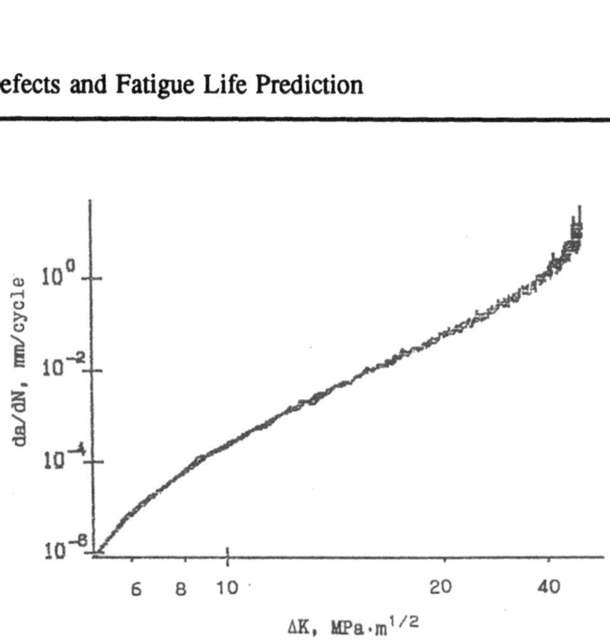

Figure 14: Sample functions of crack growth rate diagrammes (the same conditions as in Figure 12).

device as well as efficiency of the operator. Major parameters of this probability are the detection threshold a_* and the minimal size a_s of almost sure detection. For example, the following analytical representation is suitable ($\beta > 0$)

$$P[\,D \mid a\,] = \begin{cases} 0, & a \leq a_*, \\ (\frac{a-a_*}{a_s-a_*})^\beta, & a_* < a < a_s, \\ 1, & a \geq a_s. \end{cases} \tag{61}$$

Other convenient analytical forms are Pareto and three-parameter Weibull distributions.

Probability $P_D(a)$ to detect an arbitrary crack with the size larger than a placed in th field of an inspection device depends on the size a. Denote by $F(a)$ the cumulative distribution function of the size a, and by $p(a)$ the corresponding probability density function. Using Bayes' formula, we obtain

$$P_D(a) = \frac{1}{1-F(a)} \int_a^\infty P[\,D \mid a_1\,]\, p(a_1)\, da_1. \tag{62}$$

Evidently, $P_D(a) \geq P[\,D \mid a\,]$. If the size a is almost sure detectable, $P_D(a) \approx P[\,D \mid a\,] \approx 1$. If the size $a = a_d$ is deterministic, so $p(a) = \delta(a - a_d)$, and Equation (62) results into $P_D(a_d) = P[\,D \mid a\,]$. For $a < a_*$ we take $P_D(a) = P_D(a_*) > 0$.

In the general case the relation between $P_D(a)$ and $P[\,D \mid a\,]$ depends on the apriori size distribution. It is illustrated in Figure 19 where diagrammes of $P[\,D \mid a\,]$ and $P_D(a)$ are shown for the three cases of the probability density $p(a)$. In the case 1 the size of most cracks is less than the detection threshold a_*. In the case 2 the mode of the size

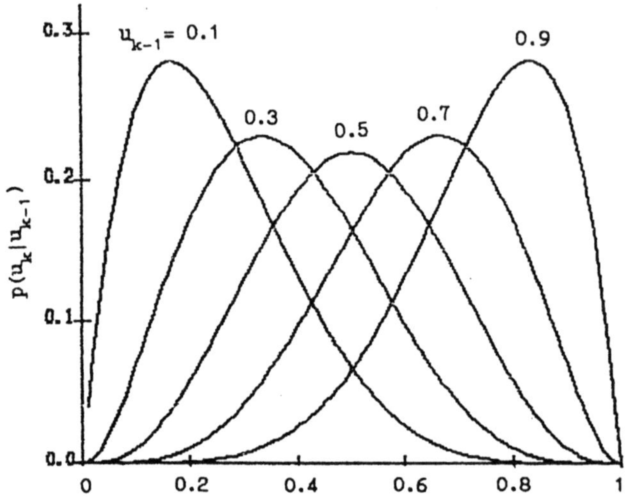

Figure 15: Transition probability $p(u_k, x_k \mid u_{k-1}, x_{k-1})$ at various u_{k-1} and $x_k - x_{k-1} =$ const.

distribution is approximately equal to a_*. In the case 3 almost all cracks are larger than a_*, and the inspection seemed to be most reliable although the probability of detection $P[\,D \mid a\,]$ is the same in all the considered cases.

To estimate the probability $P_D(a)$, special tests are required using samples of specimens with cracks distributed according to he function $F(a)$. Most records concerning post-mortem analysis of non-detected defects in pressure vessels, pipings, etc., in fact relate to the probability $P_D(a)$. As these records depend rather strongly on the crack distribution in a certain structure, they can not be considered as a "pure" assessment of the reliability of an inspection device and/or an operator.

Let consider regin M_0 with a homogeneous Poisson spatial distribution of cracks. If the expected number of all cracks in M_0 is μ_0, the expected number of cracks which size is larger than a is $\mu(a) = \mu_0(1 - F(a))$. The expected value of undetected cracks is

$$\mu_{nd}(a) = \mu(a)(1 - P_D(a)). \tag{63}$$

Denote $Q(a)$ the risk function equal to probability to miss at least one crack of the size larger than a. As $Q(a) = 1 - Q_0(a)$ where $Q_0(a)$ is determined by Equation (58), we obtain

$$Q(a) = 1 - \exp^{-\mu(a)(1-P_D(a))}. \tag{64}$$

If cracks are distributed non-homogeneously upon the region M, Equation (64) should be generalized in the following way

$$Q(a) = 1 - \exp\Bigl(-\int_M \mu(a)(1 - P_D(a))\frac{dM}{M_0}\Bigr). \tag{65}$$

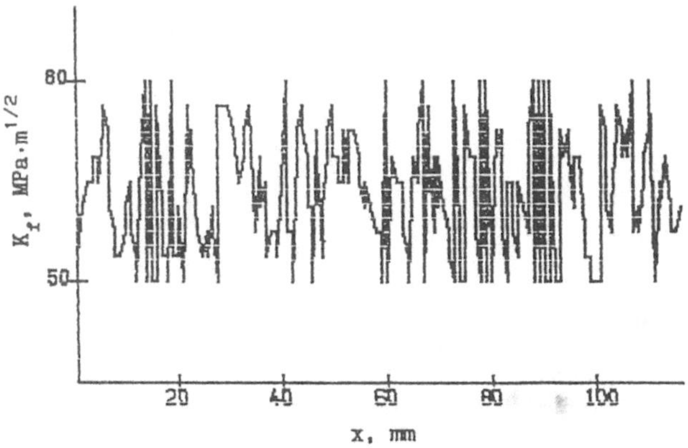

Figure 16: A sample function of fatigue thoughness along the crack path simulated with a Markov process (explanations in text).

All variables entering into the right-hand side of Equation (65), generally, depend on the position of a point in M. For example, if only a part M_0 of the region M is inspected, we put in Equation (65) $P_D(a) = 0$ in $M \setminus M_c$. Supposing that in the other aspects the properties in M are homogeneous, we obtain

$$Q(a) = 1 - \exp\Big(-\frac{M}{M_0}\mu(a)\Big(1 - \frac{M_c}{M}P_D(a)\Big)\Big). \tag{66}$$

The ratio M_c/M entering into Equation (66) corresponds to the part of non-reliability originated from the partial inspection of the area.

Some numerical results are plotted in Figure 20. The probability $P[\,D \mid a\,]$ is taken according to Equation (59), and the apriori size distribution is Weibull distribution

$$F(a) = 1 - e^{-(a/a_c)^\alpha}. \tag{67}$$

The solid lines correspond to $P[\,D \mid a\,]$ and the light ones to $P_D(a)$. Magnitudes of power indices α and β in Equation (61) and (67) are shown near the curves. In Figure 21 the probabilities $Q_k(a)$ to miss k cracks are plotted for the case $\mu = \alpha = 1$, and $k = 1$, 2 and 3. Since $Q_1(a)$ is approximately an order of magnitude larger than $Q_2(a)$, and the latter is larger than $Q_3(a)$, etc., the risk function $Q(a) = Q_1(a) + Q_2(a) + \cdots$ plotted in a logarithmic scale is rather close to $Q_1(a)$.

Using Equations (62) and (63), it is easy to find cumutative distribution function $F_{nd}(a)$ for missed cracks, and $F_d(a)$ for detected cracks. As $F_{nd}(a) = 1 - \mu_{nd}(a)/\mu_{nd}(a_*)$, we obtain

$$F_{nd} = 1 - \Big(1 - F(a)\Big)\Big(1 - P_D(a)\Big)\Big(1 - P_D(a_*)\Big)^{-1}. \tag{68}$$

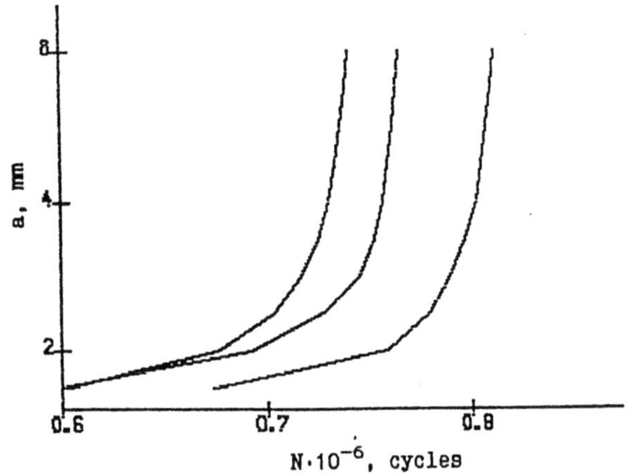

Figure 17: Sample functions of fatigue crack growth (the same conditions as in Figure 16).

Similarly $F_d(a) = 1 - \mu_d(a)/\mu_d(a_*)$ where $\mu_d(a) = \mu(a) - \mu_{nd}(a)$. Hence

$$F_d(a) = 1 - \Big(1 - F(a)\Big)\Big(P_D(a)/P_D(a_*)\Big). \tag{69}$$

Differentiating Equations (68) and (69) with respect to a, we obtain the corresponding probability density functions. They coincide with those given in [19]. Some data on the distribution of initial cracks and crack-like defect in steel structures can be found in [19–21].

4.4. Some asymptotic estimates of risk indices

The total umber of cracks and crack-like flaws in a large structures, say, in piping systems of power plants, transmission gas and oil pipelines, off-shore platforms, etc., might be very large. Meanwhile, reliability and safety requirements state that the admissible risk indices quantified as probabilities of structural failure ar to be very small compared with unity, e.g. $Q_* = 10^{-6}$ per year and even less. It means that we must deal with extremely small probabilities and, respectively, with very seldom events. Some possibilities to obtain more reliable results in this vague area are discussed in the Section.

Consider, for brevity, Equation (56) for a single set of one-parameter cracks with the random size a situated in the domain M. Equation (56) takes the form

$$R(t) = \exp\Big(-\frac{1}{M_0}\int_M \mu_0(t)\, h(t)\, dM\Big) \tag{70}$$

where μ_0 is the expected number of cracks with $a > a_*$ situated in the domain M at the time t, $h(t)$ is the probability that a randomly chosen crack happen to be non-

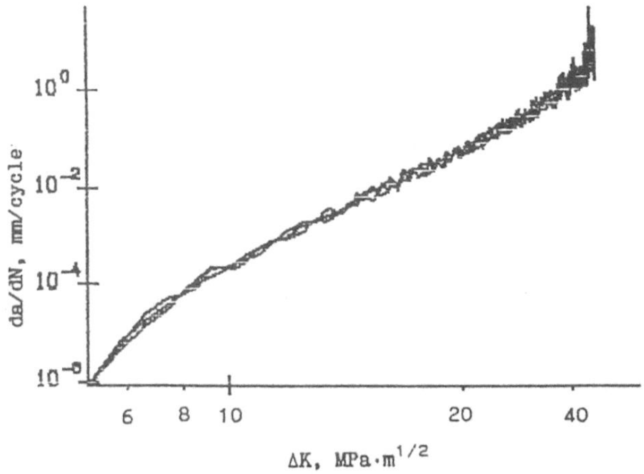

Figure 18: Sample functions of crack growth rate diagrammes (the same conditions as in Figure 16).

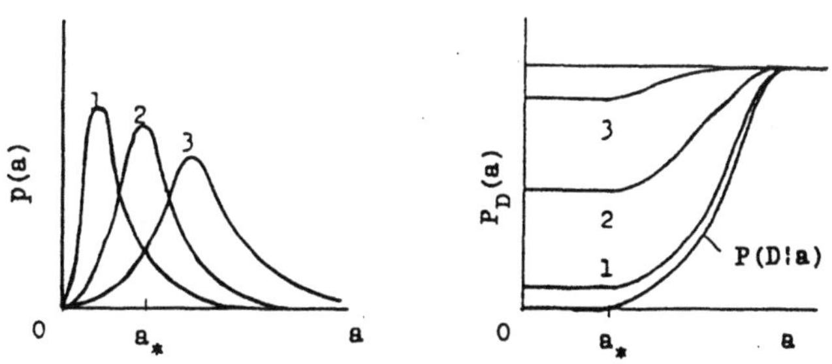

Figure 19: Probabilities of crack detection $P[\,D \mid a\,]$ and $P[\,D\,]$ for three cases of initial cracks distribution.

admissible. This probability is in fact a risk function related to one crack. This measure includes the risk connected with initial cracks as well the risk as result of the initiation and growth of new-born cracks.

When the limiting size of cracks a^* is specified, $h(t) = P[\,a(t) \geq a^*\,]$, or $h(t) = 1 - F(a^*, t)$. Here $F(a,t)$ is the cumulative distribution function of the size a at the time t. This function accounts for all types of cracks: initial, new-born, detected, etc. Since the amount of experimental data is limited, their statistical treatment does not allow to obtain results of proper confidence in the area of very small probabilities. Hence, the problem arises of extrapolation of results in the area of seldom events. To make this extrapolation more confidentable, a wide use of asymptotic distributions of extremes is to be recommended [7,8]. Their "tails" are, generally, more reliable than those of the distributions commonly used by engineers such as normal or log-normal

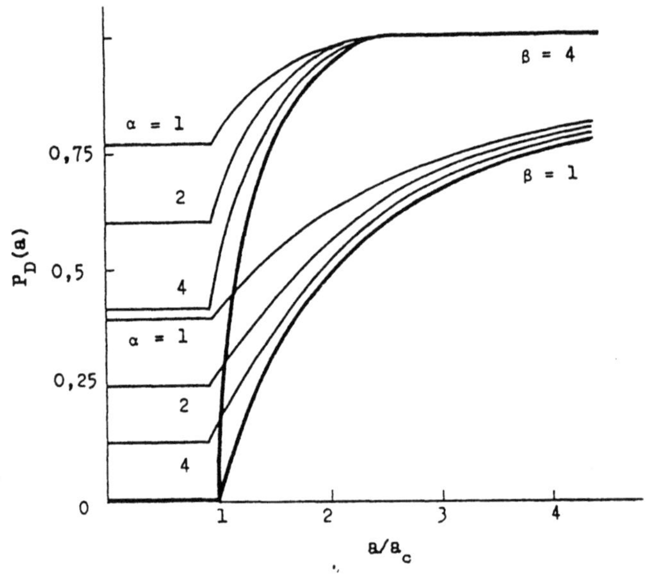

Figure 20: Probabilities of crack detection $P[\,D \mid a\,]$ and $P_D(a)$ versus a, α and β.

distributions. Additional simplifactions without loss of confidence can be achieved if an asymptotic approach [9] is applied to the calculation of integral convolutions entering into equations for $h(t)$.

As an illustrative example consider a set of one-parameter cracks in conditions when the risk function $h(t)$ from Equation (70) is given as

$$h(t) = P[\,K(t) \geq K_{IC}\,]. \tag{71}$$

Here $K(t)$ is the maximum of stress intensity factor on the segment $[\,t_0, t\,]$, and K_{Ic} is the corresponding fracture toughness assumed independent of time. If $K = Y\sigma\sqrt{a\pi}$, then

$$h(t) = \iint_{Y^2\sigma^2 a\pi \geq K_{Ic}^2} p_\sigma(\sigma, t)\, p_a(a, t)\, d\sigma\, da. \tag{72}$$

Here $p_\sigma(\sigma, t)$ and $p_a(a, t)$ are probability density functions of applied stress $\sigma(t)$ and crack size $a(t)$, respectively, and the integration area corresponds to the random event occurring into the right-hand side of Equation (71). Generally, the time instants corresponding to maxima of process $\sigma(t)$ and $a(t)$ do not coincide. The position of maximal $\sigma(t)$ is arbitrary, and maximum of $a(t)$ is attained to the end of the segment $[\,t_0, t\,]$. Since the mutual dependence of process $\sigma(t)$ and $a(t)$ on sufficiently long time segments

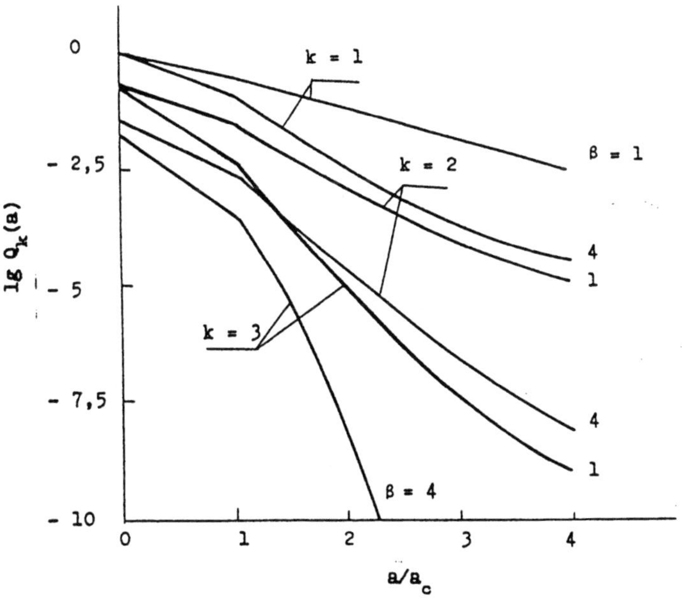

Figure 21: Probabilities $Q_k(a)$ to miss k cracks at $\mu = \alpha = 1$ and various β.

is small, one can rewrite Equation (72) in one of the two equivalent forms:

$$h(t) = \int_0^\infty F_a\Big(\frac{K_{Ic}^2}{Y^2\pi\sigma^2};t\Big)p_\sigma(\sigma;t)\,d\sigma,$$
$$h(t) = 1 - \int_0^\infty F_\sigma(\sigma;t)p_a\Big(\frac{K_{Ic}^2}{Y^2\pi\sigma^2};t\Big)\,d\sigma. \qquad (73)$$

The first of the equations contains the cumulative distribution function $F_a(a;t)$ of the crack size and the second one the cumulative distribution function $F_\sigma(\sigma;t)$ of maximal stress. The corresponding density functions are denoted $p_a(a;t)$ and $p_\sigma(\sigma;t)$.

The right-hand sides of Equations (73) are small compared with unity. On the other hand, when the risk is small, the main input into the integral is carried with the "tails" of distributions. It is illustrated in Figure 22 where the case (a) corresponds to the first, the case (b) to the second of Equations (73). The dotted lines show the ways to simplify calculations substituting exact functions with their estimates valid in the "tail" area.

Let the statistical treatment of experimental data allow to suggest that both maximal stresses and maximal crack sizes follow Fisher-Frechet distribution:

$$F_a(a;t) = e^{-(a_c/a)^\alpha}, \quad a \gg a_c,$$
$$F_\sigma(\sigma;t) = e^{-(\sigma_c/\sigma)^\beta}, \quad \sigma \gg \sigma_c. \qquad (74)$$

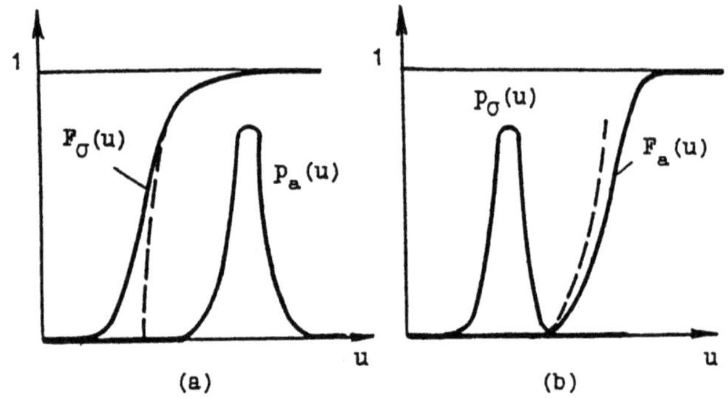

Figure 22: Two approaches to asymptotic estimation of the risk index.

Parameters a_c, σ_c, $\alpha > 0$ and $\beta > 0$, generally, depend on time t. With account of Equation (74), Equations (74) take the form

$$h = 1 - \beta \int_0^\infty u^{-(\beta+1)} e^{-(\eta u)^{2\alpha} - u^{-\beta}} du,$$
$$h = 1 - 2\alpha\eta^{2\alpha} \int_0^\infty u^{2\alpha-1} e^{-(\eta u)^{2\alpha} - u^{-\beta}} du. \tag{75}$$

Here nondimensional parameter $\eta = a_c \pi Y^2 \sigma_c^2 / K_{Ic}^2$ is introduced characterizing the relationship between the nominal stress level, defectness of material and its fracture toughness.

At $\eta \ll 1$, $2\alpha > \beta$ the exponential function in the second of Equations (75) can be replaced with the first terms of series, $e^{-u^{-\beta}} \approx 1 - u^{-\beta}$. It results into estimate $h \approx \eta^\beta \Gamma(1 - \beta/2\alpha)$. This way is illustrated in Figure 22a. If $\eta \ll 1$, $2\alpha < \beta$, the first of Equations (75) is expedient to apply with replacement $e^{-(\eta u)^{2\alpha}} \approx 1 - (\eta u)^{2\alpha}$ (see Figure 22b). Then $h \approx \eta^{2\alpha}\Gamma(1 - 2\alpha/\beta)$. At $2\alpha = \beta$ integrals in Equations (75) result, without any simplification, in $h = 1 - 2\eta^\alpha K_1(2\eta^\alpha)$ where $K_1(\cdot)$ is MacDonald function. To keep the same level of approximation, replace the latter with its asymptotic estimates. Summarizing the obtained results, we get [9]

$$h \approx \begin{cases} \eta^\beta \Gamma(1 - \beta/2\alpha), & 2\alpha > \beta, \\ \eta^{2\alpha}(2\ln(\eta^{-\alpha}) - 2\gamma + 1), & 2\alpha = \beta, \\ \eta^{2\alpha}\Gamma(1 - 2\alpha/\beta), & 2\alpha < \beta. \end{cases} \tag{76}$$

Euler's constant $\gamma = 0.577\ldots$ enter in one of these equations.

Computational results with the use of Equations (75) and (76) are presented in Figure 23. The solid lines correspond to the results obtained with the direct numerical integration, the dashed ones – those by means of the symptotic estimates. Except for small regions of η close to unity, the asymptotic estimates yields a good agreement with exact results. The less the magnitude of η, the less is the error. Equation (76) presents

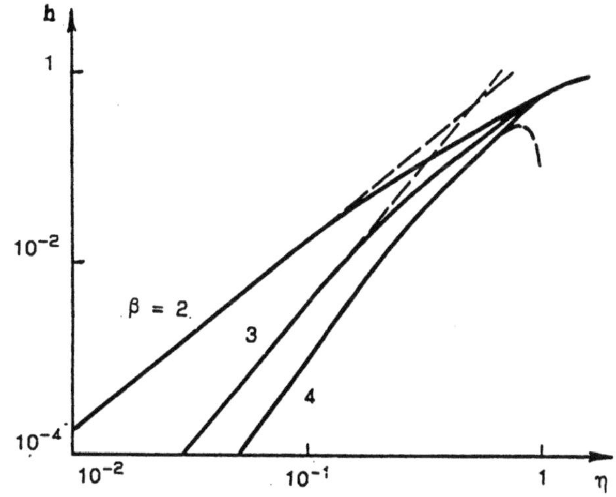

Figure 23: Risk indices obtained with the direct numerical integration (solid lines) and their asymptotic estimates (dashed lines).

very good results at $\eta \leq 10^{-4}$ when direct results happen to become unreliable because of the computational instability. The asymptotic estimates similar to Equation (76) were obtained for another combinations of distributions of extremes values, e.g. when $K_{\max}$ and K_{IK} follow the Gumbel's (double exponential) distribution [22].

4.5. Decision making

Various approaches to the estimation of structural reliability in presence of fatigue cracks were discussed in [3,20,23,24]. Hereafter we use Equations (55) and (56) based on the assumption that cracks form a Poisson set.

Let R_* is the specified limit magnitude of the reliability against cracks growth. The criterion for decision making on the design stage is

$$R(T) \geq R_* \tag{77}$$

where T is the specified lifetime or the time segment until the next inspection. The reliability function $R(t)$ is to be calculated with the use of apriori distributions of initial cracks and crack-like defect as well as cracks initiated on the segment $[t_0, t]$. If a structural component is inspected before the operational stage, the function $R(t)$ is to be calculated taking into account detected, missed and new-born cracks. The problem is more complicated when the restrictions upon the sizes of cracks are given by Equation (58). Methods of estimation of parameters $\lambda_{ij}(x, t)$ entering into Equation (59) were discussed in detail in [1,2]. If the loads are varying rather rapidly and the cracks grow sufficiently slowly, it is expedient to calculate the expected number of down crossings of the "slow" process $a_{ij}(t)$ with the "fast" process $a^*_{ij}(t)$ [3,25].

The problem of the maximal size of cracks admitted by inspection procedures is of special interest. Firstly consider the case of a homogeneous nominal stress-strain state and one-parameter set of cracks with the characteristic size a. Denote a_{ad} the admissible size of cracks, i.e. acception level. If $a > a_{ad}$, the structural component ought to be replaced or repaired. Under deterministic loading the criterion to find a_{ad} takes the form

$$L(T \mid a_{ad}, t_0) \leq a^*. \tag{78}$$

The left-hand side of Equation (78) is the solution of an equation of crack growth with initial condition $a(t_0) = a_{ad}$, and a^* is the limit size of cracks on the operational stage. Under stochastic loading Equation (78) is to be substituted with

$$F(a^*, T \mid a_{ad}, t_0) \leq 1 - R^*_{ad} \tag{79}$$

where R^*_{ad} is the specified reliability measure with respect to followed cracks. Equations (78) and (79) put a limit on the upper bound of cracks. If the number of cracks in an structure is large, Equations (56), (57), etc. should be applied. For example, let the two types of cracks are counted: admitted after being detected, and non-detected. The reliability function entering into Equation (77) takes the form

$$R(t) = \exp\Big(-\frac{M}{M_0}(\mu_d(a^*, T) + \mu_{nd}(a^*, T))\Big). \tag{80}$$

Here $\mu_d(a,t)$ is the expected number of cracks detected at the time t_0, then admitted and propagated at the time t until the size a. Similar to Equation (63)

$$\mu_d(a,t) = \mu_0 P_D(a_*)\Big(1 - \int_{a_*}^{a} F(a,t \mid a_0, t_0)\, p_d(a_0)\, da_0\Big). \tag{81}$$

The expected total number of detected cracks in the reference area M_0 is $\mu_0 P_D(a_*)$. The probability density function corresponding to the cumulative distribution function $F_d(a)$ is denoted $p_d(a)$. The expected number of non-detected cracks under the same circumstances is

$$\mu_{nd}(a,t) = \mu_0\Big(1 - P_D(a_*)\Big)\Big(1 - \int_{a_*}^{a} F(a,t \mid a_0, t_0)\, p_{nd}(a_0)\, da_0\Big). \tag{82}$$

The expected total number of non-detected cracks is calculated from Equation (63). The probability density function $p_{nd}(a)$ corresponds to the cumulative distribution function $F_{nd}(a)$ given by Equation (68).

Substituting Equation (79) into Equation (77) results

$$\mu_d(a^*, t) \leq \frac{M_0}{M} \ln \frac{1}{R_*} - \mu_{nd}(a^*, t). \tag{83}$$

The threshold value a_{ad} for cracks which may be admitted at the time t_0 enter into Equation (83) through the density $p_d(a)$ truncated at $a \leq a_{ad}$. However, the value

a_{ad} may be explicitly introduced into Equation (81) as the upper integration limit. If specifying of the admittance level a_{ad} is connected with the choice of instrumentation and/or inspection procedure, the size a_{ad} enter into detection probability $P_D(a)$ and therefore into the right-hand side of Equation (83).

When all detected cracks are followed, instead of the mathematical expectation $\mu_d(a,t)$ the estimate

$$\mu_d(a,t) = \sum(1 - F(a,t \mid a_d, t_0))$$

is to be introduced into Equations (79) and (83). Summation is applied upon all detected and admitted cracks in the reference region with the scale measure M_0.

Several ways of generalization of the presented decision rules are seen from Equations (65),(66), etc. Consider the case when all fields in M are homogeneous, but only a part M_c is subjected to inspection. Instead of Equation (79) we obtain

$$R(t) = \exp\Big(-\frac{M_C}{M_0}\big(\mu_d(a^*,T) + \mu_{nd}(a^*,T)\big) - \frac{M - M_c}{M_0}\mu(a^*,T)\Big). \tag{84}$$

The expected number $\mu(a,t)$ of cracks in the non-inspected region $M \setminus M_c$ is calculated with the use of probability density $p_0(l_0)$. The inequality

$$\mu_d(a^*,T) + \mu_{nd}(a^*,T) + \frac{M - M_c}{M_c}\mu(a^*,T) \leq \frac{M_0}{M}\ln\frac{1}{R_*}$$

contains two parameters (a_{ad} and M_c). Choosing then properly, a rational sampling inspection procedure will be planned.

Equations (79)–(84) present the general methodology for planning of inspections. All the principal factors involved into inspections and decision making enter these equations. Among them there are not only the specified reliability measure R_*, but initial cracks distribution and their expected number, reliability of inspection, loading parameters, specified lifetime T, etc.

The problems of decision making during intermediate inspection are stated Similarly. Instead of Equation (77) we use the criterion

$$R(t_{k+1} \mid t_k) \geq R_*. \tag{85}$$

The left-hand side is the reliability function on the prediction time segment $[t_k, t_{k+1}]$ calculated with account of information collected on the previous segment $[t_0, t_k]$. All types of cracks ought to be taken into consideration: detected followed and missed cracks, initial cracks in replaced and repaired components, cracks which are initiated between two sequential inspections. One of the approaches for decision making is to fix the time t_{k+1} for the next inspection putting in Equation (85) the sign of equality. This criterion can be applied together with an economic criterion. The economic criterion usually of optimization nature, and the reliability criterion enters in it as a constant.

References

[1] Bolotin, V.V. *Statistical Methods in Structural Mechanics*, Stroyizdat, Moscow 1961 (1st ed.), 1965 (2nd ed.), Engl. transl.: Holden Day, San Francisco 1969; German transl.: Verlag für Bauwesen, Berlin 1981.

[2] Bolotin, V.V. *Random Vibration of Elastic System*, Nauka, Moscow; Engl. transl.: The Hague et al. 1984.

[3] Bolotin, V.V. *Prediction of Service Life for Machines and Stuctures*, Mashinostroyeniye, Moscow 1984 (1st ed.), 1990 (2nd ed.); Engl. transl.: ASME press, New York 1989.

[4] Kogayev, V.P. *Strength Analysis under Stresses Varying in Time*, Mashinostroyeniye, Moscow 1977 (in Russian).

[5] Bogdanoff, J.L. and Kozin, F. *Probabilistic Models of Cumulative Damage*, John Wiley, New York, 1985.

[6] Sobczyk, K. and Spencer, B.F. *Random Fatigue: From Data to Theory*, Academic Press, New York et al., 1991.

[7] Bolotin, V.V. Theory of extremes and its applications to mechanics, in *Proc. of 1st World Congress of Bernoulli Society*, VNU Press, Utrecht, **2** (1987), 443–459.

[8] Castillo, E. *Extreme Value Theory in Engineering*, Academic Press, New York, 1988.

[9] Bolotin, V.V. Reliability of non-destructive monitoring systems, in: *Methods of Diagnostics of Composite Structures*, Zinatne, Riga, 1986, 5–18 (in Russian).

[10] Bolotin, V.V. Sets of multiple cracks in solids with application to the fatigue life and reliability prediction, in: *Probabilistic Methods in Mechanics of Solids and Structures*, IUTAM Symposium, Stockholm 1984 (Eds. Eggwertz, S. and Lind, N.C.), Springer, Berlin et al. 1985, 177–198.

[11] Kachnov, L.M. *Introduction to Continuum Damage Mechanics*, Martinus Nijhoff, The Hague et al. 1986.

[12] Bolotin, V.V. On the theory of delayed fracture, Trans USSR Acad. Sci., Mechanics of Solids, 1981, **N1**, 137–146, (in Russian).

[13] Bolotin, V.V. A unified model of fracture of composite materials under sustained loads, *Mechanics of Composite Materials*, 1981, **N3**, 405–420, (in Russian).

[14] Bolotin, V.V. Reliability of composite structures, in: *Handbook of Composites*, (Eds. Kelly, A and Rabotnov, Yu.N.), **2**, *Structures and Design*, North-Holland, Amsterdam, 1989. 263–349.

[15] Ditlevsen, O. and Sobczyk, K. Random fatigue crack growth with retardation, *Engineering Fracture Mechanics*, **29** (1986), 861–878.

[16] Bolotin, V.V. Equations of fatigue cracks growth, Trans. USSR Acad. Sci., *Mechanics of Solids*, 1983, **N4**, 153–160, (in Russian).

[17] Bolotin, V.V. Mechanics of fatigue fracture, in: *Nonlinear Fracture Mechanics*, CISM course, **N314**, (Ed. Wnuk, M.), Springer, Berlin, 1990, 1–69.

[18] Virkler, D.A., Hillberry, B.A. and Goel, P.K. The statistical nature of fatigue crack propagation, *Journal of Engineering Mechanics and Technology*, Trans. ASME 101 (1979), 148–152.

[19] Heller, R.A. and Stevens, G.H. Bayesian estimation of crack initiation times from service data. *Journal of Aircraft*, **15** (1978), N11, 794–798.

[20] Engesvik, K.N. and Moan, T. Probabilistic analysis of the uncertainty in the fatigue capacity of welded joints, *Engineering Fracture Mechanics*, **18** (1983), 743–762.

[21] Payne, A.O., Mallison, G.D. and Graham, A.D. Reliability approach to structural safety and safety criteria, in: *Probabilistic Methods in the Mechanics of Solids and Structures*, Proc. IUTAM Symposium, Stockholm 1984 (Eds. Eggwertz, S. and Lind, N.C.), Springer, Berlin et al. 1985, 529–557.

[22] Bolotin V.V. and Chirkov, V.P. Asymptotic estimates of non-failure probabilities for action-resistance models, *Problems of Mechanical Engineering and Reliability of Machines*, 1992, **N6**, 3–12, (in Russian).

[23] Yang, J.N. and Chen, S. Fatigue reliability of structural components under scheduled inspection and repair maintenance, in: *Probabilistic Methods in the Mechanics of Solids and Stuctures*, Proc. IUTAM Symposium, Stockholm 1984, (Eds. Eggwertz, S. and Lind, N.C.), Springer, Berlin et al. 1985, 559–568.

[24] Palmberg, B., Blom, A.F. and Eggwertz, S. Probabilistic Damage Tolerance Analysis of Aircraft Structures, FFA, Stockholm, 1985.

[25] Bolotin, V.V. On safe crack sizes under random loading, Trans. USSR Acad. Sci., Mechanics of Solids, 1980, **N2**, 124–130, (in Russian).

STOCHASTIC CUMULATIVE MODELS FOR FATIGUE

K. Sobczyk
Polish Academy of Sciences, Warsaw, Poland

ABSTRACT

This chapter is devoted to modelling of fatigue crack growth via cumulative jump stochastic processes. A wide class of such processes can be represented in the form of a stochastic integral with respect to counting Poisson measure in particular, such processes take a form of a random sum of random components. The cumulative jump processes are adopted as a model-process for fatigue crack growth in engineering materials subjected to random loading and other uncertainties. After brief presentation of the previous author' results, the extension of the approach to a more general situation where elementary crack increments are mutually correlated is shown. The use of Morgenstern model for the joint probability distribution leads to approximate characterization of the crack size. The results quantify the effect of the correlation in crack's increments on statistics of a crack size and show the features analogous to those in existing experimental data. In the last part of the chapter a stochastic cumulative jump model of fatigue crack growth with retardation is presented.

1. INTRODUCTION

According to many experimental investigations fatigue crack grows intermitently and the growth itself consists of active and dormant period. It is known that fatigue crack propagates successively creating strations which, as it is observed, are most often randomly spaced with considerable statical dispersion (cf. Itaaki, Shinozuka [1]). This is supposed to be mainly due to the fact that the damping stresses experienced by the material and causing crack growth are primarily generated by such "factors" as peaks of the loading process, its rises and falls, etc., which are random and discrete events at time.

In spite of this general mechanism, shocks are a common case of time-varying loads which occur in separate, discrete instants of time. They can, for instance, be generated by explosion-type loads. But also the loads acting on offshore structures due to sea waves during storm weather conditions can be regarded as sequences of random shocks.

It is, therefore, justified to regard the crack growth process as a discontinuous random process consisting of random number of jumps. Although some elements of such a reasoning are also present in the Markov chain models, it seems to be important to treat random elementary crack increments (jumps) in fatigue growth in more explicit way. The approach to modelling of fatigue crack growth by random sums of random components has recently been elaborated in the papers by Sobczyk and Trębicki [2], [3] and for random crack growth with retardation by Ditlevsen and Sobczyk [4]. In this paper we will expound the basic features of the cumulative jump models along with the results relevant to fatigue life prediction.

2. DESCRIPTION OF THE MODEL

Fatigue process is characterized here by a stochastic process $A(t,\gamma)$ which is regarded as the length of a dominant crack at time t. The letter γ symbolizes the elementary event belonging to the sample space Γ ; for each $\gamma \in \Gamma, A(t,\gamma)$ represents a possible sample function of the crack growth. Process $A(t,\gamma)$ is represented as a random sum of random components:

$$A(t,\gamma) = A_0 + \sum_{i=1}^{N(t)} Y_i(\gamma), \qquad Y_i(\gamma) = \Delta A_i \tag{1}$$

where A_0 is the initial crack length of sufficient size to propagate (assumed to be known from experimental data); it can be regarded as a deterministic constant or as a random variable. $Y_i(\gamma)$ are the random variables characterizing the magnitudes of elementary crack increments in crack length and $N(t)$ is an integer-valued stochastic process (a counting stochastic process) characterizing the number of crack increments in the interval $[0,t]$.

A a matter of fact, the model-process $A(t,\gamma)$ constitutes a class of cumulative models; particular members of this class are specified by the hypotheses posed on the counting process $N(t)$ and on the random variables $Y_i(\gamma), i = 1,2,\ldots$.

Although the counting process $N(t)$ could be taken in quite general form, in order to make the model effective we have to assume the appropriate specific process (for example, inhomogenoeus Poisson process with suitable intensity function, birth process, etc.). Since in modelling of fatigue the growth intensity should be state-dependent, it seems that the birth process could be a possible candidate for $N(t)$.

As far as components $Y_i(\gamma)$, $i = 1, 2, \ldots$ are concerned, we shall assume first that they are independent and identically distributed, non-negative random variables, and then extension will be presented when $Y_i(\gamma)$ are correlated random variables. Through all analysis we additionally assume that $N(t)$ is independent of $\{Y_i(\gamma)\}$.

In the model (2.1) randomness of fatigue crack growth process is accounted for via the probabilistic mechanism of the transition from one fatigue state (crack's length) to another (with pure birth intensity) and by the probability distribution of the magnitude of elementary crack increments.

3. CUMULATIVE MODEL WITH UNERLYING BIRTH PROCESS

Let us assume that the counting process $N(t)$ is a linear birth process in which the intensity, or birth rate, is $\lambda = \lambda^0 k$, where λ^0 is a positive constant. In this case

$$\begin{aligned} P_k(t) = \; & P\{N(t) = k\} = e^{-\lambda^0 t}(1 - e^{-\lambda^0 t})^{k-1}, \quad t < 0 \\ P_0(t) = \; & 0 \qquad\qquad\qquad\qquad\qquad\qquad\qquad\qquad k \leq 1 \end{aligned} \tag{2}$$

Although it is not necessary to make restriction to specific probability distribution of random variables $Y_i(\gamma)$, here we assume that they are independednt and exponentially distributed with parameter α (i.e. $\frac{1}{\alpha}$ is a mean of elementary jump magnitude).

As it was shown in paper [2] the probability density of a crack size is

$$f_{A(t)}(a) = \beta_t e^{-\beta_t(a - A_0)}, \quad \beta_t = \alpha e^{-\lambda^0 t} \tag{3}$$

Using the definitions of mean and variance and performing the appropriate integration we obtain

$$\begin{aligned} \langle A(t,\gamma) \rangle &= A_0 + \frac{1}{\alpha} e^{\lambda^0 t}, \\ var A(t,\gamma) &= \frac{1}{\alpha^2} e^{2\lambda^0 t}. \end{aligned} \tag{4}$$

The distribution of random variable T , the time to reaching by process $A(t,\gamma)$ the critical value ξ , is easily obtained by use of the formula

$$P(T < t) = F_T(t) = 1 - F_{A(t)}(a) \mid_{a=\xi} \tag{5}$$

The result is:

$$F_T(t) = exp[-\alpha(\xi - A_0)e^{-\lambda^0 t}] \tag{6}$$

Differentiation of the above with respect to t gives the following expression for the probability density of the life-time

$$\begin{aligned} f_T(t) &= \alpha\lambda^0(\xi - A_0)exp[-\lambda^0 t - \alpha(\xi - A_0)e^{-\lambda^0 t}] \\ &= \beta\lambda^0 \exp[-\lambda^0 t - \beta e^{-\lambda^0 t}] \end{aligned} \tag{7}$$

where $\beta = \alpha(\xi - A_0)$.

This result for the distribution of the life-time derived from the cumulative birth process, supports the conjencture of Kozin and Bogdanoff (cf. Proc. of ICOSSAR'85, Kobe, Japan (edited by I. Konishi, A.H.S. Ang and M. Shinozuka) vol. 2, pp. 331-340, 1985) formulated on the basis of the Virkler data that the extreme-type distribution is a possible candidate for the first-passage time of fatigue crack growth process.

According to general methodology of stochastic modelling of real phenomena an important further step, after choosing an appropriate model - process and determining its probabilistic characteristics, consists in relating the model parameters to measurable physical parameters. This stage can be regarded as a part of model validation, which is usually performed by comparison of the model predictions with data form repeated experiments.

The parameter of the model-process (1), i.e. statistic of a single crack increment (probability distribution of random variables $Y_i(\gamma)$, or-in particular-parameter α in assumed exponential distribution) and properties of the counting process $N(t)$, e.g. the growth intensity λ^0 in assumed birth process, should be estimated from empirical data. However, to do this a new type of fatigue experiments is necessary in which fatigue crack growth is regarded as a cumulative random process. In designing such fatigue experiments a care should be taken to create possibility for identifying the underlying counting process $N(t)$ including the estimation of its intensity. Also, one should prepare experimental conditions for estimations the statistics of random elementary crack increments. Since such experimental studies seem to be of future concern, the model parameters can be related to the traditional experimental results (cf.[2]).

A possible way of approximation of parameter α characterizing the average magnitude of the crack increments (in assumed exponential distribution) is the use of the following mean-square criterion. Namely, we can determine α from the following condition

$$F(\alpha) = \int_0^{\bar{t}} < [A_p(t) - A(t,\gamma)]^2 > dt = minimum \tag{8}$$

where $A_p(t)$ is obtained from the Paris equation (approximately modified-when random loading is considered) and $A(t,\gamma)$ is form cumulative model (2.1); $\bar{t}$ is the time in which $A_p(t)$ reaches critical level ξ . Criterion (3.7) gives an effective procedure for relating parameter α to physical constants contained in $A_p(t)$. It should be noticed, however that the value of α determined by use of (3.7) is the same for all considered time interval. To overcome this deficiency one can determine many values of α according to the criterion

$$F_i(\alpha) = \int_0^{t_i} < [A_p(t) - A(t,\gamma)]^2 > dt = minimum \tag{9}$$

where

$$t_i = t_{i-1} + \Delta t, \quad \Delta t = \frac{\bar{t}}{n}, \quad t_0 = 0, \quad i = 1, 2, \ldots, n \tag{10}$$

In this way we obtain different values α_i of parameter α for different elementary time intervals on the interval $[0, \bar{t}]$.

Much weaker criterion for determination of α is as follows. We require that the global uncertainties (expressed by the entropies) of the process $A(t,\gamma)$ defined by (1) and the random crack growth governed by other stochastic model should be equal. So, the criterion has the form

$$\int_0^{\bar{t}} H_A(t)dt = \int_0^{\bar{t}} \tilde{H}_A(t)dt \tag{11}$$

where $H_A(t)$ is the one-dimensional (informational) entropy of the process $A(t,\gamma)$ defined by model (2.1) and $\tilde{H}_A(t)$ is the entropy of the known stochastic fatigue crack growth model e.g. $\tilde{H}_A(t)$ can be taken as the entropy of the process obtained from a randomized Paris equation. One-dimensional entropy of the process $A(t,\gamma)$ is defined as usual

$$H_A(t) = -\int_0^{\infty} f_{A(t)}(a) ln f_{A(t)}(a) da \tag{12}$$

More detailed analysis of the estimation of parameters in model (2.1) can be found in paper [2].

4. CUMULATIVE JUMP–CORRELATED MODEL

4.1. Description of the model

Let us assume that the fatigue crack growth is characterized by stochastic process $A(t,\gamma)$ represented analytically by formula (1) where $N(t)$ is a homogeneous Poisson process and $Y_i(\gamma)$ are dependent random variables.

Though in general the dependent random variables are characterized by the multivariate distributions, the available information on the dependence (usually estimated from incomplete data) is most often limited to knowledge of covariances between random variables in question. On the other hand, in many situation marginal distributions of single variables can be prescribed on the basis of empirical data or on the physical grounds. It is, therefore, a need for appropriate constructions (models) of multivariate distributions in terms of known one-dimensional (marginal) distributions and covariances (or, correlation coefficients). A construction which we adopt here is known as the Morgenstern model (cf. [5]); its generalization has been suggested by Kotz [6] whereas the conditions for validity of the model are expounded in the paper [7] by Liu and Kiureghian.

Let $Y_1, Y_2, \ldots, Y_n$ be dependent random variables with given marginal densities $f_{Y_I}(y_i)$ and joint correlational moments. The joint probability density can be represented by the formula (cf. [7])

$$f(y_1, y_2, \ldots, y_n) = \prod_i f_{Y_i}(y_i)\{1 + \sum_{i<j} \alpha_{ij}[1 - 2F_{Y_i}(y_i)][1 - 2F_{Y_j}(y_j)] + \sum_{i<j<k} \alpha_{ijk}[1 - 2F_{Y_i}(y_i)][1 - 2F_{Y_j}(y_j)][1 - 2F_{Y_k}(y_k)] + \ldots\} \tag{13}$$

where $F_{Y_i}(y_i)$ are the marginal cumulative distributions,

$$\alpha_{12\ldots k} = \frac{\rho_{12\ldots k}}{(-2)^k Q_1 Q_2 \ldots Q_k} \tag{14}$$

and $\rho_{12\ldots k}$ is the k-dimensional, normalized, joint central moment

$$\rho_{12\ldots k} = \langle\left[\left(\frac{Y_1 - m_1}{\sigma_1}\right)\left(\frac{Y_2 - m_2}{\sigma_2}\right)\ldots\left(\frac{Y_k - m_k}{\sigma_k}\right)\right]\rangle \tag{15}$$

m_i and σ_i are the mean and standard deviation of Y_i , and

$$Q_i = \int \left(\frac{Y_i - m_i}{\sigma_i}\right) f_{Y_i}(y_i) F_{Y-i}(y_i) dy_i \tag{16}$$

Of course, since the joint density (13) has to be nonnegative the Morgenstern model is valid under the restrictions on $\alpha_{12\ldots k}$. For instance, to assure the nonnegativeness of the bivariate density $f(u_i, y_j)$ we need $\mid \alpha_{ij} \mid \leq 1$ which implies the appropriate condition for $\mid \rho_{ij} \mid$.

Formula (13) includes in general all higher-order moments (14). However, they are seldom available in practical applications. On the other hand, using formula (13) with account for many moments $\alpha_{12\ldots k}$ makes essential computational difficulties. Therefore, the analysis performed here will use a truncated series preserving only binary correlation α_{ij}.

4.2. Moments of the Crack Size

We shall concentrate here on deriving the formulae for the statistical moments of the crack size represented by the jump-correlated model (1). This model will represented (for convenience) in the form

$$A(t, \gamma) = A_0 + A_1(t, \gamma) \tag{17}$$

$$A_1(t,\gamma) = \sum_{i=1}^{N(t)} Y_i(\gamma) \tag{18}$$

and we will look for the statistics of process $A_1(t,\gamma)$; accounting for A_0 does not make any essential trouble (cf.[2]).

If $N(t)$ is a homogeneous Poisson process with intensity λ_0 , then the moment of order m of $A_1(t,\gamma)$

$$\begin{aligned}\langle A(t,\gamma)\rangle &= \sum_{n=0}^{\infty} \langle A_i^m \mid N(t)=n\rangle P_n(t) = \\ &= exp(-\lambda_0 t)\sum_{n=0}^{\infty}\langle \left(\sum_{i=1}^{N(t)} Y_i(\gamma)\right)^m \rangle \frac{(\lambda_0 t)^n}{n!} = \\ &= exp(-\lambda_0 t)\sum_{n=0}^{\infty} E_n^m \frac{(\lambda_0 t)^n}{n!}\end{aligned} \tag{19}$$

where E_n^m is a conditional moment of order m of $A_1(t,\gamma)$, that is moment of sum (18) when the number of crack increments in the interval $(0,t]$ is fixed $N(t)=n$. Making use of the Newton formula we have the following expression for E_n^m

$$\begin{aligned}E_n^m &= \langle \left[\sum_{i=1}^{n} Y_i(\gamma)\right]^m \rangle = \\ &= \sum_{h_1+\ldots+h_n=m} \frac{m!}{h_1!\ldots h_n!}\langle Y_1^{h_1}\ldots Y_n^{h_n}\rangle\end{aligned} \tag{20}$$

where summation is extended over all possible combinations of n elements giving the sum equal to m .

In order to make theabove formulae entirely explicit one has to evaluate the joint moments of random variables $Y_i(\gamma)$. To do this we make use of basic formula (13) for the joint probability density. When only binary correlations are accounted for, the joint moments occurring in (20) are

$$\begin{aligned}\langle Y_1^{h_1}\ldots Y_n^{h_n}\rangle &= \prod_{i=1}^{n}\langle Y_i^{h_i}\rangle + \\ &+\sum_{i<j}\alpha_{ij} \prod_{\substack{r\neq i\\ r\neq j}} \langle Y_r^{h_r}\rangle M_{h_i} M_{h_j}\end{aligned} \tag{21}$$

where

$$M_{h_r} = \int y_r^{h_r} f_{Y_r}(y_r)[1 - 2F_{Y_r}(y_r)]dy_r \tag{22}$$

Let us assume that the crack increments $Y_i(\gamma) = \Delta A_i$ have exponential distribution with parameters η_i . In this case

$$\langle Y_1^{h_1} \dots Y_n^{h_n} \rangle = \prod_{i=1}^{n} \frac{h_i!}{(\eta_i)^{h_i}} \{1 + \sum_{i<j} \alpha_{ij} \tilde{M}_{h_i} \tilde{M}_{h_j}\} \tag{23}$$

where

$$\tilde{M}_{h_r} = \frac{1}{2^{h_r}} - 1 \tag{24}$$

The conditional moments E_n^m given by general formula (20) in our case are

$$E_n^m = \sum_{h_1+\dots+h_n=m} m! \left[\prod_{i=1}^{n} \eta_i^{h_i} \right]^{-1} \{1 + \sum_{i<j} \alpha_{ij} \tilde{M}_{h_i} \tilde{M}_{h_j}\} \tag{25}$$

To make further analysis more effective a special form of the binary correlation between the crack increments has to be assumed. Also, the calculations can be made less cumbersome of the parameters in the exponential distribution of crack elementary increments assume a special from.

Since the correlation between i-th and j-th increments is a decreasing function of the distance between jumps Y_i and Y_j (cf. analysis of fractographical patterns of fatigue – Itagaki, Shinozuka [1]), and due to the numerous observations indicating that the crack jumps increase (in average) as a number of cycles increases we assumed that:

$$\alpha_{ij} = B\left[1 - \frac{|\, j - i \,|}{max(i, j)}\right], \quad \begin{matrix} i, j = 1, \dots, n \\ i \neq j \\ B \in [0, 1] \end{matrix}$$

$$\eta_i = \frac{\eta}{i^\nu}, \eta > 0, \nu - \text{natural number} \tag{26}$$

The assumption (26) means that $< Y_i >= \frac{i^\nu}{\eta}$ increases with a number of jumps i ; the parameter ν should be calibrated to data.

Under hypotheses (26) the moments of random crack growth process take the form:

$$\langle A_1^m(t, \gamma) \rangle = \sum_{k=1}^{m(\nu+1)} C_k(B) D_k(\lambda_0 t) \tag{27}$$

where $D_k(z)$ are the Stirling polynomials of degree k defined by the recurrence formula

$$D_0(x) = 1,\ D_k(z) = z\left[\frac{dD_{k-1}(z)}{dz} + D_{k-1}(z)\right],\ k = 1, 2, \dots \tag{28}$$

and $C_k(B)$ are coefficients (algebraically depending on B) in representation of E_n^m in polynomial form. For example,

$$\langle A_1(t,\gamma)\rangle = \frac{\lambda_0 t}{6\eta}\{2(\lambda_0 t)^2 + 9\lambda_0 t + 6\},$$

$$\langle A_1^2(t,\gamma)\rangle = \frac{\lambda_0 t}{720\eta^2}\{5(\lambda_0 t)^5(16+3B) + 18(\lambda_0 t)^4(88+13B)\}$$

$$+345(\lambda_0 t)^3(28+3B) + 60(\lambda_0 t)^2 \times (348+23B) \tag{29}$$

$$+360\lambda_0 t(38+B) + 1440)\},$$

Assuming the same material and loading characteristics as in paper [2] and performing numerical calculations we obtained a quantitative effect of the correlation of crack increments on statistics of fatigue crack growth. For example, Fig. 1 – 2 illustrate the dependence of the variance (or, standard deviation σ) of the crack size on the strength of correlation B between the crack jumps. The predictions agree with the features of fatigue crack growth inferred from the Virkler et al. empirical data (cf. Virkler et al. [8]); for example, the standard deviation of the crack size strongly increases, like its empirical counterpart, with the number of cycles.

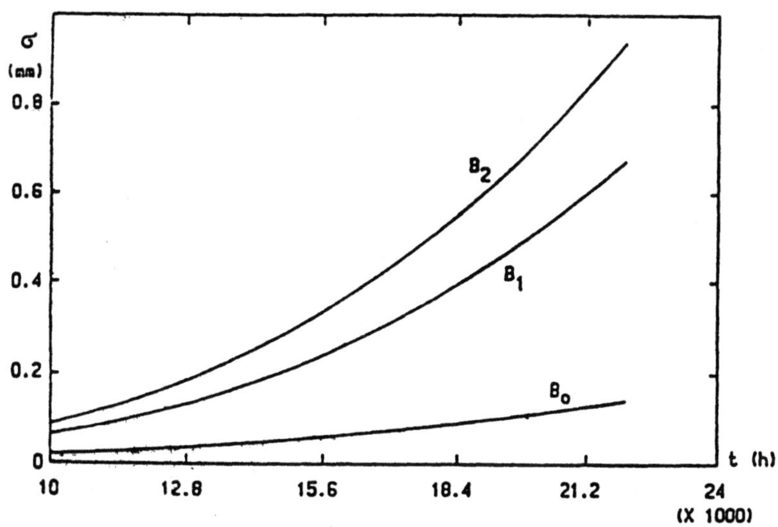

Figure 1. Effect of the correlation strength of crack increments on the variance of crack size.

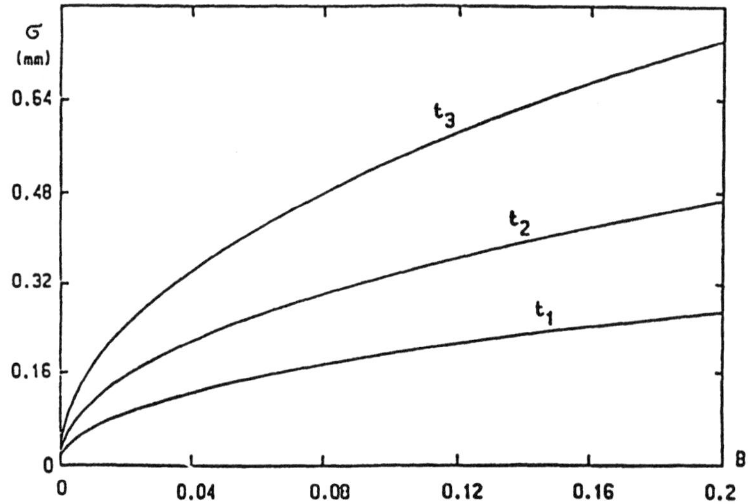

Figure 2. Variance of the crack size versus correlation strength parameter B, for different times.

Remarks

It does not seem to be possible to obtain an exact analytical formula for the probability distribution of a crack size at arbitrary time t . However, having the moments of arbitrary finite order one can construct an approximate crack size distribution using the information contained in moments. A possible approach to do it is the usage of the maximum entropy principle (cf. Sobczyk, Trębicki [9]).

5. STOCHASTIC MODELLING OF FATIGUE CRACK GROWTH WITH RETARDATION

5.1. Introduction

It has been recognized that an important mode of structural failure due to time-varying loading is fatigue damage. The mechanical fatigue appears also as significant deteriorating effect of random vibrations (Crandall, Mark [10]). It occurs because of the successive incremental reduction of the response measures as a results of crack growth in each load repetition. Today, the modelling and analysis of fatigue cracks in structural elements is one of the central issues of ongoing research in engineering mechanics.

However, the traditional fatigue crack propagation laws are based on fixed stress level fatigue experiments (mostly, a constant amplitude homogeneous cyclic loading) and they do not take into account so called interaction effects due to variable-amplitude loading (single or repeated overloads, step loading, programmed block loading, random loading etc.). Of particular interest is the decrease in growth rate (crack retardation) which normally follows a high overload. As pointed out by many researchers this retardation can have a significant influence on the fatigue life of a structure. When the retarding effect of peak overload on crack growth is disregarded, the prediction of material life is usually very conservative. So, as it was also underlined in (Scharton, Crandall [11]) it is very desirable to understand and mathematically describe the peak overload mechanisms and the transient interaction effect. However, all researchers investigating the retarding effects agree that retardation in fatigue crack growth is a highly complex phenomenon and that no satisfactory model has been developed to account for all the observed behaviour (cf. Wei, Shih [12], Stouffer, Williams [13]).

In this report we wish to describe the possible probabilistic modelling of the retarded crack growth. First, the basic features of the retardation phenomenon shall be briefly summarized and then (Section 3) a modelling by use of a stochastic reasoning will be presented together with analytical results concerning the statistics of a crack size.

5.2. Retardation in fatigue crack growth

Although the effects accompanying load interactions are of permanent importance in the prediction of fatigue life of various structural components, the micromechanisms of the retardation phenomenon are not yet clearly understood. In this situation the basic knowledge is drawn from the experimental results which provide an essential information about the influence of overload on the fatigue crack growth in real materials.

The most important conclusions drawn from the observations reported in recent papers on fatigue crack growth retardation can be verbalized as follows (cf. Bernard, Lindley, Richards [14]; Jones [15]; Schijve [16]; Ditlevsen, Sobczyk [4].

1. Almost all investigations have revealed that retardation occurs above a threshold overload value in all materials and specimen geometries. The threshold level, the lowest overload level at which significant retardation occur, is generally around 40 to 60% overload relative to the baseline K_{max} , where K_{max} is the maximum stress intensity in periodic loading.

2. Positive overloads introduce significant crack growth retardation, but in general, longer retardation is generated by: (i) increasing the magnitude of the overload, (ii) repeating the overload during the crack propagation period, (iii) application of blocks of overloads instead of single overload.

3. Negative overloads (compressive stress conditions) have a relatively small effect

on crack growth. But, when a tensile (i.e. positive) overload is followed immediately by a compressive overload, the effect of crack growth retardation is greatly diminished. Most of the prediction methods is based on the assumption that only rising tensile load rangers cause growth.

4. Most often, crack growth retardation does not occur immediately after an overload. This is usually called as delayed retardation. Delayed retardation does not always occur, and some authors have provided evidence of an immediately decrease in propagate rate.

5. Retardation in crack growth is affected by a broad range of metallurgical, loading and environmental variables and it is very difficult to separate contribution of each. In particular, the retardation clearly depend on the ductility of the material; if ductility of an alloy is controlled by heat treatment, a lower yield strength will produce longer retardation. As loading factors are concerned, the experiments show a permanent reduction in growth rate with increasing stress biaxiality.

Numerous investigations have made effort to treat retardation phenomenon quantitatively. Since the stress intensity factor range, ΔK , turned out to be very useful in theoretical prediction of fatigue crack growth under irregular loading (including single or multiple overloads) is usually also based on various modification of ΔK . However, most of the analyses of retardation, to date, used data fitting techniques in conjunction with the effective plastic zone size or the effective crack opening. In this context the Wheeler model and the crack closure model of Elber are the most notable. It should be emphasized, however, that there is no agreement concerning the ability of satisfactory prediction of real fatigue crack growth by these models (cf. Robin, Louah, Pluvinage [17], Schijve [16]).

5.3. Probabilistic modelling

a. Stochastic cumulative birth model

It is likely that one reason that the existing quantitative descriptions of retarded crack growth are not satisfactory is that these models are deterministic while the fatigue crack growth process was recognized as highly complicated random phenomenon. So, it is essential to look at crack growth processes from a more general point of view taking into account an inherent randomness of the phenomenon. Such approach leads to the representation of the crack growth by a suitable stochastic process. In the paper by Ditlevsen and Sobczyk [4] the authors have presented a stochastic cumulative model with underlying birth process.

The sample of the material is regarded as a system whose fatigue states are described by the stochastic process $L(t,\gamma)$, where $L(t,\gamma)$ is interpreted as a length of a dominant crack at time t; $\gamma \in \Gamma$ where Γ is a space of elementary events (sample space) on which the probability is defined (for each $\gamma \in \Gamma$, $L(t,\gamma)$ represents a possible sample function of the crack length). Since crack grows mainly due to a sequence of events occurring randomly in time and since the fatigue is a cumulative process we represent the crack size at time t as

$$L(t,\gamma) = L_0 + Y_1(\gamma) + Y_2(\gamma) + \ldots + Y_{N(t,\gamma)}(\gamma) \tag{30}$$

where L_0 is the initial crack growth length (sufficient to propagate), $Y_i(\gamma) = \Delta L_i(\gamma)$ are the random partial crack increments and $N(t,\gamma)$ is an integer-valued stochastic process characterizing a number of crack increments in the interval $[0,t]$ .It is assumed that $Y_i(\gamma)$ are independent and identically distributed non-negative random variables with the common distribution $G(\gamma)$. Randomness of the fatigue crack growth process is, therefore, taken into account through the probabilistic mechanism of the transition from one state to another (process $N(t)$) and by the fact that elementary increments are allowed to be random. Since in modelling of fatigue it is essential the growth intensity be state-dependent the process $N(t,\gamma)$ is assumed to be the birth process (originating in the theory of population growth).

In the case of homogeneous periodic loading the infinitesimal intensity of $N(t)$ is assumed as

$$\lambda_k = \lambda^0 k \qquad (k = 1,2,\ldots,\lambda^0 > 0) \tag{31}$$

which means that the probability to transition from state k to $k+1$ in the interval $(t, t+\Delta t)$ is proportional to state k . The averaged crack size grows exponentially with time and the rate of this exponential growth can be related to the basic characteristics of the loading and material properties.

Let us consider a more general case when the crack growth is generated by periodic loading with n overloads of different magnitude which are sufficiently separated to assure that retardation is effectively over after each overload. Although some sequence effects can, in general, occur the influence of the peak overload sequence will be neglected; this sequence effect is often considered secondary as compared to the stress magnitude effects (Porter [19]). In this case the growth intensity is postulated to have the form

$$\lambda_k = \lambda_k(t) = k[\lambda^0 - \lambda_{OL}(t; t_1, 1_2, \ldots, t_n)] \tag{32}$$

with

$$\lambda_{OL}(t; t_1, \ldots, t_n) = \sum_{i=1}^{n} \mu_i(t,\zeta) e^{-\alpha(t-t_1)} H_i(t - t_i) \tag{33}$$

$$H_i(t - t_i) = \begin{cases} 1, & t_i + \theta_i < t < t_i + \theta_i + \tau_i \\ 0, & \text{otherwise} \end{cases} \tag{34}$$

where $t_{i+1} - t_i > \tau_i$ and $0 < \lambda_{0L} < \lambda^0$.

The factor $\mu_i(t, \zeta)$, suitably bounded, describes the retarding effect of an overload at $t = t_i (i = 1, 2, \ldots, n); \alpha_i$ is a decay parameter associated with i th overload. A delay θ_i in retardation can likely be assumed to be the same for each overload, i.e. $\theta_i = \theta$. The retardation magnitude function $\mu(t, \zeta)$ depends on time t and a collection of relevant variables denoted symbolically by ζ (e.g. the overload ratio, the stress biaxiality etc.). The parameter τ_i is the retardation time associated with i th overload; when $t = t_i + \tau_i$ the growth rate comes back to its level just prior to the overload.

A situation which is often investigated experimentally is concerned with the blocks of overloads. it has been observed that when the blocks of overloads are applied instead of a single overload a longer retardation is obtained. If the n overload blocks are separated enough it seems to be justified to characterize this case by the retarding intensity (eq. 33 and 34) where t_i is a characteristic instant for an i – – th overload block and $\mu_i(t, \zeta)$ is replaced by $\mu_i(t, \zeta) = \beta_i \mu_i(t, \zeta)$ where $\beta_i > 1$ describes the block effect. Also the retardation times τ_i should be modified adequately.

If the time intervals between overloadings are not long enough to assure independence of the retarding effects the situation is much more complicated. In this case the retardation are less obvious as seen at the crack growth curve. In fact, the effects of interaction of overloadings can be either positive (retardation) or negative (accelerated growth). The existing experimental results give no clear indication of how to characterize quantitatively the interaction effects of overloading. Many mechanisms seems to contribute to these effects and it is difficult to recognize and separate them. It is observed that reapplication of a high overload after a previous one early enough before the retardation from the previous overload has become effective, may considerably reduce the crack growth retardation. A possible way of modelling the crack growth under periodic loading with irregularly occurring overloads (within the cumulative birth model described above) has also been proposed in (Ditlevsen, Sobczyk, [4]); the "birth" growth intensity has been represented as

$$\begin{aligned} \lambda_k(t) &= \lambda_{int}(t)\lambda_k \\ &= \psi(t)X(t, \gamma)\lambda_k \end{aligned} \tag{35}$$

where $\lambda_{int}(t)$ is an interaction factor and characterizes an overall effect of irregularity of peak overloads and the load history. Since these effects are random $\lambda_{int}(t)$ is represented according to eq. (35) where $X(t, \gamma)$ describes random effect of overload interaction.

b. Statistics of crack size

The probability distribution of the crack size can the obtained by calculating first the characteristic function of process $L(t)$. The first two moments are given by the formulae ($\langle \cdot \rangle$ denotes the average value)

$$\begin{aligned} \langle L(t)\rangle &= L_0 + \langle N(t,\gamma)\rangle\langle Y(\gamma)\rangle \\ &= L_0 + \langle Y\rangle e^{\eta(t)}, \end{aligned} \tag{36}$$

$$\text{var } L(t) = e^{\eta(t)} \text{ var } Y + \langle Y\rangle^2 e^{\eta(t)}[e^{\eta(t)} - 1] \tag{37}$$

where (cf. Ditlevsen, Sobczyk, [4]);

$$\eta(t) = \int_0^t \lambda(s)ds \tag{38}$$

and $\lambda(t)$ is the time-varying factor in the birth intensity. In the case of homogeneous periodic loading

$$\lambda(t) = \lambda^0 = \text{const.} \tag{39}$$

and in the case of multiple (separated) overloads

$$\lambda(t) = \lambda^0 - \lambda_{0L}(t; t_1, \ldots, t_n) \tag{40}$$

where $\lambda_{0L}(t; t_1, \ldots, t_n)$ is given by eq. 33 and 34. Of course, in case of eq. 39, $\eta(t) = \lambda^0 t$ and

$$\begin{aligned} \langle L(t)\rangle &= L_0 + \langle Y\rangle e^{\lambda^0 t} \\ \text{var} L(t) &= e^{\lambda^0 t}\{\text{var} Y + \langle Y\rangle^2 (e^{\lambda^0 t} - 1)\}; \end{aligned} \tag{41}$$

in the case of eq. 40

$$\eta(t) = \lambda^0 t - \sum_{i=1}^{n} \mu_i(\zeta)\frac{1}{\alpha_i}[e^{-\alpha_i\theta_i} - e^{-\alpha_i(t-t_i)}]H_i(t-t_i) \tag{42}$$

where it has been assumed that $\mu(t,\zeta) = \mu(\zeta)$.

The model described above needs appropriate empirical verification by comparison with data from the experiment in which the crack growth is treated as random cumulative process. The characteristics of the model, such as statistics of random elementary increments as well as the properties of the infinitesimal growth intensity λ^0 should than be estimated from the data. Since such a methodology is a future matter, the parameters of the model should be related to the existing experimental predictions (cf. Ditlevsen, Sobczyk [4]). In particular, the retardation time τ occurring in the model (and equal to: $\tau = \frac{1}{\omega} N_{ret}$)can be expressed in terms of parameters occurring in the Wheeler empirical model according to which

$$\left(\frac{dL}{dN}\right)_{OL} = C_p \left(\frac{dL}{dN}\right)_{C\cdot A} \tag{43}$$

where

$$\begin{aligned} C_P &= \left(\frac{r_{pi}}{L_* + r_{po} - L_i}\right)^m = \left(\frac{r_{pi}}{s - L_i}\right)^m, \quad r_{pi} < s - L_i \\ C_P &= 1 \qquad\qquad\qquad\qquad\qquad\qquad\quad r_{pi} \geq s - L_i \end{aligned} \tag{44}$$

and r_{pi} = current plastic zone (in the i th cycle); L_i = current crack size; r_{po} = plastic zone size caused by overload; L_* = crack size at which overload occurred; m = empirical (data fitting) constants; $s = L_* + r_{po}$. The number of cycles, N_{ret} , over which retardation occurs (after each overload) has been estimated from the above formulae. In particular (since $L_i \approx L_*$ (cf. Broek, Smith [19]), $C_P \approx \left(\frac{r_i}{r_{po}}\right)^m$) it can be shown that

$$N_{ret} = \frac{\rho^{2m}}{C\alpha\sigma_{ys}^2} \frac{\rho^2 - 1}{(\sigma_{max})^{n-2}(\pi L_*)^{\frac{n}{2}-1}} \tag{45}$$

where $= \frac{K_{max,0}}{K_{max,i}}$ is the overload ratio. K_{max} is the maximum stress intensity in a given cycle. $K_{max,0}$ is the stress intensity at the overload. $\alpha = 2\pi$ for plane stress and $\alpha = 6\pi$ for plane strain.

c. Stochastic cumulative model with exponential decay

Let us assume that in the absence of overloading the crack is growing according to constant amplitude fatigue crack growth low at time i the crack size is characterized by $L_{CA}(t)$; $L_{CA}(t)$ can be regarded as deterministic or random.

The retardation due to overload can be accounted for by assuming that the net crack size is considered. If we suppose that the retardation due to overloads occurs with a delay θ_i and that the magnitude of the retardation due to a particular overload is random, the crack size at time t can be represented as:

$$L(t,\gamma) = L_{CA}(t) - L_{OL}(t,\gamma;t_1,t_2\dots t_n), \tag{46}$$

$$L_{OL}(t;\gamma;t_1,\dots,t_n) = \sum_{i=1}^{n} X_i(t,\zeta;\gamma)e^{-\alpha(t-t_1)}H_i(t-t_i) \tag{47}$$

where n denotes a number of overloads and t_i, $(i = 1,2,\dots,n)$ expresses the instant of occurance of i th overload. $X_i(t,\zeta;\gamma)$ is random factor characterizing the magnitude of retardation at time t due to i th overload which occurred at $t = t_i$; it depends on time t and collection of variable denoted symbolically by ζ . The step function

$H_i(t-t_i)$ is defined by eq. 34. For each fixed t and ζ the retarding factors $X_i(t,\zeta;\gamma)$ are assumed to be mutually independent, non-negative, random variables. To obtain the probability distribution of a crack size at arbitrary time t is convenient to calculate the characteristic function $\phi_{L(t)}(u)$ of the process $L(t,\gamma)$ given by eq. 46 and 47.

If $L_{CA}(t)$ is deterministic, then

$$\phi_{L(t)}(u) = e^{iuL_{CA}(t)}\phi_{0L}(-u) \tag{48}$$

where

$$\phi_{0L}(-u) = \prod_{i=1}^{n} \phi_{X_i}(-ue^{-\alpha_i(t-t_i)}H_i(t-t_1)) \tag{49}$$

Differentiation of eq. 49 with respect to u and use of the well known relation between the moments and derivatives of the characteristic function leads to the following expressions for the mean and the variance:

$$\begin{aligned} \langle L(t,\gamma)\rangle &= L_{CA}(t) + \sum_{i=1}^{n}\langle X_i(t,\zeta;\gamma)\rangle e^{-\alpha(t-t_1)}H_i(t-t_i) \\ \text{var } L(t,\gamma)\rangle &= \sum_{i=1}^{n} \text{var } [X_i(t,\zeta;\gamma)]e^{-\alpha(t-t_1)}H_i(t-t_i) \end{aligned} \tag{50}$$

To make the model described above effective one needs to have statistics of the random factors $X_i(t,\zeta;\gamma)$ estimated from experimental data. The retardation parameters α_i are inversely proportional to the retardation times τ_i discussed in the previous sub-section and expressed in terms of the existing data (associated with the Wheeler empirical model).

It is clear that the analysis can be extended to the more general situation where $X_i(t,\zeta;\gamma)$ are dependent random variables and $L_{CA}(t)$ is a stochastic process. It seems that the model can also be modified to include the effects of interaction of irregularly occurring overloads.

References

1. Itagaki H. and Shinozuka M., Application of the Monte Carlo Technique to fatigue Analysis under Random Loading, in: Probabilistic Aspects of Fatigue (Ed. Heller R.A.) Amer. Soc. for Testing and Materials, STP511, Philadelphia, PA, 1972.

2. Sobczyk K. and Trębicki J., Modelling of Random Fatigue by Cumulative Jump Processes, Eng. Fract. Mech., Vol. 34, No 2, pp. 477-493, 1989.

3. Sobczyk K. and Trębicki J., Cumulative Jump–Correlated Model for Fatigue, Eng. Fract. Mech., Vol. 40, No 1, pp. 201-210, 1991.

4. Ditlevsen O. and Sobczyk K., Random Fatigue Crack Growth with Retardation, Eng. Fract. Mech., Vol. 34, No 2, pp. 861-878, 1986.

5. Morgenstern D., Einfache Beispiele Zweidimensionaler Verteilungen, Miteilingsblatt fur Math. Statik, Vol.8, pp. 234-235, 1956.

6. Kotz S., Multivariate Distribution at a Cross Road, in: Statistical Distributions in Scientific Work (eds. Patial, G.P., et al.) Reidel Publ. Comp., Dordrecht, Vol. 1, pp. 247-270, 1975.

7. Liu P.L. and Der Kiureghian A., Multivariate Distribution Model with Prescribed marginals and Covariances, Probab. Eng, Mech., Vol. 1, No 2, pp. 105-112, 1986.

8. Virkler D.A., Hillberry B.M., and Goel P.K., The Statistical Nature of Fatigue Crack Propagation, Journal Eng. Material and Techn., ASME, Vol. 101, pp. 148-153, 1979.

9. Sobczyk K. and Trębicki J., Maximum Entropy Principle in Stochastic Dynamics, Prob. Eng. Mech., Vol. 5, No 3, 1990.

10. Crandall S.H. and Mark W.D., Random vibration in mechanical systems, Academic Press, New York, 1963.

11. Scharton T.D. and Crandall S.H., Fatigue failure under complex stress histories, Trans. ASME J. Basic Eng., Vol. 88, Ser. D., No 1, p. 247-251, 1966.

12. Wei R.P. and Shih T.T., Delay in fatigue crack growth, Trans. ASME. J.Basic Eng., Ser. D., Vol. 94, No 1., p. 181-186, 1972.

13. Stouffer D.C. and Williams J.F., A model for fatigue crack growth with a variables stress intensity factor, Eng. Frac. Mech, Vol. 11 p. 525-536, 1979.

14. Bernard P.J., Lindley T.C. and Richards C.E., Mechanisms of overload retardation during fatigue crack propagation, in: Fatigue Crack Growth under Spectrum Loads, ASTM STP 595, 1976.

15. Jones R.E., Fatigue crack growth retardation after single-cycle peak overload in Ti/6Al/4V titanium alloy, Eng. Fract. Mech., Vol. 5 pp. 585-604, 1973.

16. Schijve J., Observations on the prediction of fatigue crack growth propagation under variable-amplitude loading, in: Fatigue Crack Growth under Spectrum Loads. ASTM STP 595, 1976.

17. Robin C., Louah M. and Pluvinage G., Influence of an overload on fatigue crack growth in steels, fatigue Eng. Mat. Struct., Vol. 6 p. 1-13, 1983.

18. Porter T.R., Method of analysis and prediction for variable amplitude fatigue crack growth, Eng. Fract. Mech. Vol. 4 pp. 717-736, 1972.

19. Broek D. and Smith S.H., The prediction of fatigue crack growth under flight-by--flight loading, Eng. Fracture Mech., Vol. 11, p.123-141, 1979.

STOCHASTIC DIFFUSION MODELS FOR FATIGUE CRACK GROWTH AND RELIABILITY ESTIMATION

B.F. Spencer, Jr.
University of Notre Dame, Notre Dame, IN, USA

Abstract

Crack propagation is a major task in the design and life prediction of fatigue-critical structures such as aircraft, offshore platforms, bridges, etc. Experimental data indicate that fatigue crack propagation involves a large amount of statistical variation and is not adequately modeled deterministically. The lectures presented herein discuss the basic analysis and use of fracture mechanics-based random process fatigue crack growth models that can be represented by Markov diffusion processes. For completeness, the random variable models are presented as a special case of the random process models. The use of the models in fatigue reliability estimation is also discussed.

1. Introduction

Failures due to fatigue loading, especially of high performance structures such as aircraft and space vehicles, have always been a main concern of structural engineers. By now it is widely accepted that fatigue failure of a structure involves three stages: (a) crack initiation; (b) crack propagation; and (c) unstable crack growth leading to catastrophic failure [1], [2]. Because crack propagation often constitutes a large portion of the fatigue life of a structure, researches have proposed many fatigue crack growth models. Unfortunately, at this time the knowledge of the behavior of engineering materials is insufficient for fatigue crack growth models to be based solely upon fundamental physical laws. Generally, phenomenological crack growth models based on the principles of fracture mechanics are of the form (e.g., Bluham [3], Miller and Gallagher [4])

$$\frac{da(t)}{dt} = Q(\Delta K, K_{\max}, K_c, S, a, R), \tag{1.1}$$

where $a(t)$ is the deterministic crack size, t is the time or cycle number, $Q(\cdot)$ is a non-negative function, $\Delta K = Y(a)\Delta s\sqrt{\pi a}$ is the stress intensity factor range, $Y(a)$ is the geometry factor, Δs is the stress range, $K_{\max}$ is the maximum stress intensity factor, K_c is the fracture toughness, S is the maximum stress amplitude in the loading spectrum, and R is the stress ratio.

Some commonly used fatigue crack growth models, such as the Paris-Erdogan model [5], the Forman model [6], the hyperbolic sine model [7], and the cubic polynomial model [8], are given respectively by

$$\frac{da(t)}{dt} = C(\Delta K)^m, \tag{1.2}$$

$$\frac{da(t)}{dt} = \frac{C(\Delta K)^m}{(1-R)K_c - \Delta K}, \tag{1.3}$$

$$\frac{da(t)}{dt} = 10^{C_1 \sinh[C_2(\log \Delta K + C_3)] + C_4}, \tag{1.4}$$

$$\frac{da(t)}{dt} = \exp[C_0 + C_1(\ln \Delta K) + C_2(\ln \Delta K)^2 + C_3(\ln \Delta K)^3], \tag{1.5}$$

where the constants C, m, C_0, C_1, C_2, C_3 and C_4 are determined from experimental crack growth data.

Data obtained from experimental tests provide the main source of information regarding fatigue of engineering materials. Figure 1.1 shows the sample functions of crack length versus cycle number resulting from a statistical investigation of fatigue crack growth reported by Virkler, et al. [9]. A more recent collection of fatigue crack growth data reported by Ghonem and Dore [10] is given in Fig. 1.2. Both of these data sets resulted from replicate constant-amplitude crack growth tests conducted in tightly controlled laboratory settings. As shown here, crack growth data exhibit considerable statistical variability, even when obtained using deterministic loading in the laboratory settings. Consequently, one concludes that deterministic

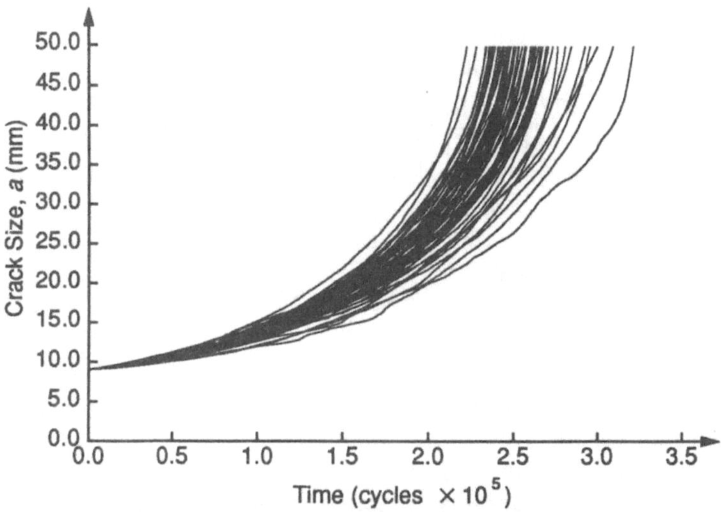

Figure 1.1. Crack length versus number of cycles for the Virkler data [9].

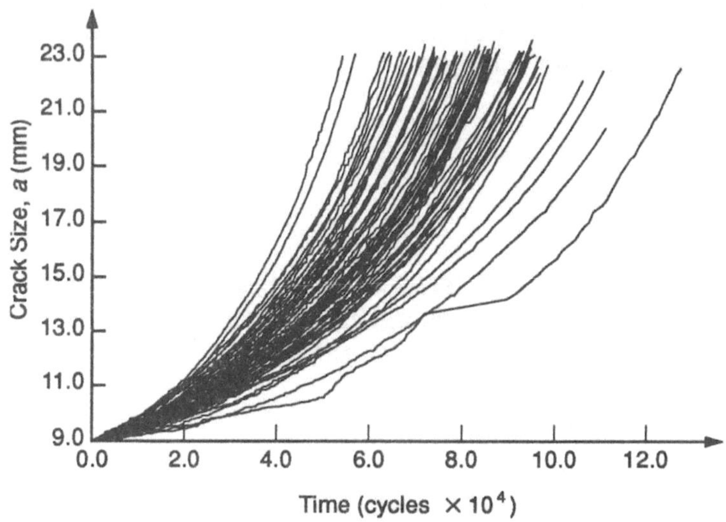

Figure 1.2. Crack length versus number of cycles for the Ghonem and Dore data [10].

models cannot adequately describe crack growth behavior, and that the variability in the data should be appropriately taken into account in the design and analysis of fatigue-critical structures.

To account for the random nature of the fatigue crack growth phenomena, several researchers have suggested that Eq. (1.1) be *randomized* through the introduction of random variables and/or random processes to appropriately characterize the statistical variability in, for example, the stress range, structural geometry, material properties, environment, etc. (e.g., see Bolotin [11], [12], Ditlevsen [13], Ditlevsen and Olesen [14], Dolinski [15], [16], Guers and Rackwitz [17], Kozin and Bogdanoff [18], Lin and Yang [19], [20], Madsen [21], [22], Madsen, et al. [23], Ortiz and Kiremidjian [24], [25], Sobczyk [26], [27], [28], Solomos [29], Spencer and coworkers [30], [31], [32], Tanaka and Tsurui [33], Tsurui and Ishikawa [34], Tsurui, et al. [35], Veers [36], Veers, et al. [37], and Winterstein and Veers [38], and Yang, et al. [39], [40], [41]). The randomized form of the fatigue crack growth model in Eq. (1.1) can be written as

$$\frac{dA(t)}{dt} = Q(A(t), \mathbf{X}), \tag{1.6}$$

where $A(t)$ is the random crack size, $\mathbf{X}$ is a vector of random variables and/or random processes characterizing the variability in the crack growth rate, and A_0 is the random variable representing the initial crack size.

The following sections discuss the basic analysis and use of the fracture mechanics-based stochastic differential equation model described by Eq. (1.6) that can be represented by Markov diffusion processes. For completeness, random variable models are presented as a special case of the random process models. The use of the models in fatigue reliability estimation is also discussed.

The reader is referred to Sobczyk and Spencer [42] for a more general exposition of the methods of stochastic fatigue.

2. Random Variable Models

2.1 General Formulation

A special case of Eq. (1.6) of interest is when the vector $\mathbf{X}$ contains only random variables. These types of models are termed *random variable* models. The random variable fatigue crack growth model is then

$$\begin{aligned} \frac{dA(t)}{dt} &= Q(A(t), X_1, \ldots, X_m), \\ &= Q(A(t), \mathbf{X}) \\ A(t_0) &= A_0, \end{aligned} \tag{2.1}$$

where $X_1, \ldots, X_m$ are the random variables in the model, A_0 is the random initial crack size, and $f_0(a_0, x_1, \ldots, x_m)$ is the joint density function of $X_1, \ldots, X_m$ and A_0.

If Eq. (2.1) has a solution given by

$$A(t) = g_t(A_0, X_1, \ldots, X_m), \tag{2.2}$$

which for each t possesses a one-to-one inverse transformation

$$A_0 = g_t^{-1}(A, X_1, \ldots, X_m), \tag{2.3}$$

then the joint probability density for the crack length as a function of time is given by

$$f(a, x_1, \ldots, x_m; t) = f_0(g_t^{-1}(a, x_1, \ldots, x_m), x_1, \ldots, x_m) \frac{\partial g_t^{-1}}{\partial a}. \tag{2.4}$$

The marginal density function of crack size can be obtained from

$$f_A(a;t) = \int \ldots \int f(a, x_1, \ldots, x_m; t)\, dx_1 \ldots dx_m, \tag{2.5}$$

and the associated marginal cumulative distribution function is then

$$F_A(a;t) = \int_0^a f_A(\xi;t)\, d\xi. \tag{2.6}$$

To illustrate the above concepts, consider a power-law random variable model proposed by Yang, et al. [41] for crack growth in fastener holes under spectrum loading,

$$\frac{dA(t)}{dt} = C_0 [A(t)]^b, \tag{2.7}$$

where C_0 and b are constants depending on the characteristics of the spectrum loading and on the material and geometry of the fastener hole specimens. The solution of Eq. (2.7) as a function of the random variables is

$$A(t) = \begin{cases} [A_0^\beta + \beta C_0 (t - t_0)]^{1/\beta}, & \beta \neq 0 \\ A_0 e^{C_0 (t - t_0)}, & \beta = 0 \end{cases} \tag{2.8}$$

where $\beta = 1 - b$.

Eq. (2.7) and its solution Eq. (2.8) contain three parameters, A_0, C_0 and b (or equivalently, β). The material parameter b is often assumed to be a deterministic constant and will be considered as such in this discussion. However, the material parameter C_0 and the initial crack size A_0 do exhibit significant variability and should be modeled as random variables.

To avoid the possibility of negative crack growth rates, C_0 is taken to be a non-negative random variable with probability density $f_{C_0}(c_0)$. Further, it is assumed that A_0 has a probability density $f_{A_0}(a_0)$ that C_0 and A_0 are statistically independent. Thus, the joint probability density function is given by

$$f_{A_0,C_0}(a_0,c_0) = f_{A_0}(a_0)\, f_{C_0}(c_0)\,. \tag{2.9}$$

Solving Eq. (2.8) for A_0 and substituting into Eq. (2.9) and subsequently into Eq. (2.4) yields joint density function

$$f_{A,C_0}(a,c_0;t) = \begin{cases} f_{A_0}\left([a^\beta - \beta c_0(t-t_0)]^{\frac{1}{\beta}}\right) f_{C_0}(c_0)\, a^\beta [a^\beta - \beta c_0(t-t_0)]^{\frac{1}{\beta}-1}, & \beta \neq 0 \\ f_{A_0}(a e^{-c_0(t-t_0)})\, f_{C_0}(c_0)\, e^{-c_0(t-t_0)}, & \beta = 0 \end{cases} \tag{2.10}$$

The marginal probability density and cumulative distribution functions of crack size as a function of time are given respectively by

$$f_A(a;t) = \int_0^\infty f_{A,C_0}(a,\xi;t)\, d\xi, \tag{2.11}$$

$$F_A(a;t) = \int_0^a f_A(\eta;t)\, d\eta = \int_0^a \int_0^\infty f_{A,C_0}(\eta,\xi;t)\, d\xi d\eta. \tag{2.12}$$

An alternative formulation is obtained by first considering the distribution of crack sizes, conditional on the initial crack size. Solving Eq. (2.8) for C_0 yields

$$C_0 = h_t = \begin{cases} \dfrac{a^\beta - A_0^\beta}{\beta(t-t_0)}, & \beta \neq 0 \\ \dfrac{1}{t-t_0} \ln \dfrac{a}{A_0}, & \beta = 0 \end{cases} \tag{2.13}$$

The probability distribution of A conditional on A_0 is then

$$f_A(a;t|A_0;t_0) = \begin{cases} 0, & a \le A_0 \\ f_{C_0}(h_t)\dfrac{dh_t}{da}, & a > A_0 \end{cases}, \tag{2.14}$$

$$F_A(a;t|A_0;t_0) = \begin{cases} 0, & a \le A_0 \\ F_{C_0}(h_t), & a > A_0 \end{cases}. \tag{2.15}$$

The unconditional probability distribution of crack size analogous to Eqs. (2.11)–(2.12) becomes

$$f_A(a;t) = \int_0^\infty f_A(a;t|a_0;t_0)\, f_{A_0}(a_0)\, da_0, \tag{2.16}$$

$$F_A(a;t) = \int_0^\infty F_A(a;t|a_0;t_0)\, f_{A_0}(a_0)\, da_0. \tag{2.17}$$

Eqs. (2.16)–(2.17) are often more useful for problems in probabilistic inspection updating.

The distribution of time to reach a given crack size a^* can also be derived. Let $T(a^*)$ be a random variable denoting the time to reach the crack size a^*. Solving Eq. (2.8) for $T(a^*)$ yields

$$T(a^*) = \begin{cases} \dfrac{1}{\beta C_0}(a^{*\beta} - A_0^\beta) + t_0, & \beta \ne 0 \\ \dfrac{1}{C_0}\ln\dfrac{a^*}{A_0} + t_0, & \beta = 0 \end{cases}, \tag{2.18}$$

and the cumulative distribution of the random time to reach the crack size a^* is

$$F_T(t;a^*) = P\{T(a^*) \le t\} = \int_{t_0}^{t}\int_0^{a^*} (1 - F_{C_0}(h_\tau))\, f_{A_0}(a_0)\, da_0 d\tau, \tag{2.19}$$

where h_t is given in Eq. (2.13).

2.2 Numerical Illustration

Comparison with experimental data for fatigue cracks in fastener holes [43] indicates that the exponent in the crack growth law is $b = 1.01$ and that C_0 is lognormally distributed with $\lambda = \langle \ln C_0 \rangle = -8.91$ and $\zeta = \sigma_{\ln C_0} = 0.1276$. The probability distribution of C_0 is

$$f_{C_0}(c_0) = \frac{1}{c_0 \zeta}\phi\left(\frac{\ln c_0 - \lambda}{\zeta}\right), \quad c_0 \geq 0, \tag{2.20}$$

$$F_{C_0}(c_0) = P\{C_0 \leq c_0\} = \Phi\left(\frac{\ln c_0 - \lambda}{\zeta}\right), \quad c_0 \geq 0, \tag{2.21}$$

where $\Phi(\cdot)$ and $\phi(\cdot)$ are the standard normal distribution and density functions, respectively.

A probability distribution for A_0 that well represents the initial fatigue quality of fastener holes is given by (see Yang and Manning [44])

$$f_{A_0}(a_0) = \begin{cases} \exp\left[-\left(\frac{\ln(a_u/a_0)}{Q_0\beta_0}\right)^{\alpha_0}\right] \frac{\alpha_0}{a_0 Q_0 \beta_0}\left(\frac{\ln(a_u/a_0)}{Q_0\beta_0}\right)^{\alpha_0 - 1}, & 0 \leq a_0 \leq a_u \\ 0, & \text{otherwise} \end{cases} \tag{2.22}$$

$$F_{A_0}(a_0) = \begin{cases} \exp\left[-\left(\frac{\ln(a_u/a_0)}{Q_0\beta_0}\right)^{\alpha_0}\right], & 0 \leq a_0 \leq a_u \\ 1.0, & a_0 > a_u, \end{cases} \tag{2.23}$$

where $a_u = 0.03$ in. is the upper bound on the initial crack size, and $\alpha_0 = 1.823$ and $Q_0\beta_0 = 1.928$ are shape parameters for the distribution.

Substituting Eqs. (2.20) and (2.22) into Eq. (2.14) and subsequently into Eq. (2.16) yields

$$f(a;t) = \int_0^{\min(a,a_u)} \left\{ \frac{1}{h_t \zeta}\phi\left(\frac{\ln h_t - \lambda}{\zeta}\right) \frac{a^{\beta-1}}{t} \right. \tag{2.24}$$
$$\left. \times \exp\left[-\left(\frac{\ln(a_u/a_0)}{Q_0\beta_0}\right)^{\alpha_0}\right] \frac{\alpha_0}{a_0 Q_0\beta_0}\left(\frac{\ln(a_u/a_0)}{Q_0\beta_0}\right)^{\alpha_0-1} \right\} da_0 ,$$

where $h_t = (a^\beta - A_0^\beta)/(\beta(t-t_0))$, $\beta = -0.01$, and $t_0 = 0$. Eq. (2.24) was integrated numerically to obtain the distributions of crack size at time $T_1 = 8{,}000$ flight hours. The results are compared to the initial crack size distribution in Fig. 2.1. Notice that the mean and dispersion of the crack size have increased with time.

Only two random variables, C_0 and A_0 were considered in the example just presented. Even for this case, analytical solution was not possible and direct nu-

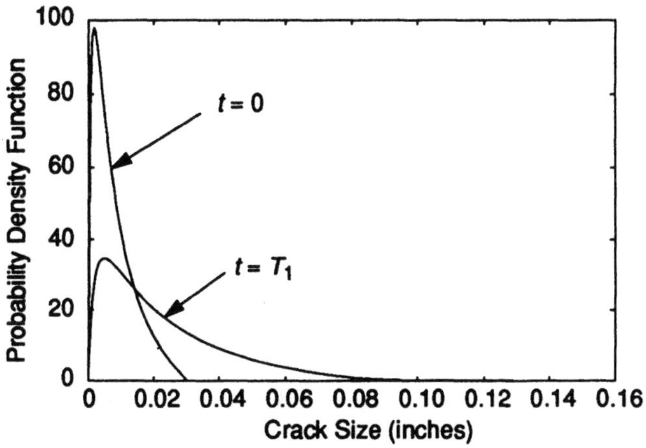

Figure 2.1. Evolution of the crack size probability distribution before the inspection at T_1 = 8,000 flight hours.

merical integration was employed. As the number of random variables in the model increases, this approach becomes intractable. The next section discusses an alternative method that is appropriate for complex models involving many random variables.

2.3 First and Second Order Reliability Approach

For many problems of practical importance, the manner in which random vector **X** enters Eq. (2.1) is too complex to analytically determine the probability distribution of crack size. First and Second Order Reliability Methods (FORM/SORM) can be employed to efficiently analyze the crack growth problem (e.g., see Madsen, et al. [23]). This approach has been extensively employed in inspection and maintenance problems for offshore structures [22]. The formulation of appropriate limit state functions for the FORM/SORM analysis is illustrated in this section by example.

Consider the Paris–Erdogan equation given by

$$\frac{dA}{dt} = C(\Delta K)^m, \tag{2.25}$$

$$A(t_0) = A_0, \tag{2.26}$$

where t is the number of cycles. The material constants C and m and the initial crack size A_0 are assumed to be random variables; thus the vector **X** in Eq. (2.1) is a vector containing the random variables A_0, C and m. Substituting the stress intensity factor range,

$$\Delta K = Y(A)\,\Delta S\sqrt{\pi A}, \tag{2.27}$$

into Eq. (2.25) yields

$$\frac{dA}{dt} = C[Y(A)\,\Delta S\sqrt{\pi A}]^{m}, \tag{2.28}$$

where $Y(a)$ is the geometry function, and ΔS is the stress range. In general, uncertainties are present in $Y(a)$ due to idealizations in the structural geometry and can be modeled by assigning random variables to the parameters in $Y(a)$. The form of the geometry function will depend upon the specific problem under consideration.

For constant-amplitude loading, Eq. (2.28) can be separated and integrated to obtain

$$\Psi(A) = C(\Delta S)^{m}t, \tag{2.29}$$

where $\Psi(A)$ is denoted the damage function, and is given by

$$\Psi(A) = \int_{a_0}^{A} \frac{dz}{Y(z)^{m}(z\pi)^{m/2}}. \tag{2.30}$$

If the failure of the specimen is defined as the point at which the crack size reaches the critical crack size a_c, then a limit state can be defined as

$$g(\mathbf{X}) = \Psi(a_c) - \Psi(a) = \int_{a_0}^{a_c} \frac{dz}{Y(z)^{m}(z\pi)^{m/2}} - C(\Delta S)^{m}t. \tag{2.31}$$

The probability of failure p_f is given by

$$\begin{aligned} p_f &= P\{g(\mathbf{X}) \le 0\} = P\{\Psi(a_c) \le \Psi(a)\} \\ &= P\left\{\int_{a_0}^{a_c} \frac{dz}{Y(z)^{m}(z\pi)^{m/2}} - C(\Delta S)^{m}t \le 0\right\}. \end{aligned} \tag{2.32}$$

The limit state function for fatigue crack growth under variable-amplitude loading can also be obtained for the case when sequence effects are neglected. The differential equation for crack growth (2.25) can be written in an incremental form given by

$$\Delta A_i = CY(A)^m (\Delta S_i)^m (A_i \pi)^{m/2}, \tag{2.33}$$

where ΔA_i is the increase in the crack length due to the ith applied stress cycle, and A_i is the crack length in the ith stress cycle. Equation (2.29) then becomes

$$\Psi(a) = C \sum_{i=1}^{N} \Delta S_i^m. \tag{2.34}$$

For high cycle fatigue where the structural member experiences a large number of stress cycles, the uncertainty in the sum on the right hand side of Eq. (2.34) can be neglected, and the sum replaced by its expected value, i.e.,

$$C \sum_{i=1}^{t} \Delta S_i^m = C\nu_0 t \langle \Delta S^m \rangle, \tag{2.35}$$

where $\langle \cdot \rangle$ is the expected value operator.

The limit state for the probability that the crack size reaches the critical crack size a_c is then formulated as

$$g(\mathbf{X}) = \Psi(a_c) - \Psi(a) = \int_{a_0}^{a_c} \frac{dz}{Y(z)^m (z\pi)^{m/2}} - C \sum_{i=1}^{t} \Delta S_i^m \tag{2.36}$$

The probability of failure p_f is given by

$$\begin{aligned} p_f &= P\{g(\mathbf{X}) \le 0\} = P\{\Psi(a_c) \le \Psi(a)\} \\ &= P\left\{ \int_{a_0}^{a} \frac{dz}{Y(z)^m (z\pi)^{m/2}} - C \sum_{i=1}^{N} \Delta S_i^m \le 0 \right\}. \end{aligned} \tag{2.37}$$

As discussed above, these probabilities are not easily determined analytically, however, they can be effectively calculated by first or second order reliability methods using computer programs such as PROBAN [45]. A more detailed discussion can be found in Madsen, et al. [23].

3. Scalar Diffusion Models

Careful examination of Figs. 1.1–1.2 indicates that the sample functions of crack length versus time are non-smooth functions with considerable intermingling.

This type of behavior cannot be portrayed by random variable fatigue crack growth models. A more general approach to stochastic modeling of fatigue crack growth is to explicitly introduce appropriate random processes into the model. When a single random process is employed to characterize the combined random interactions affecting crack growth, the following model has been considered

$$\frac{dA(t)}{dt} = Q(\Delta K, K_{max}, K_c, S, A, R)\, X(t)\,, \tag{3.1}$$

where $X(t)$ is a non-negative random process. Unfortunately, analytical solution of Eq. (3.1) for a arbitrary random process $X(t)$ is not generally available. The next sections considers a special case of Eq. (3.1) for which analytical solution can be obtained.

3.1 White Noise Model

The case in which $X(t)$ is assumed to be a Gaussian white noise will be initially considered. Such a process is nonphysical and, due to its Gaussian nature, admits the possibility of negative crack growth rates. However, this idealization proves useful because effective analytical methods are available to determine the sought probability distributions.

Because the quantities ΔK, K_{max}, K_c, S, R, μ_X, etc., are assumed to be known, Eq. (3.1) is rewritten as

$$\frac{dA(t)}{dt} = Q(A)\, X(t)\,. \tag{3.2}$$

The random process $X(t)$ is then modeled as

$$X(t) = \mu_X + \sqrt{2D}\; \xi(t)\,, \tag{3.3}$$

$$\langle \xi(t_1)\, \xi(t_2) \rangle = \delta(t_2 - t_1)\,, \tag{3.4}$$

where μ_X is the constant mean value of $X(t)$, $2D$ is the intensity of the random fluctuations of $X(t)$, and $\xi(t)$ is a unit intensity white noise. For fatigue crack growth problems, the fluctuating part of the process $X(t)$ is generally small in comparison with the process mean μ_X. Substituting Eq. (3.3) into Eq. (3.2) yields

$$\frac{dA(t)}{dt} = m(A) + \sigma(A)\, \xi(t)\,, \tag{3.5}$$

where

$$m(A) = Q(A)\, \mu_X, \tag{3.6}$$

$$\sigma^2(A) = 2DQ^2(A). \tag{3.7}$$

There are two well-known interpretations of the stochastic differential equation in Eq. (3.5), the Stratonovich and the Itô interpretations (see Sobczyk [46]). The Stratonovich interpretation is adopted here. The Itô stochastic differential equation that is equivalent to Eq. (3.5) is then given by

$$dA(t) = m^*(A)\,dt + \sigma(A)\,dW(t), \tag{3.8}$$

where $W(t)$ is a standard Wiener process, and

$$m^*(A) = m(A) + \frac{1}{2}\sigma(A)\frac{\partial\sigma}{\partial A}. \tag{3.9}$$

Under appropriate assumptions regarding $m^*(A)$ and $\sigma(A)$, the solution of Eq. (3.8) (i.e., $A(t)$), is a diffusion Markov process whose transition density function satisfies the Fokker–Planck–Kolmogorov equation given by [46]

$$\frac{\partial p}{\partial t} + \frac{\partial}{\partial a}[m^*(a)\,p] - \frac{1}{2}\frac{\partial^2}{\partial a^2}[\sigma^2(a)\,p] = 0. \tag{3.10}$$

or

$$\frac{\partial p}{\partial t} + \frac{\partial}{\partial a}\left[Q(a)\,\{\mu_X + D\frac{\partial Q(a)}{\partial a}\}\,p\right] - D\frac{\partial^2}{\partial a^2}[Q^2(a)\,p] = 0. \tag{3.11}$$

Here, $p = p(a, t| a_0, t_0)$ is the transition probability density function for the crack length $A(t)$. The initial condition for the partial differential equation in Eq. (3.10) is given by

$$p(a, t_0| a_0, t_0) = \delta(a - a_0). \tag{3.12}$$

Boundary conditions are also necessary for Eq. (3.10). Crack growth is considered on the finite interval $[a_0, a_c] \subset [a_0, \infty)$, where a_c denotes a critical crack size. The conditions that seem to be physically appropriate are: a reflecting boundary condition at $a = a_0$, and a zero derivative boundary condition at $a = a_c$.

A reflecting boundary condition indicates that flow of the probability mass through $a = a_0$ is not allowed; if a trajectory reaches the state $a = a_0$, it is instantaneously reflected to the interior of the interval where it is again governed by Eq. (3.10). This type of boundary condition guarantees that no cracks are smaller than a_0 and that

$$\int_{a_0}^{\infty} p(a,t)\,dt = 1, \tag{3.13}$$

i.e., all probability mass is concentrated on the open interval $[a_0, \infty)$. Analytically, this condition takes the form [47],

$$m^*(a)\,p(a,t_0|\,a_0,t_0) - \frac{1}{2}\frac{\partial}{\partial a}[\sigma^2(a)\,p(a,t_0|\,a_0,t_0)] = 0, \quad \text{at } a = a_0. \tag{3.14}$$

The boundary condition at $a = a_c$ is given on physical grounds by Oh [48]. When the crack length a reaches the critical crack size a_c, it is assumed to have an infinite velocity (i.e., unstable crack growth). Therefore, one can argue that

$$p(a_c + \Delta a, t) = p(a_c, t), \tag{3.15}$$

which implies

$$\left.\frac{\partial p}{\partial a}\right|_{a_c} = \lim_{\Delta a \to 0} \frac{p(a_c+\Delta a, t) - p(a_c, t)}{\Delta a} = 0. \tag{3.16}$$

The solution of the Fokker–Planck–Kolmogorov equation, i.e., Eq. (3.10), satisfying the prescribed initial and boundary conditions provides the basic information concerning the probabilistic evolution of the process $A(t)$. From the transition density $p(a,t|\,a_0,t_0)$, one can obtain quantities that are of interest to fatigue life prediction. For example, the probability distribution of the crack size at time t conditional on the initial crack size is given by

$$F_A(a;t|\,a_0;t_0) = P\{A(t)<a|A(t_0)=a_0\} = \int_{a_0}^{a} p(a,t|a_0,t_0)\,da. \tag{3.17}$$

The probability that the crack size exceeds a given critical level a_c is

$$P\{A(t) > a_c|A(t_0)=a_0\} = 1 - F_A(a_c;t|\,a_0;t_0), \tag{3.18}$$

and the probability distribution of random time T_{a_c} when the crack size $A(t)$ reaches a given value a_c is

$$F_T(t) = P\{T_{a_c} \le t\} = 1 - P\{A(t) < a_c\}$$
$$= 1 - \int_{a_0}^{a_c} p(a, t| a_0, t_0)\, da\,. \tag{3.19}$$

An alternative boundary condition may be considered at $a = a_c$. If a_c is such that cracks larger than a_c do not exist (e.g., if a_c is the width of the plate), then an absorbing boundary condition should be prescribed at $a = a_c$. This indicates that trajectories that reach this state are absorbed and will remain there forever. In this situation, the probability $p(a, t| a_0, t_0)$ should vanish, i.e.,

$$p(a_c, t| a_0, t_0) = 0. \tag{3.20}$$

Additionally, if the initial crack length is a random variable, then the unconditional probability density function $f_A(a;t)$ may be of interest, i.e.,

$$f_A(a;t) = \int_0^\infty p(a, t| a_0, t_0)\, f_{A_0}(a_0)\, da_0, \tag{3.21}$$

where $f_{A_0}(a_0)$ is the distribution of initial crack sizes. Multiplying Eq. (3.10) by $f_{A_0}(a_0)$ and integrating from 0 to ∞ shows that $f(a;t)$ also satisfies the Fokker–Planck–Kolmogorov equation, i.e., Eq. (3.10). The appropriate initial condition is then

$$f_A(a_0;t_0) = f_{A_0}(a_0)\,. \tag{3.22}$$

Here, the boundary condition given in Eq. (3.14) is applied at the minimum realizable initial crack size (e.g., at $a_0 = 0$), whereas the boundary condition in Eq. (3.16) (or Eq. (3.20)) remains unchanged.

3.2 Solution of Fokker–Planck–Kolmogorov Equation

To obtain the solution of Eq. (3.11), the following transformation is introduced (Sobczyk [26], [28], Lin and Yang [20]):

$$z(t) = \int_{a_0}^{a(t)} \frac{dv}{Q(v)}. \tag{3.23}$$

Substituting Eq. (3.23) into Eq. (3.11), one obtains

$$\frac{\partial p_z}{\partial t} + \mu_X \frac{\partial p_z}{\partial z} - D\frac{\partial^2 p_z}{\partial z^2} = 0, \tag{3.24}$$

where $p_z = p_z(z, t| z_0, t_0)$. The initial condition for Eq. (3.24) is

$$p_z(z, t_0| 0, t_0) = \delta(z). \tag{3.25}$$

The transition probability density for the random crack length A is related to the solution of Eq. (3.24) via

$$p_z(z, t| z_0, t_0) = Q(a)\, p(a, t| a_0, t_0), \tag{3.26}$$

or

$$p(a, t| a_0, t_0) = \frac{1}{Q(a)} p_z\left(\int_{a_0}^{a} \frac{dv}{Q(v)}, t \middle| z_0, t_0\right). \tag{3.27}$$

Assuming the boundary conditions for Eq. (3.24) are at infinity, the solution that satisfies the initial condition in Eq. (3.25) is given by

$$p_z(z, t| 0, t_0) = \frac{1}{\sqrt{2\pi}\sqrt{2D(t-t_0)}} \exp\left\{-\frac{[z-\mu_X(t-t_0)]^2}{4D(t-t_0)}\right\}. \tag{3.28}$$

The corresponding solution to Eq. (3.11) is

$$p(a, t| a_0, t_0) = \frac{1}{Q(a)\sqrt{2\pi}\sqrt{2D(t-t_0)}} \exp\left\{-\frac{\left[\int_{a_0}^{a} \frac{dv}{Q(v)} - \mu_X(t-t_0)\right]^2}{4D(t-t_0)}\right\}. \tag{3.29}$$

Notice that the solution in Eq. (3.29) must also admit values of a smaller than a_0, a physical impossibility. This situation arises because the diffusive Markov process described in Eq. (3.11) is only an approximation to the physical crack length process.

A better approximation can be obtained by enforcing the appropriate boundary conditions. The boundary conditions for Eq. (3.24) that correspond to those given in Eqs. (3.14) and (3.16) are given respectively by

$$D\frac{\partial p_z}{\partial z} - \mu_X p_z = 0, \text{ at } z = 0, \tag{3.30}$$

$$\frac{\partial p_z}{\partial z} - \left.\frac{\partial Q}{\partial a}\right|_{a_c} p_z = 0, \text{ at } z = z_c, \tag{3.31}$$

where

$$z_c = \int_{a_0}^{a_c} \frac{dv}{Q(v)}. \tag{3.32}$$

The boundary conditions for Eq. (3.24) that is equivalent to the absorbing boundary condition in Eq. (3.18) is

$$p_z(z_c, t|0, t_0) = 0. \tag{3.33}$$

The solution of Eq. (3.24) subject to appropriate boundary conditions can be obtained by the method of separation of variables. Thus, substituting

$$p_z(z, t) = p_z(z, t|0, t_0) = Z(z)\,T(t) \tag{3.34}$$

into Eq. (3.24), the differential equations for Z and T are

$$\frac{dT}{dt} + \lambda DT = 0, \tag{3.35}$$

$$\frac{d^2 Z}{dz^2} - 2\eta\frac{dZ}{dz} + \lambda Z = 0, \tag{3.36}$$

where λ is the separation constant, and $\eta = \mu_X/(2D)$. The solution to Eq. (3.35) is given by

$$T(t) = T_0 e^{-\lambda D(t-t_0)}. \tag{3.37}$$

Rewriting Eq. (3.36) with its boundary conditions constitutes the following Sturm–Liouville problem:

$$\frac{1}{e^{-2\eta z}}\{\frac{d}{dz}(e^{-2\eta z}\frac{dZ}{dz})\} + \lambda Z = 0, \tag{3.38}$$

$$\frac{dZ}{dz} - 2\eta Z = 0, \text{ at } z = 0, \tag{3.39}$$

$$\frac{dZ}{dz} - \left.\frac{\partial Q}{\partial a}\right|_{a_c} Z = 0, \text{ at } z = z_c. \tag{3.40}$$

The solution of this Sturm–Liouville problem satisfies the orthogonality condition

$$\int_0^{z_c} e^{-2\eta z} Z_i Z_j = 0, \qquad i \neq j, \tag{3.41}$$

where Z_i is the eigenfunction corresponding to λ_i.

Two case must be considered: (i) $\eta^2 - \lambda \geq 0$ and $\eta^2 - \lambda < 0$. In the first case, the solution is of the form

$$Z(z) = e^{\eta z}\{c_1 e^{\sqrt{\eta^2 - \lambda z}} + c_2 e^{-\sqrt{\eta - \lambda z}}\}, \tag{3.42}$$

where c_1 and c_2 are constants. Applying the boundary conditions in Eqs. (3.39) and (3.40) yields

$$\frac{c_1}{c_2} = \frac{\eta + \alpha}{\eta - \alpha}, \tag{3.43}$$

$$\frac{c_1}{c_2} = \frac{\alpha - \eta + \gamma}{\alpha + \eta - \gamma} e^{-2\alpha z_c}, \tag{3.44}$$

and the associated characteristic equations

$$\frac{(\eta - \alpha)(\alpha - \eta + \gamma)}{(\eta + \alpha)(\alpha + \eta - \gamma)} e^{-2\alpha z_c} = 1, \tag{3.45}$$

where $\alpha = \sqrt{\eta^2 - \lambda}$ and $\gamma = \partial Q/\partial a|_{a_c}$. The single eigenvalue associated with Eq. (3.45) is designated λ_0, and the corresponding eigenfunction is given by

$$Z_0(z) = C_0 e^{\eta z}\left(\frac{\eta \alpha_0}{\eta - \alpha_0} e^{\alpha_0 z} + e^{-\alpha_0 z}\right). \tag{3.46}$$

where $\alpha_0 = \sqrt{\eta^2 - \lambda_0}$.

For the case where $\eta^2 - \lambda < 0$, the eigenfunctions are oscillatory and given by

$$Z(z) = e^{\eta z}\{c_3 \cos(\sqrt{\lambda - \eta^2}\, z) + c_4 \sin(\sqrt{\lambda - \eta^2}\, z)\}, \tag{3.47}$$

where c_3 and c_4 are constants. Defining $\delta = \sqrt{\lambda - \eta^2}$ and applying the boundary conditions in Eqs. (3.39) and (3.40) yields

$$\frac{c_3}{c_4} = \frac{\delta}{\eta}, \tag{3.48}$$

$$\frac{c_3}{c_4} = \frac{\delta + (\eta - \gamma)\tan\delta z_c}{\gamma - \eta + \delta\tan\delta z_c}. \tag{3.49}$$

The characteristic equation for the eigenvalues λ_i, $i = 1, 2, \ldots$, is obtained as

$$\frac{\delta}{\eta} = \frac{\delta + (\eta - \gamma)\tan\delta z_c}{\gamma - \eta + \delta\tan\delta z_c}. \tag{3.50}$$

Substituting Eq. (3.48) into Eq. (3.47) and rearranging yields

$$Z_i(z) = C_i e^{\eta z}\{\sin(\delta_i z + \phi_i)\}, \qquad i = 1, 2, \ldots, \tag{3.51}$$

where C_i is a constant, $\delta_i = \sqrt{\lambda_i - \eta^2}$, and $\phi_i = \tan^{-1}\sqrt{\lambda_i/\eta^2 - 1}$. The general solution to Eq. (3.24) subject to the boundary conditions in Eqs. (3.30) and (3.31) is then

$$\begin{aligned} p(z, t|0, t_0) &= \sum_{i=0}^{\infty} T_i(t) Z_i(z) \\ &= K_0 e^{\eta z - \lambda_0 D(t - t_0)} \left(\frac{\eta\alpha_0}{\eta - \alpha_0} e^{\alpha_0 z} + e^{-\alpha_0 z} \right) \\ &\quad + \sum_{i=1}^{\infty} K_i e^{\eta z - \lambda_i D(t - t_0)} \sin(\delta_i z + \phi_i). \end{aligned} \tag{3.52}$$

The constants K_i are obtained using the orthogonality relation in Eq. (3.41) and the initial condition in Eq. (3.25) and are given by

$$K_0 = \frac{\dfrac{\eta\alpha_0}{\eta - \alpha_0} + 1}{\left(\dfrac{\eta\alpha_0}{\eta - \alpha_0}\right)^2 \dfrac{(e^{2\alpha_0 z_c} - 1)}{2\alpha_0} - \dfrac{(e^{-2\alpha_0 z_c} - 1)}{2\alpha_0} + \dfrac{2\eta\alpha_0 z_c}{\eta - \alpha_0}}, \tag{3.53}$$

$$K_i = \frac{2\delta_i^2}{\sqrt{\lambda_i}\{\delta_i\left(z_c + \frac{\delta_i}{\eta\sqrt{\lambda_i}}\right) - \cos(\delta_i z_c + \phi_i)\sin(\delta_i z_c + \phi_i)\}}. \tag{3.54}$$

The sought probability density function $p(a, t| a_0, t_0)$ is recovered by substituting the series in Eq. (3.52) into Eq. (3.27). The distribution of times to reach a given crack size can then be evaluated using Eq. (3.19), and the marginal distribution of crack sizes as a function of time is obtained via (3.21).

The problem has also been formulated and solved by Solomos [29] for the absorbing boundary condition at z_c, i.e., $Z(z_c) = 0$.

3.3 Correlated Noise Models

The white noise model presented in the preceding section provides an illustrative case where the fatigue crack length is modeled as a diffusive Markov process. However, as discussed previously, randomization of the crack growth model using a white noise process admits the possibility that the model process takes negative crack growth rates, which are physically unrealizable.

Another difficulty in the white noise model is that the values of the process $X(t)$ are uncorrelated in time. If $X(t)$ is considered to be perfectly correlated (i.e., a random variable), then the largest statistical dispersion is obtained; whereas if $X(t)$ is uncorrelated at each instant of time (i.e., a Gaussian white noise), then the smallest statistical dispersion of crack size is obtained. Experimental evidence indicates that the correct situation is somewhere in between. In this situation, an important problem arises: Is it possible to relax the Gaussian white noise assumption to include the necessary correlation information without losing the analysis advantages of Markov process theory?

The first paper to pose and discuss this question was Sobczyk [26], in which two approaches were suggested to alleviate these difficulties yet still remain within the scope of Markov processes theory. The first approach uses the concept of stochastic averaging to address the problem of correlation in the white noise model and will be discussed in the next few paragraphs. The second approach is to construct vector Markov process models such that the process has an appropriate correlation structure and the crack growth rate is always positive. Section 4 studies this problem.

Consider the crack growth equation given by

$$\frac{dA}{dt} = Q[A(t), X(t)]. \tag{3.55}$$

If the correlation time of $X(t)$ is sufficiently short as compared with the characteristic time of $A(t)$, the process $A(t)$ governed by Eq. (3.55) can be approximated by a diffusive Markov process that is governed by the Itô stochastic differential equation,

$$dA = m^* (A)\, dt + \sigma(A)\, dW, \tag{3.56}$$

with the following drift and diffusion coefficients:

$$\begin{aligned} m^* (A) &= \langle Q(A, X) \rangle + \int K_{\tilde{Q}'\tilde{Q}}(A, \tau)\, d\tau, \\ \sigma^2 (A) &= 2\int \langle \tilde{Q}(A, X(\tau))\, \tilde{Q}(A, X(t+\tau)) \rangle d\tau, \end{aligned} \tag{3.57}$$

Here, $\langle Q(A, X) \rangle$ is the mean value of the right-hand side of Eq. (3.55) when A is fixed; $\tilde{Q}(A, X(t))$ is the fluctuating component of $Q(A, X(t))$, i.e.,

$$\tilde{Q}(A, X(t)) = Q(A, X(t)) - \langle Q(A, X(t)) \rangle, \tag{3.58}$$

and $K_{\tilde{Q}'\tilde{Q}}(A, \tau)$ is a mutual correlation function of the derivative of the fluctuation part of Q (with respect to A) and Q in different time instants. More details can be found in [26], [19], [20], and [42].

Applying this reasoning to the crack growth equation given in Eq. (3.2), i.e.,

$$\begin{aligned} \frac{dA}{dt} &= Q(A)\, X(t) \\ &= Q(A)\, [m_X + Y(t)] \\ &= \mu_X Q(A) + Q(A)\, Y(t), \end{aligned} \tag{3.59}$$

leads to the Itô differential equation whose diffusion and drift coefficients are

$$m^* (A) = m(A) + \frac{1}{2}\sigma(A)\frac{\partial\sigma}{\partial A}, \tag{3.60}$$

$$\sigma^2 (A) = I_e Q^2 (A). \tag{3.61}$$

Using Eqs. (3.60) and (3.61) corresponds to replacing the process $Y(t)$ with an *equivalent* white noise process. Here, I_e is the intensity of the *equivalent* white noise process in the analysis given by

$$I_e = 2\int_0^\infty K_Y(\tau)\, d\tau. \tag{3.62}$$

Eqs. (3.60) and (3.61) correspond to Eqs. (3.9) and (3.7), respectively, where the intensity of the white noise, $2D$, is replaced by the intensity of the equivalent white noise, I_e. As a result, solution of the associated Fokker–Planck–Kolmogorov equation also can be obtained as shown in Section 3.2.

Lin and Yang [19], [20] have performed a detailed analysis of the random fatigue problem using this approach giving specific examples and comparisons with experimental data. More recently, Solomos [29] reported the analysis of the first-passage time distribution for reflecting and absorbing boundary conditions at the ends of interval the $[a_0, a_c]$.

As indicated above, although the use of the concept of averaging described before provides an avenue to account for the correlation of the processes affecting crack growth, it still includes the Gaussian distribution assumption, which in general results in the possibility of negative value of the crack growth rate.

3.4 Other Approaches

A number of researchers have suggested that the multiplicative random process in Eq. (3.1) be a function of crack size a, rather than time or cycle number. For constant-amplitude tests in tightly controlled laboratory settings, this is physically reasonable, as the majority of the variability found in fatigue crack growth is attributed to the inhomogeneity of the material, i.e., the crack growth rate varies as a function of the resistance of the material through which it is propagating. In these approaches, the randomness in crack growth rates is broken down into (i) specimen-to-specimen variation, and (ii) within-specimen variation. The specimen-to-specimen variation is represented by random variables, whereas the within-specimen variation is modeled using a random process. For more details of these approaches, the interested reader is referred to Ortiz and Kiremidjian [24], [25], Madsen, et al. [23], Ditlevsen [13], and Ditlevsen and Olesen [14].

4. Two-State Models for Stochastic Crack Growth

This section presents a model formulation that augments the fatigue crack growth rate model given in Eq. (3.1) with an auxiliary equation to accurately model the correlation structure of the crack growth process and to eliminate the possibility of a negative crack growth rate. The formulation of the boundary value problem governing the evolution of the crack size probability distribution is consistent with physical phenomena, does not require stochastic averaging and gives the solution as a function of initial crack size, a_0.

Consider the auxiliary random process $Z(t)$, which is a filtered white noise process, and a deterministic function $g(z)$, which transforms the process $Z(t)$ into a positive process $X(t)$ (e.g., $g(z)$ could be taken as an exponential function). Then the random process $X(t)$ in Eq. (3.1) can be represented by $X(t) = g(Z(t))$, and a coupled system of stochastic differential equations is obtained,

$$\frac{dA}{dt} = Q[A(t)]\,g[Z(t)]$$
$$\frac{dZ}{dt} = F[Z(t)] + \sqrt{2D}\,\xi(t) \qquad (4.1)$$
$$A(0) = A_0, \quad Z(0) = Z_0,$$

where $\xi(t)$ is a unit intensity Gaussian white noise process, and A_0, Z_0 are general random variables characterizing the initial crack size and the initial state of the auxiliary random process, respectively. Because both Q and g are defined as positive functions of their respective arguments, the possibility of negative crack growth rates encountered in the white noise models therefore has been eliminated. Q can then be appropriately constructed such that the random fluctuations in crack growth rate have the proper correlation structure.

If the system of equations in Eq. (4.1) are interpreted in the Itô sense, the vector process $[A(t), Z(t)]^T$ is a diffusion Markov process, and all of the attendant theory is available for the description of the random crack size $A(t)$. Such an approach has been considered by Spencer and co-workers (see [30], [31] and [32]).

4.1 Reliability Problem Formulation

The transition probability density function $p = p(a, z, t | a_0, z_0, t_0)$ is the joint probability density function for the random crack size $A(t)$ and the auxiliary random process $Z(t)$, conditional on their initial values, and is governed by the Fokker–Planck–Kolmogorov equation,

$$\frac{\partial p}{\partial t} = -\frac{\partial}{\partial a}[Q(a)\,g(z)\,p] - \frac{\partial}{\partial z}[F(z)\,p] + D\frac{\partial^2 p}{\partial z^2}. \qquad (4.2)$$

The transition density is also governed by the formal adjoint of Eq. (4.2), the backward Kolmogorov equation, given by

$$\frac{\partial p}{\partial t} = Q(a_0)\,g(z_0)\frac{\partial p}{\partial a_0} + F(z_0)\frac{\partial p}{\partial z_0} + D\frac{\partial^2 p}{\partial z_0^2}. \qquad (4.3)$$

The cumulative distribution of times to reach a given crack size a_c, or conversely, the probability that a crack will not reach a critical crack size in a given time, is of great interest in the design of fatigue-critical structures. Denoting the probability that the crack size is less than the critical crack size (conditional on the initial crack size and the initial value of $Z(t)$) as

$$R = R(t|a_0, z_0) = P\{A(t) < a_c; t|a_0, z_0, t_0\}$$
$$= \int_{a_0}^{a_c}\int_{-\infty}^{\infty} p(a, z, t|a_0, z_0, t_0)\, dz da, \tag{4.4}$$

one see that $R(t|a_0, z_0)$ also satisfies the backward Kolmogorov equation (4.3) and can write

$$\frac{\partial R}{\partial t} = Q(a_0)\, g(z_0)\frac{\partial R}{\partial a_0} + F(z_0)\frac{\partial R}{\partial z_0} + D\frac{\partial^2 R}{\partial z_0^2}. \tag{4.5}$$

Notice that, due to the lack of a second derivative with respect to a_0 in Eq. (4.5), only one boundary condition needs to be specified in the a_0 direction in order that the initial-boundary value problem be well-posed (see Fichera [49]). The initial and boundary conditions can be stated on physical grounds as follows:

$$R(0|a_0, z_0) = 1, \qquad 0 \le a_0 < a_c, \tag{4.6}$$

$$R(t|a_c, z_0) = 0, \qquad \forall\, z_0, \tag{4.7}$$

$$\lim_{z_0 \to \infty} R(t|a_0, z_0) = 0, \qquad 0 \le a_0 < a_c, \tag{4.8}$$

$$\lim_{z_0 \to -\infty} \frac{\partial R(t|a_0, z_0)}{\partial z_0} = 0, \qquad 0 \le a_0 < a_c. \tag{4.9}$$

Further discussion of the initial and boundary conditions follows.

The initial condition given in Eq. (4.6) is rather evident; the reliability of the structural component is assumed to be equal to one before being put into service (i.e., the initial crack size is considered to be below the critical crack size at the time the structural member is put into service).

The boundary condition given by Eq. (4.7) is also easy to envision and simply indicates that the reliability of the component is zero for all z_0 when the initial crack size is the critical crack size (i.e., $a_0 = a_c$).

Assume that the function $F(z)$ is such that the domain of $Z(t)$ is the interval $(-\infty, \infty)$, that $g(z)$ is a one-to-one mapping onto the interval $(0, \infty)$, and that $\partial g/\partial z \ge 0$. Because $Q(\cdot)$ is positive for all physically realizable arguments, the boundary condition specified in Eq. (4.8) can be justified by examining Eq. (4.1) and noting that as $z_0 \to \infty$, so also does $dA/dt \to \infty$. Since the random process $Z(t)$ is continuous, $Z(t)$ and thus dA/dt cannot instantaneously make large changes. Therefore, as $z_0 \to \infty$, $A(t)$ will grow very rapidly, and in the limit, the reliability will go to zero (i.e., $R(t|a_0, z_0) \to 0$).

The third boundary condition is more difficult to envision. Because of the assumptions on $g(\cdot)$ stated in the previous paragraph, as $z_0 \to -\infty$, then $dA/dt \to 0$. Thus, there is no crack growth. As a result, one might expect the reliability to go to one (i.e., the time to reach the critical crack size goes to infinity) as the initial value of $Z(t)$ goes to negative infinity. However, there is a limiting factor. Once the time to reach the critical crack size a_c becomes much larger than the correlation time for $Z(t)$, the initial value of $Z(t)$ will have little effect on the reliability; that is, the increment of the reliability approaches zero as the initial value for $Z(t)$ goes to negative infinity, implying that the derivative of the reliability goes to zero. This boundary condition has also been demonstrated through Monte Carlo simulation [30].

The initial-boundary value problem given by Eqs. (4.5)–(4.9) is therefore defined over a semi-infinite two-dimensional domain. Here, the problem has been constructed based upon a two-dimensional Markov process such that negative crack growth rates are impossible and the correlation structure of the random crack growth rate can be included without employing stochastic averaging.

In this problem formulation, the solution is a function of the initial value of the auxiliary random process, $Z(t)$. To eliminate the dependence on z_0, one can assume a stationary start condition for $Z(t)$ and make use of the theory of total probability to obtain

$$R(t|a_0) = \int_{-\infty}^{\infty} R(t|a_0, z_0)\, f_Z(z_0)\, dz_0, \tag{4.10}$$

where $f_Z(z_0)$ is the stationary probability density function for $Z(t)$. In this final step, the formulation is completed to obtain the reliability of a component against fatigue crack growth conditional on the initial flaw size.

Remark: One of the most attractive features of this approach is that the solution is obtained as a function of the initial crack size. This allows one to readily assess the effect of initial fatigue quality on the reliability calculations through the use of initial crack size distributions. For example, if the distribution of initial crack sizes is given by $f_{A_0}(a_0)$, then the unconditional reliability can be determined as a function of time via

$$R(t) = \int_{0}^{a_c} R(t|a_0)\, f_{A_0}(a_0)\, da_0, \tag{4.11}$$

and thus the effect of initial flaw size distribution is readily assessed.

4.2 Moments of Time to Reach a Critical Crack Size

A problem of interest to many researchers is determination of the statistical moments of a random process, not only because they are more readily obtained than the complete probability distribution function, but because they generally give the main features of a random process. In this section, the boundary value problem for the statistical moments of the time to reach a critical crack size is formulated.

Denoting the ordinary moments of the random time to reach a critical crack size a_c as

$$T^n(a_0, z_0) = \int_0^\infty t^n \frac{\partial R(t|a_0, z_0)}{\partial t} dt, \tag{4.12}$$

Eq. (4.5) is multiplied by t^{n-1} and integrated with respect to time to obtain a recursive set of boundary value problems given by

$$-nT^{n-1} = Q(a_0)g(a_0)\frac{\partial T^n}{\partial a_0} + F(z_0)\frac{\partial T^n}{\partial z_0} + D\frac{\partial^2 T^n}{\partial z_0^2}, \quad n = 1,2,\dots, \tag{4.13}$$

where, by definition, $T^0(a_0, z_0) = 1$, and the boundary conditions are given respectively by

$$T^n(a_c, z_0) = 0, \qquad \forall\ z_0, \tag{4.14}$$

$$\lim_{z_0 \to \infty} T^n(a_0, z_0) = 0, \qquad 0 \le a_0 < a_c, \tag{4.15}$$

$$\lim_{z_0 \to -\infty} \frac{\partial T^n(a_0, z_0)}{\partial z_0} = 0, \qquad 0 \le a_0 < a_c. \tag{4.16}$$

The boundary conditions were obtained directly from Eq. (4.12) and the boundary conditions for the reliability problem.

The central moments can be of more interest than the ordinary moments. The second central moment (i.e., the variance V) is given in terms of the ordinary moments as

$$V = T^2 - (T^1)^2. \tag{4.17}$$

The boundary value problem can then be formulated directly in terms of the second central moments as

$$-2\pi S_0\left(\frac{\partial T^1}{\partial z_0}\right)^2 = Q(a_0)g(a_0)\frac{\partial V}{\partial a_0} + F(z_0)\frac{\partial V}{\partial z_0} + D\frac{\partial^2 V}{\partial z_0^2}, \tag{4.18}$$

where the boundary conditions are

$$V(a_c, z_0) = 0, \qquad \forall\ z_0, \tag{4.19}$$

$$\lim_{z_0 \to \infty} V(a_0, z_0) = 0, \qquad 0 \le a_0 < a_c, \tag{4.20}$$

$$\lim_{z_0 \to -\infty} \frac{\partial V(a_0, z_0)}{\partial z_0} = 0, \qquad 0 \le a_0 < a_c. \tag{4.21}$$

The boundary conditions for the equations governing the central moments are also obtained directly from Eq. (4.12) and the boundary conditions for the reliability problem. The preceding manipulations could be carried out for the higher-order central moments in a similar manner.

As stated in the previous section, the dependence of the solution on z_0 can be eliminated by assuming a stationary start condition for the random process $Z(t)$ and making use of the theorem of total probability. The solutions, however, are also a function of the initial crack size. Thus, either the actual initial crack size or the initial crack size distribution can be used to obtain the unconditional moments of the random time to reach a critical crack size.

Although analytical solution of the boundary value problems presented in the preceding is unavailable at this time, a simple and efficient numerical solution has been reported by Enneking, et al. [50] and will be used for the examples presented in the next section.

4.3 Correlation with Experimental Data

Data are essential for validation of any mathematical model. The fatigue crack growth data set reported by Virkler [9] is one of high replication under tightly controlled laboratory conditions and is well-suited for validation of probabilistic models.

The test specimens used in the experiment were 0.10 in. (2.54 mm) thick center cracked panels of 2024–T3 aluminum alloy measuring 2.2 in. (558.8 mm) long and 6 in. (152.4 mm) wide. To minimize statistical uncertainty in the tests, each of the specimens was taken from a single lot provided by the Aluminum Company of America, and each test was performed by the same operator on the same machine. The tests were run at a constant stress range of $\Delta s = 48.28$ MPa and with a stress ratio $R = 0.2$ to facilitate replication of test conditions and to provide a wide range of ΔK data. The load was sinusoidal with a frequency of 20 Hz. Fatigue crack growth was monitored with a 150x zoom stereo microscope that was mounted on a traversing system and had a resolution of 0.001 mm. The average experimental measurement error of this setup was found to be only 0.00141 mm. The accumulated number of cycles was recorded at specified crack lengths between 9 mm and 49.8 mm. Measurements were taken every 0.2 mm in the range $9.00 \le a \le 36.20$ mm, every 0.4 mm in the range $36.20 \le a \le 44.2$ mm, and every 0.8 mm in the range $44.2 \le a \le 49.8$ mm. A total of 164 measurements were taken for each of the 68 replicate specimens. This exacting data will be used to validate the two-state stochastic model and to demonstrate its accuracy and versatility.

For the study of fatigue crack growth under constant-amplitude loading, the crack growth rate is commonly considered as a function of the elastic stress intensity range, given for a finite-width plate by

$$\Delta K = \Delta s \sqrt{\frac{\pi a}{\cos(\pi a/b)}}, \qquad \frac{a}{b} < 0.7, \tag{4.22}$$

with no more than 0.3% error, where ΔS is the stress range, and b is the width of the plate.

A plot of $\ln(dA/dt)$ versus $\ln(\Delta K)$ for the Virkler data is depicted in Fig. 4.1. While the predominantly linear trend in Fig. 4.1 indicates that the Paris–Erdogan-type model would well represent the data, Spencer, et al. [31] has shown that the nonlinear trends are significant and can introduce systematic error into predicted results. To eliminate this systematic error, Kung and Ortiz [8] have suggested that a cubic polynomial fatigue crack growth law be used. In addition, the function $F(\cdot)$ in Eq. (4.1) is chosen to be linear, and $X(t)$ is chosen to be an exponentially correlated lognormal random process. The system of equations governing crack growth then becomes

$$\begin{aligned} \frac{dA}{dt} &= Q[A(t)] \exp[Z(t)], \\ \frac{dZ}{dt} &= -\zeta\, Z(t) + \sqrt{2D}\xi(t), \end{aligned} \tag{4.23}$$

where ζ is the correlation parameter for the process $Z(t)$. For the cubic polynomial fatigue crack growth laws, $Q[a(t)]$ is given by

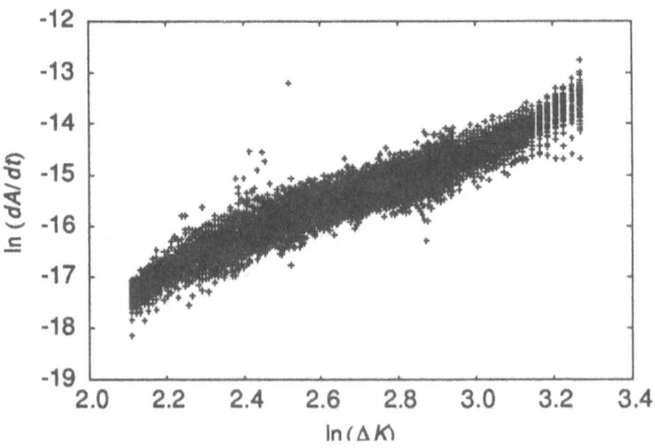

Figure 4.1. Ln(dA/dt) versus ln(ΔK) for the Virkler data.

$$Q[a(t)] = \exp[\beta_0 + \beta_1(\ln \Delta K) + \beta_2(\ln \Delta K)^2 + \beta_3(\ln \Delta K)^3], \qquad (4.24)$$

where $\Delta K = \Delta K(a)$ is given by Eq. (4.22). The parameters determined using the modified finite integral approach are (see [31]): $\beta_0 = -74.060$, $\beta_1 = 60.265$, $\beta_2 = -21.390$, $\beta_3 = 2.6348$, and $\langle X(t) \rangle = 1$.

The random residuals Z_i for this model are depicted in Fig. 4.2. One sees that the residuals are zero mean and have a constant variance. Therefore, the multiplicative process $X(t) = \exp[Z(t)]$ is also stationary. The standard deviation of X is found via Eq. (5.16) to be $\sigma_X = 0.227$ and the correlation parameter is chosen to be $\zeta = 9\times10^{-5}$.

The boundary value problem in Eqs. (4.13)–(4.9) is solved for the statistical moments of time to reach a critical crack size a_c as a function of the initial crack size a_0. Figures 4.3–4.4 present the solutions for the mean and mean square, respectively, of the time to reach a critical crack size of 49.8 mm. The Virkler data is superimposed for comparison, and as can be seen, the predicted results are indiscernible from the data over the entire range of initial crack sizes.

Solving the boundary value problem in Eqs. (4.18)–(4.9) yields the variance of the time to reach a critical crack size a_c as a function of the initial crack size a_0. The square root of this result, the standard deviation, is depicted in Fig. 4.5. Again, the Virkler data is superimposed for comparison, and very good agreement is seen ith the results predicted by the model.

The boundary value problem for the reliability function given in Eqs. (4.5)–(4.9) is also solved to yield the distribution of the random time to reach a critical crack size. The results are given in Figs. 4.6–4.7. In Fig. 4.6, the initial crack size is held constant and the distribution of times to reach several crack sizes is presented. The distribution functions for several initial crack sizes, holding the critical crack size, are presented in Fig. 4.7. In all cases, the results fit the data very well, and no bias is observed.

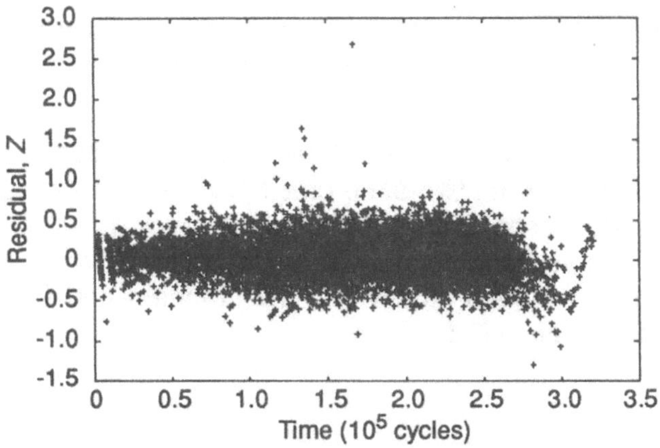

Figure 4.2. Residuals for the cubic polynomial model.

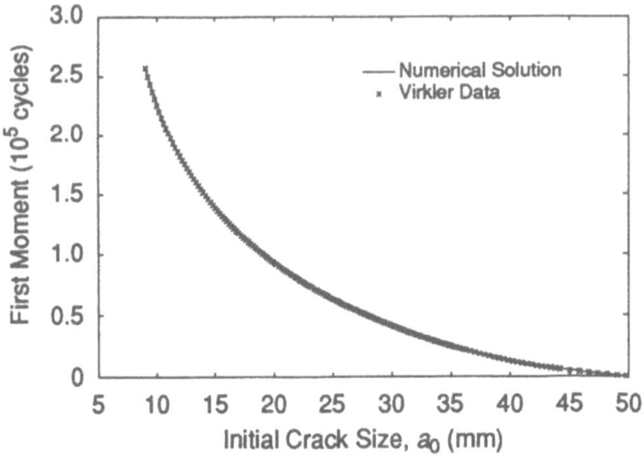

Figure 4.3. Comparison between the predicted mean time to reach a_c = 49.8 mm and the experimental data.

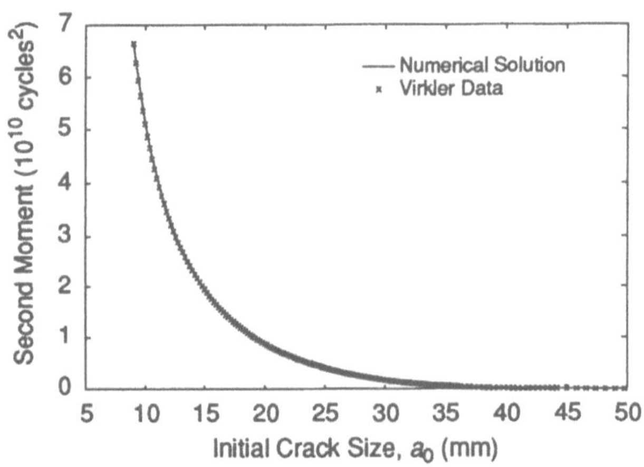

Figure 4.4. Comparison between the predicted mean-square time to reach a_c = 49.8 mm and the experimental data.

Because of the excellent agreement of the results predicted by the model and the data, it is concluded that the two-state crack growth model is capable of accurately describing the stochastic nature of the fatigue crack growth phenomenon for a wide variety of initial crack lengths, as well as critical crack lengths.

Comparison of the two-state model with the Ghonem and Dore data [10] for several stress ratios is given in [32].

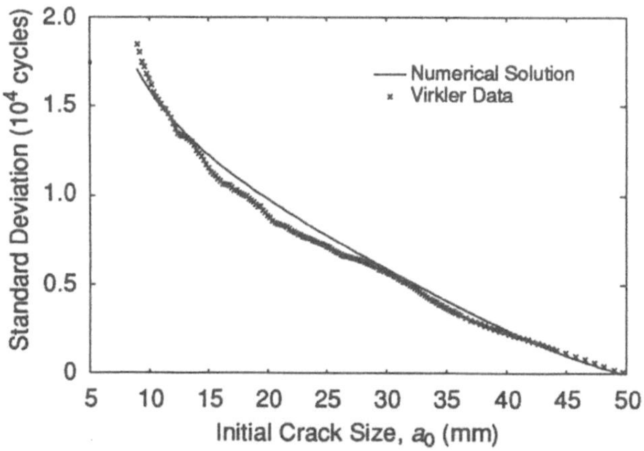

Figure 4.5. Comparison between the predicted standard deviation of the time to reach $a_c = 49.8$ mm and the experimental data.

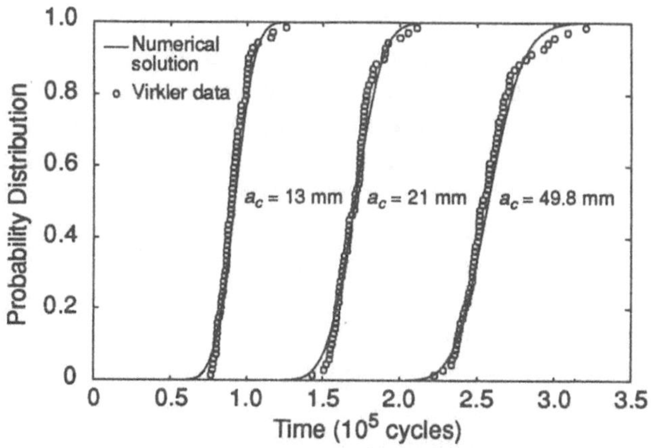

Figure 4.6. Comparison between the predicted distribution function of the time to reach $a_c = 13$, 21, and 49.8 mm, and the experimental data ($a_0 = 9$ mm).

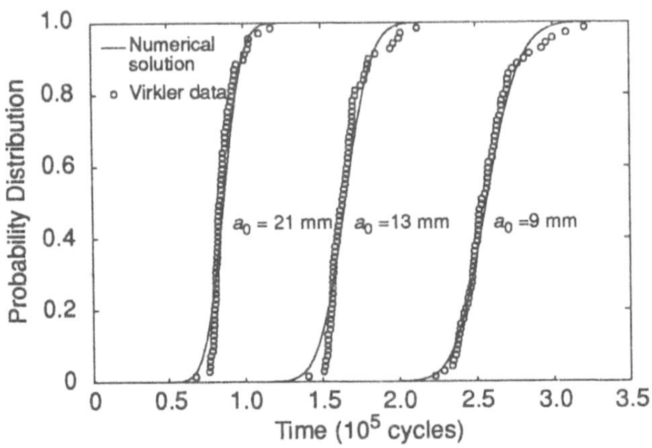

Figure 4.7. Comparison between the predicted distribution function of the time to reach a_c = 49.8 mm and the experimental data (a_0 = 9, 13, and 21 mm).

5. Data Reduction for Stochastic Analysis

The methods of data reduction and parameter estimation for deterministic fatigue crack growth analysis are not necessarily the most appropriate for use in stochastic analysis. The available data are usually in the form of crack length measurements A at the associated points in time t (or cycle n). In contrast, fracture mechanics models for fatigue crack growth such as the Paris–Erdogan equation don't directly describe the crack length, but rather the crack growth rate, dA/dt. As a result, the first step in determining the fatigue crack growth model parameters usually involves estimating the crack growth rate from the experimental data, i.e., numerically differentiating the discrete data. These estimates can then be used in a least squares regression analysis to determine the model parameters.

Considerable error can be introduced in the process of reducing the experimental A vs. t data to the sought dA/dt data, depending on the particular numerical differentiation scheme utilized and the measurement increment, Δa. In fact, regression analysis of differentiated data followed by subsequent integration of the resulting crack growth rate equations leads to inconsistent reproduction of the original data (see Ostergaard and Hillberry [51]).

For completeness and to better understand the difficulties involved in fatigue data reduction, the next section discusses the least squares regression procedure along with several methods for estimating the crack growth rate from experimental data. Subsequently, a modified finite integral method discussed in Spencer, et al. [31] is presented that can be employed to avoid the numerical differentiation of the experimental data.

5.1 Least-Squares Regression Analysis

To effectively discuss parameter estimation procedures, attention is first focused on estimating the parameters of the randomized Paris–Erdogan equation,

$$\frac{dA}{dt} = C(\Delta K)^m X(t), \tag{5.1}$$

where, as is customary in the literature, $X(t)$ is assumed to be a stationary lognormal random process. The parameters that need to be estimated in this model are then C, m, and the characteristics of $X(t)$. The data are assumed to be available in the form of observed values of crack length a_i at a given time instant t_i.

Taking the logarithm of both sides of Eq. (5.1) yields

$$\ln\left(\frac{dA}{dt}\right) = m \ln \Delta K + \ln C + Z(t), \tag{5.2}$$

where $Z(t) = \ln X(t)$ is a stationary Gaussian process. For each sample point (a_i, t_i) in the experimental data, a random residual can be defined as

$$z_i = \ln\left(\frac{da}{dt}\right)_i - m \ln \Delta K_i - \ln C. \tag{5.3}$$

One criterion for choosing m and C is to minimize the sum of the squares of Z_i. This naturally leads to a least-square regression analysis to estimate m and $\ln C$. Defining the quantities $x = \ln \Delta K$ and $y = \ln(dA/dt)$, the parameter estimates are

$$m = \frac{\Sigma x_i y_i - n\bar{x}\,\bar{y}}{\Sigma x_i^2 - n\bar{x}} \qquad \ln C = \bar{y} - m\bar{x}, \tag{5.4}$$

where the subscript i refers to a particular data point (a_i, t_i), $\Sigma = \Sigma_{i=1}^{n}$, and $\bar{x}, \bar{y}$ are the sample means of x and y, respectively.

Because $Z(t)$ is assumed to be stationary, σ_Z^2 is constant and can be estimated by taking the sample variance of the difference between the actual data and the regression line. The stationary variance of $Z(t)$ is then given by

$$\sigma_Z^2 = \frac{1}{n-2} \sum_{i=1}^{n} z_i^2. \tag{5.5}$$

The factor of $1/(n-2)$ is required to give an unbiased estimate of the sample variance of $Z(t)$ obtained in the regression analysis.

An inherent bias is introduced by the log–log regression, or least-squares, approach discussed in this section. As pointed out by Ferguson [52], the arithmetic mean, instead of the geometric mean, is obtained for dA/dt using this method.

To use the least-squares method to estimate the parameters, calculation of the derivative of the crack length with respect to time is required. Several approaches can be employed to determine $(da/dt)_i$ for use in the preceding analysis. For deterministic analyses, the American Society for Testing and Materials has advocated two techniques for these calculations: the secant method and the incremental polynomial method (see [53]). These two methods will be elaborated upon in the next few paragraphs.

The secant method approximates da/dt by passing a line through the points (a_i, t_i) and (a_{i+1}, t_{i+1}) and using the slope of the secant line as the crack growth rate at the crack length $(a_i + a_{i+1})/2$, i.e.,

$$\left.\frac{da}{dt}\right|_{\bar{a}} \approx \frac{a_{i+1} - a_i}{t_{i+1} - t_i}, \tag{5.6}$$

where $\bar{a} = (a_{i+1} + a_i)/2$ (see Fig. 5.1).

There are difficulties associated with the secant method. Using the secant estimates in Eq. (5.4) to determine the parameters in a stochastic model will generally result in a model that overpredicts the structural fatigue life. This effect is illustrated in Fig. 5.1, where it can be seen that the slope of the secant line will generally be smaller than the slope of the true tangent line at $\bar{a}$. This implies that the secant estimates of (da/dt) will underestimate the true crack growth rates, thus accounting for the overprediction of specimen fatigue life.

Alternatively, Ortiz and Kiremidjian [24] suggested that the slope of the secant line be used to represent the crack growth rate at a_i, i.e.,

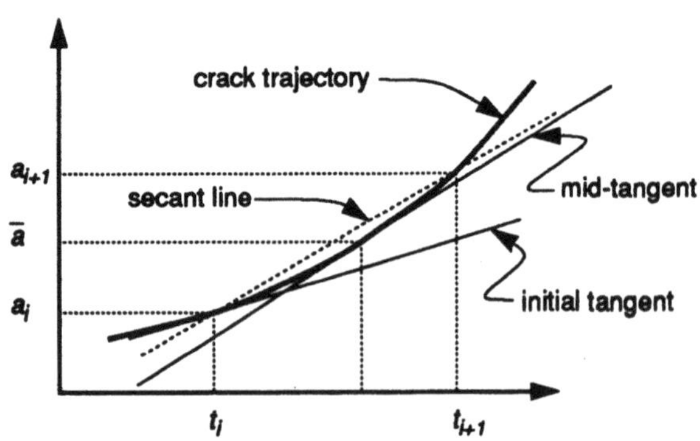

Figure 5.1. Illustration of secant method.

$$\left.\frac{da}{dt}\right|_{a_i} \approx \frac{a_{i+1} - a_i}{t_{i+1} - t_i}. \tag{5.7}$$

As shown in Fig. 5.1, this approximation will usually overpredict the crack growth rate at point a_i, resulting in an underprediction of fatigue life.

The other technique recommended by ASTM for estimating da/dt from the a versus t data is the incremental polynomial method. In this approach, a second-order polynomial is fit to three, five, seven, or nine successive data points using least-squares. The rate of crack growth is then obtained from the derivative of the fitted polynomial at the middle data point. The more data points employed, the more *smoothed* the estimates of the da/dt become, i.e., the dispersion in the data is artificially reduced by fitting the polynomial to the data. As a result, employing the incremental polynomial method to estimate the parameters in a stochastic model can produce poor results [31].

As indicated previously, methods that differentiate the a versus t data to obtain the crack growth rate da/dt can introduce considerable error, depending on the particular numerical differentiation scheme utilized and the measurement increment, Δa. In fact, regression analysis of the differentiated data followed by subsequent integration of the resulting crack growth rate equations leads to inconsistent reproduction of the actual data [51]. The next section will present an approach to alleviate some of these difficulties.

5.2 Modified Finite Integral Approach

Ostergaard and Hillberry [51] developed a finite integral method to avoid the numerical differentiation of the original experimental data in the estimation of parameters in the deterministic fatigue crack growth law. In this approach, the deterministic analog of Eq. (5.1) is separated and integrated to obtain

$$\int_{a_0}^{a_i} \frac{da}{C(\Delta K)^m} = \int_0^{T_{ri}} dt = T_{ri}, \tag{5.8}$$

where a_0 is the initial crack size, and T_{ri} is the time predicted by the model at which the crack will have a length a_i.

A sum of the squared errors $\bar{e}$ can then be defined as

$$\bar{e} = \sum_{i=1}^{N_{\Delta a}} (T_{ai} - T_{ri})^2, \tag{5.9}$$

where T_{ai} is the cumulative cycle count from the crack growth data, and $N_{\Delta a}$ is the number of crack intervals into which the data are broken. The optimal set of model parameters are defined as those that minimize the squared error $\bar{e}$.

This approach can be modified for parameter estimation in stochastic fatigue crack growth models [31]. Here, Eq. (5.1) is separated, integrated, and averaged over the ensemble to obtain

$$\int_{a_0}^{a_l} \frac{da}{C(\Delta K)^m} = \left\langle \int_0^{T_{rl}} X(t)\,dt \right\rangle, \tag{5.10}$$

where a_l is fixed, thus making the left-hand side of the equation a constant, and $\langle\cdot\rangle$ denotes expectation with respect to the ensemble. Employing the mean value theorem, the integral on the right-hand side of Eq. (5.10) can be rewritten as

$$\int_0^{T_{rl}} X(t)\,dt = X(s)\,T_{ri}, \tag{5.11}$$

where s is on the interval $[0, T_{ri}]$. Thus,

$$\left\langle \int_0^{T_{rl}} X(t)\,dt \right\rangle = \langle X(s)\,T_{ri}\rangle = \langle X(s)\rangle\langle T_{ri}\rangle + \rho\sigma_X\sigma_T, \tag{5.12}$$

where ρ is the correlation between $X(s)$ and T_{ri} (*N.B.*, $|\rho| \le 1$), σ_X is the standard deviation of the stationary process $X(t)$, and σ_T is the standard deviation of the random variable T_{ri}.

One can argue that the term $\rho\sigma_X\sigma_T$ is negligible as compared to $\langle X(s)\rangle\langle T_{ri}\rangle$ by noting that $\rho \to 0$ at T_{ri} increases and that practical experience indicates that $\sigma_X \approx 1$ and $\sigma_T \ll \langle T_{ri}\rangle$. Therefore, Eq. (5.10) can be well-approximated as

$$\int_{a_0}^{a_l} \frac{da}{C(\Delta K)^m} = \langle X(t)\rangle\langle T_{ri}\rangle. \tag{5.13}$$

This approximation has been verified with several extensive data sets in [31] and [32].

The sum of the squared errors for the stochastic parameter estimation problem can then be defined as

$$\bar{e} = \sum_{i=1}^{N_{\Delta a}} (\langle T_{ai}\rangle - \langle T_{ri}\rangle)^2, \tag{5.14}$$

where $\langle T_{ai}\rangle$ is the mean cumulative cycle count obtained from the data, and $\langle T_{ri}\rangle$ is calculated from Eq. (5.13).

The optimum values of the parameters in the stochastic model are obtained by minimizing the sum of the squared errors $\bar{e}$. This can be effectively done by using a numerical search algorithm such as the IMSL routine ZXSSQ [54].

In the case of the randomized Paris–Erdogan model (5.1), the parameters C, m, and $\langle X(s)\rangle$ are determined from the modified finite integral method. However, there are two more parameters to be determined in the model, σ_X and a correlation parameter for the random process $X(t)$. While numerical differentiation of the fatigue crack growth data is not required in the modified finite integral method to determine the material constants, it is required to obtain an estimate of the variance σ_X and can be calculated as follows. Write $X(t)$ as

$$X(t) = \frac{dA/dt}{C(\Delta K)^m}, \tag{5.15}$$

with the variance given by

$$\sigma_X^2 = \frac{1}{n-2}\sum_{i=1}^{N_{\Delta a}}\left(\frac{(dA/dt)_i}{C(\Delta K_i)^m} - \langle X(t)\rangle\right). \tag{5.16}$$

The secant method will give the largest estimation of σ_X^2 among those methods recommended by ASTM [53]; therefore, to be conservative, the secant method is recommended for calculation of $(dA/dt)_i$. Determination of the correlation parameter is dependent upon the particular form of the function $F(\cdot)$ in Eq. (4.1) (see Spencer, et al. [31]).

6. Vector Diffusion Models for Random Loading

This section presents a recently developed multi-state fatigue crack growth model that includes load sequence effects through the introduction of a history-dependent parameter, the *reset stress*. The reset stress provides an indication of the state of residual compression at the crack tip (see Veers [36], Veers, et al. [37], and Winterstein and Veers [38]).

6.1 General Properties

Consider the Paris–Erdogan-type crack growth law modified for random loading and given by

$$\frac{dA}{dt} = C(\Delta K_{\text{eff}})^m, \tag{6.1}$$

where ΔK_{eff} is the effective stress intensity factor range defined as in the crack closure model of Elber (see Section 4.3), i.e.,

$$\Delta K_{\text{eff}} = Y(a)\,\Delta S_{\text{eff}}\,\sqrt{\pi a}, \tag{6.2}$$

where $Y(a)$ is the geometry factor and ΔS_{eff} is the effective stress intensity range. The basic idea here is that the crack will propagate only when the stress is great enough to cause the crack faces to fully separate or open. The effective stress range is then defined as

$$\Delta S_{\text{eff}} = \begin{cases} S_{\max} - S_{\text{op}}, & S_{\max} > S_{\text{op}},\ S_{\min} < S_{\text{op}} \\ S_{\max} - S_{\min}, & S_{\max} > S_{\text{op}},\ S_{\min} \geq S_{\text{op}} \\ 0, & S_{\max} \leq S_{\text{op}}, \end{cases} \tag{6.3}$$

where S_{op} is the opening stress, which is determined experimentally. The effective stress intensity factor range then becomes

$$\Delta K_{\text{eff}} = \begin{cases} Y(a)\sqrt{\pi a}\,(S_{\max} - S_{\text{op}}), & S_{\max} > S_{\text{op}},\ S_{\min} < S_{\text{op}} \\ Y(a)\sqrt{\pi a}\,(S_{\max} - S_{\min}), & S_{\max} > S_{\text{op}},\ S_{\min} \geq S_{\text{op}} \\ 0, & S_{\max} \leq S_{\text{op}}. \end{cases} \tag{6.4}$$

In the case of constant-amplitude loading, Elber proposed that the opening stress is related to the maximum stress as

$$S_{\text{op}} = c_f\, S_{\max}, \tag{6.5}$$

where c_f is usually considered to be a function of the stress ratio R.

For random loading, Eq. (6.5) is generalized as

$$S_{\text{op}} = c_f\, S_r, \tag{6.6}$$

where S_r, termed the *reset stress,* is defined as the stress level necessary to *reset* the maximum extent of the overload-affected zone at the crack tip. In this manner, sequence effects are included, and the magnitude of the crack growth rate retardation is correlated with the magnitude of the overload.

To better understand the concept of reset stress, a few quantities are introduced. A tensile overload applied to a cracked specimen is designated S_{ol} and the associated stress intensity factor is given by

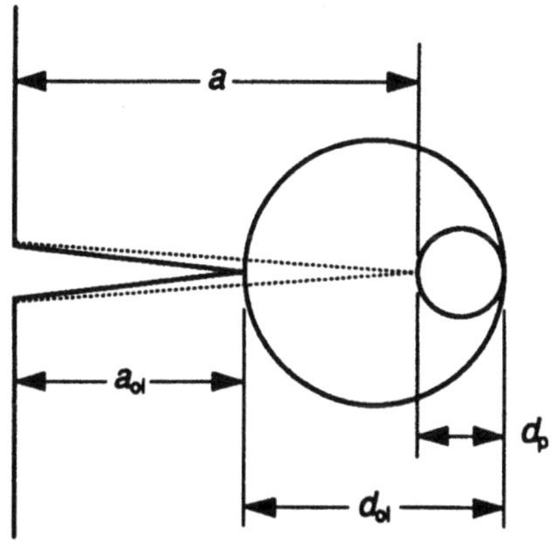

Figure 6.1. Overload-affected zone.

$$K_{ol} = Y(a)\sqrt{\pi a}\; S_{ol}. \tag{6.7}$$

Here, the maximum extent of the influence of the overload d_{ol} is approximated as twice the radius of the plastic zone (see Johnson [55]) and given by

$$d_{ol} = \frac{K_{ol}^2}{\lambda \pi S_y^2}, \tag{6.8}$$

where λ is a constraint factor ($\lambda = 1$ for plane stress and $\lambda = 3$ for plane strain), and S_y is the yield stress. The reset stress S_r can then be approximated as

$$S_r = \frac{S_y}{Y(a)}\left(\frac{\lambda d_p}{a}\right)^{1/2}, \tag{6.9}$$

where d_p is defined as the distance from the crack tip to the maximum extent of the overload influence zone. Figure 6.1 provides a graphical depiction of the quantity d_p and the concept of the reset stress; the reset stress is the stress necessary to create an overload influence zone that has a magnitude greater than d_p.

6.2 Four-State Model

To develop a model of crack growth that is a vectored diffusion process, an expression for the time derivative of the reset stress is needed. Thus, taking the derivative of S_r with respect to time yields

$$\frac{dS_r}{dt} = \frac{\partial S_r}{\partial a}\frac{da}{dt} + \dot{S}H(\dot{S})\,H(S - S_r), \tag{6.10}$$

where S and $\dot{S}$ are the applied stress and its time derivative, respectively, and the $H(\cdot)$ is the Heaviside function. The first term in Eq. (6.10) is due to the gradual reduction in the reset stress as the crack grows through the plastic zone, and the second term is due to the instantaneous increase in the reset stress when the applied stress exceeds the reset stress. Substituting appropriate terms into Eq. (6.10) yields

$$\frac{dS_r}{dt} = -\frac{S_r}{Y(a)}\left[\frac{dY(a)}{da} + \frac{1}{2a}\left(Y(a) + \frac{\lambda S_y^2}{Y(a)\,S_r^2}\right)\right]\frac{da}{dt} + \dot{S}H(\dot{S})\,H(S - S_r)\,. \tag{6.11}$$

Upon examination of the previous expression, it is seen that the applied stress and stress rate must be states variables in the model. The applied stress can be modeled as a white noise passed through a linear second-order filter. Therefore, the state vector for this model is defined as $\mathbf{X} = [A, S_r, S, \dot{S}]$, which is governed by the vector stochastic differential equation,

$$\dot{\mathbf{X}} = \boldsymbol{\eta}(\mathbf{X}) + \mathbf{D}\xi(t)\,, \tag{6.12}$$

where $\boldsymbol{\eta}$ and $\mathbf{D}$ are the drift and diffusivity vectors, respectively. The drift vector can be written explicitly as

$$\boldsymbol{\eta} = [\eta_A, \eta_{S_r}, \eta_S, \eta_{\dot{S}}]^T, \tag{6.13}$$

where

$$\eta_A = C[Y(a)\sqrt{\pi A}\Delta s_{eff}]^m, \tag{6.14a}$$

$$\eta_{S_r} = -\frac{s_r}{Y(a)}\left[\frac{dY(a)}{da} + \frac{1}{2a}\left(Y(a) + \frac{\lambda S_y^2}{Y(a)\,s_r^2}\right)\right]\eta_A + \dot{s}H(\dot{s})\,H(s - s_r)\,, \tag{6.14b}$$

$$\eta_S = \dot{s}, \tag{6.14c}$$

$$\eta_{\dot{S}} = -\omega_0^2(s - \mu_S) - 2\zeta\omega_0\dot{s}, \tag{6.14d}$$

μ_S is the mean of the applied stress process, and the parameters ω_0 and ζ describe the characteristics of the linear filter for the applied stress process. In addition, $D_4 = 2\sigma_S\sqrt{\zeta\omega_0^3}$ is the only nonzero element in the diffusivity vector where σ_S^2 is the stationary variance of the applied stress process. To complete the formulation of this four-state model, S_{max} must be approximated in the expression for ΔS_{eff}, Eq. (6.3). This is accomplished by modeling the maximum of the applied stress as

$$S_{max} = \mu_S + \mathcal{A}, \tag{6.15}$$

where $\mathcal{A}$ is the amplitude envelope given in terms of the states by

$$\mathcal{A} = \sqrt{(S-\mu_S)^2 + (\dot{S}/\omega_0)^2}. \tag{6.16}$$

6.3 Reliability Problem Formulation

In this section, an initial-boundary value problem is developed for the probability that the crack length is less than the critical crack length based on the four-state model (see Enneking [56]). The transition probability density function $p = p(\mathbf{x}, t|\mathbf{x}_0, t_0)$ for the four-state vector $\mathbf{x} = [A, S_r, S, \dot{S}]$ is governed by the Fokker–Planck–Kolmogorov equation,

$$\begin{aligned}\frac{\partial p}{\partial t} = &-\frac{\partial}{\partial a}[\eta_A(\mathbf{x})\,p] - \frac{\partial}{\partial s_r}[\eta_{S_r}(\mathbf{x})\,p] - \frac{\partial}{\partial s}[\eta_S(\mathbf{x})\,p] \\ &-\frac{\partial}{\partial \dot{s}}[\eta_{\dot{S}}(\mathbf{x})\,p] + 2\zeta\omega_0^3\sigma_S^2\frac{\partial^2 p}{\partial \dot{s}^2}.\end{aligned} \tag{6.17}$$

The backward Kolmogorov equation also governs the probability density function and is given by

$$\begin{aligned}\frac{\partial p}{\partial t} = &\,\eta_A(\mathbf{x}_0)\frac{\partial p}{\partial a_0} + \eta_{S_r}(\mathbf{x}_0)\frac{\partial p}{\partial s_{r0}} + \eta_S(\mathbf{x}_0)\frac{\partial p}{\partial s_0} \\ &+ \eta_{\dot{S}}(\mathbf{x}_0)\frac{\partial p}{\partial \dot{s}_0} + 2\zeta\omega_0^3\sigma_S^2\frac{\partial^2 p}{\partial \dot{s}_0^2}.\end{aligned} \tag{6.18}$$

In a development similar to that in Section 4.1, it is assumed that fatigue failure has not occurred if the crack has not reached a length of a_c. In addition, failure of the structure is defined to be when the applied stress reaches the allowable stress, s_{all}. The cumulative distribution of times to reach a given crack size a_c, or conversely, the probability that a crack will not reach a critical crack size in a given time, is then defined as (conditional on the initial value of the state vector $\mathbf{x}_0$)

$$R = R(t|\mathbf{x}_0) = P\{A(t) < a_c, S < s_{all}; t|\mathbf{x}_0; t_0\}$$
$$= \int_{a_0}^{a_c} \int_{0}^{s_{all}} \int_{-\infty}^{s_r} \int_{-\infty}^{\infty} p(a, s_r, s, \dot{s}, t|\mathbf{x}_0; t_0)\, d\dot{s}\; ds\; ds_r da\,. \tag{6.19}$$

The reliability function R also satisfies the backward Kolmogorov equation (6.18), i.e.,

$$\frac{\partial R}{\partial t} = \eta_A(\mathbf{x}_0)\frac{\partial R}{\partial a_0} + \eta_{S_r}(\mathbf{x}_0)\frac{\partial R}{\partial s_{r0}} + \eta_S(\mathbf{x}_0)\frac{\partial R}{\partial s_0}$$
$$+ \eta_{\dot{S}}(\mathbf{x}_0)\frac{\partial R}{\partial \dot{s}_0} + 2\zeta\omega_0^3\sigma_S^2\frac{\partial^2 p}{\partial \dot{s}_0^2}\,. \tag{6.20}$$

Only one boundary condition needs to be specified in the a_0, s_{r0}, and s_0 directions for the initial-boundary value problem to be well-posed (see Fichera [49]). Possible initial and boundary conditions are suggested on physical grounds and given by

$$R(0|a_c, s_{r0}, s_0, \dot{s}_0) = 1, \qquad 0 \le a_0 < a_c, \; s_0 \le s_{r0} < s_{all}, \tag{6.21}$$

$$R(t|a_c, s_{r0}, s_0, \dot{s}_0) = 0, \qquad s_0 \le s_{r0} < s_{all}, \tag{6.22}$$

$$R(t|a_0, s_{all}, s_0, \dot{s}_0) = 0, \qquad 0 \le a_0 \le a_c, \; s_0 < s_{all}, \tag{6.23}$$

$$\lim_{s_0 \to -\infty} R(t|a_0, s_{r0}, s_0, \dot{s}_0) \to 0, \qquad 0 \le a_0 \le a_c, \; s_{r0} < s_{all}, \tag{6.24}$$

$$\lim_{|\dot{s}_0| \to \infty} R(t|a_0, s_{r0}, s_0, \dot{s}_0) \to 0, \quad 0 \le a_0 \le a_c, \; s_0 \le s_{r0} < s_{all}. \tag{6.25}$$

The determination of the preceding boundary conditions can be made following logic similar to that presented for the two-state crack growth problem in Section 4.1. First, the initial stress is noted to be less than the initial reset stress. Further, both must be less than the maximum allowable stress, i.e., $s_0 \le s_{r0} \le s_{all}$. In addition, the reset stress must be greater than zero, i.e., $s_{r0} \ge 0$. The initial condition is given by Eq. (6.21) and simply states that the reliability is one at $t = 0$ as long as the initial crack length is less than the critical crack length. The boundary condition given in Eq. (6.22) indicates that the reliability is zero if the initial crack size is the critical crack size. Equation (6.23) states that the reliability should be zero when the initial reset stress is equal to the maximum allowable stress, s_{all}. The boundary condition in Eq. (6.24) indicates that as the initial stress goes to negative infinity, the stress amplitude will become very large, resulting in rapid crack growth, system failure, and zero reliability. The last boundary condition (6.25) states that the reliability will go to zero as the initial stress rate and, thus, the stress amplitude (e.g., see Eq. (6.16)) will go to infinity, again causing the crack to grow rapidly and the system to fail.

The boundary value for the statistical moments of time to reach the critical crack size can also be formulated in a manner similar to that found in Section 5. This problem may be somewhat simpler than the reliability problem to solve, since it is not time-dependent.

The model presented in this section is physically appealing, and Monte Carlo simulation indicates that many important features of the fatigue phenomena under random loading are captured (see [36]). Unfortunately, there are no analytical solutions, and more research is needed to develop numerical solutions for this problem formulation.

To alleviate this problem, analytical results for the mean life have been developed that use a set of scalar diffusion models that are on different time scales; e.g., models of S_r at various crack sizes. These models are shown to produce results that agree well with simulation (see [36] and [38]).

7. Reliability Estimation

After appropriate stochastic models are developed which accurately portray the fatigue crack growth behavior in a structural system under consideration, they can be used as a basis from which reliability estimates are made. Because of the increasing number of aging engineering structures, one aspect of the reliability problem which has drawn considerable attention is that of fatigue reliability updating through scheduled inspections. In this section, the inspection problem is considered to illustrate the use of stochastic crack growth models in reliability estimation.

7.1 Inspection Updating

To prevent catastrophic failure, fatigue-critical structures such as airframes and offshore platforms are usually subjected to scheduled inspection maintenance. For example, a typical fighter aircraft has tens of thousands of fastener holes — each of which serves as a site for crack initiation. The damage tolerance requirement in MIL-A-87221 [57] states that an "airframe shall have adequate residual strength in the presence of flaws for specified periods of usage. These flaws shall be assumed to exist initially in the structure as a result of manufacturing processes." In-service inspections are often essential to meet this requirement.

Determination of optimal inspection and repair strategies requires that the effect of scheduled maintenance on the safety and reliability of the structure be assessed. However, many uncertainties are involved, e.g., initial fatigue quality, in-service load spectrum, material characteristics, crack growth laws, nondestructive evaluation capabilities, etc.

Two types of results are considered for an inspection that occurs at time T_i: a) No crack is found during the inspection — implying that the crack size is smaller than the smallest detectable crack size A_{di}, and b) a crack of length A_{mi} is detected during the inspection. These results can be formally represented as

$$A(T_i) < A_{di} \tag{7.1}$$

and

$$A(T_i) = A_{mi}, \tag{7.2}$$

for outcomes (a) and (b), respectively. The minimum detectable crack size is usually random, because a crack longer than A_{di} is undetected with a certain probability, depending on the crack size and inspection method. The distribution of A_{di}, $F_{A_{di}}(a)$, is the distribution of undetected crack size and is identical to the probability of detection curve (POD) for the particular inspection method used (see Madsen et al. [58]). Due to uncertainties in measurement techniques, the measured crack size A_{mi} is generally also random, with distribution function $F_{A_{mi}}(a)$.

If a dominant crack exists in a structure and the failure criterion is based on this crack exceeding a critical crack length, the probability of failure can be written as

$$P_f(t) = P\{A(t) \geq a_c\}, \tag{7.3}$$

where a_c is the critical crack size. Prior to the first inspection, the probability of failure is then

$$P_f(t) = 1 - F_A(a_c;t) = 1 - \int_0^{a_c} f_A(a;t)\, da, \qquad t \leq T_1, \tag{7.4}$$

where

$$f_A(a;t) = \int_0^{\infty} f_A(a;t|x;t_0)\, f_{A_0}(x)\, dx, \tag{7.5}$$

$f_A(a;t|x;t_0)$ is the probability distribution at time $t > t_0$, conditional on $a(t_0) = x$, and $f_{A_0}(a)$ is the probability distribution of initial crack sizes. Substituting Eq. (7.5) into Eq. (7.4) and integrating, the probability of failure becomes

$$P_f(t) = 1 - \int_0^{\infty} F_A(a_c;t|x;t_0)\, f_{A_0}(x)\, dx, \tag{7.6}$$

where $F_A(a;t|x;t_0)$ is the cumulative distribution of crack sizes conditional on $A(t_0) = x$.

If an inspection is conducted at time T_1 and no crack is detected, the updated probability of failure, based on the definition of conditional probability, takes the form,

$$P_f^{upd}(t) = P\{A(t) \geq a_c | A(T_1) < A_{d1}\} = \frac{P\{A(t) \geq a_c \cap A(T_1) < A_{d1}\}}{P\{A(T_1) < A_{d1}\}}. \tag{7.7}$$

Equation (7.7) can be written explicitly in terms of the updated distribution of crack sizes as

$$P_f^{upd}(t) = 1 - F_A^{upd}(a_c;t) = 1 - \int_0^{a_c} f_A^{upd}(a;t)\, da, \qquad t > T_1, \tag{7.8}$$

where

$$f_A^{upd}(a;t)\, da = P\{a < A(t) \leq a + da | A(T_1) < A_{d1}\} = \frac{P\{a < A(t) \leq a + da \cap A(T_1) < A_{d1}\}}{P\{A(T_1) < A_{d1}\}}. \tag{7.9}$$

Assuming A_{d1} to be statistically independent of $A(T_1)$, Eq. (7.8) is found to be [42]

$$f_A^{upd}(a;t) = \frac{1}{k_{d1}} \int_0^{\infty} f_A(a;t|x;T_1)\, \{1 - F_{A_{d1}}(x)\}\, f_A(x;T_1)\, dx, \qquad t > T_1, \tag{7.10}$$

where

$$k_{d1} = \int_0^{\infty} \{1 - F_{A_{d1}}(x)\}\, f_A(x;T_1)\, dx. \tag{7.11}$$

and Eq. (7.8) becomes,

$$P_f^{upd}(t) = 1 - \frac{1}{k_{d1}} \int_0^{a_c}\int_0^{\infty} f_A(a;t|x;T_1)\,\{1 - F_{A_{d1}}(x)\}\, f_A(x;T_1)\,dx\,da$$
$$= 1 - \frac{1}{k_{d1}} \int_0^{\infty} F_A(a_c;t|x;T_1)\,\{1 - F_{A_{d1}}(x)\}\, f_A(x;T_1)\,dx, \qquad t > T_1, \tag{7.12}$$

Alternatively, when a crack is detected, the updated probability of failure becomes

$$P_f^{upd}(t) = P\{A(t) \geq a_c | A(T_1) = A_{m1}\}$$
$$= \frac{P\{A(t) \geq a_c \cap A(T_1) = A_{m1}\}}{P\{A(T_1) = A_{m1}\}}. \tag{7.13}$$
$$= 1 - \int_0^{a_c} f_A^{upd}(a;t)\,da, \qquad t > T_1$$

where the updated probability density function of crack sizes is now

$$f_A^{upd}(a;t) = \frac{1}{k_{m1}} \int_0^{\infty} f_A(a;t|x;T_1)\, f_{A_{m1}}(x)\, f_A(x;T_1)\,dx, \qquad t > T_1. \tag{7.14}$$

Here, $f_A(x;T_1)$ is given in Eq. (7.5), and

$$k_{m1} = \int_0^{\infty} f_{A_{m1}}(x)\, f_A(x;T_1)\,dx. \tag{7.15}$$

Substituting Eqs. (7.14) and (7.15) into Eq. (7.13) and integrating, the updated probability distribution becomes

$$P_f^{upd}(t) = 1 - \frac{1}{k_{m1}} \int_0^{\infty} F_A(a_c;t|x;T_1)\, f_{A_{m1}}(x)\, f_A(x;T_1)\,dx, \qquad t > T_1. \tag{7.16}$$

After the first inspection, the updated failure probability after the second inspection at a time $T_2 > T_1$ is of interest. If no crack is detected on the second inspection, the updated probability of failure is given by

$$P_f^{upd}(t) = P\{A(t) \geq a_c | A(T_1) < A_{d1} \cap A(T_2) < A_{d2}\}$$
$$= \frac{P\{A(t) \geq a_c \cap A(T_1) < A_{d1} \cap A(T_2) < A_{d2}\}}{P\{A(T_1) < A_{d1} \cap A(T_2) < A_{d2}\}}. \tag{7.17}$$

In terms of the updated distribution of crack sizes, Eq. (7.17) can be written as

$$P_f^{upd}(t) = 1 - F_A^{upd}(a_c;t) = 1 - \int_0^{a_c} f_A^{upd}(a;t)\, da, \qquad t > T_2, \tag{7.18}$$

where

$$f_A^{upd}(a;t) = \frac{1}{k_{d2}} \int_0^{\infty} f_A(a;t | x;T_2)\, \{1 - F_{A_{d2}}(x)\}\, f_A^{upd}(x;T_2)\, dx, \qquad t > T_2, \tag{7.19}$$

$$k_{d2} = \int_0^{\infty} \{1 - F_{A_{d2}}(x)\}\, f_A^{upd}(x;T_2)\, dx. \tag{7.20}$$

Finally,

$$P_f^{upd}(t) = 1 - \frac{1}{k_{d2}} \int_0^{a_c}\int_0^{\infty} f_A(a;t | x;T_2)\, \{1 - F_{A_{d2}}(x)\}\, f_A^{upd}(x;T_2)\, dx da$$
$$= 1 - \frac{1}{k_{d2}} \int_0^{\infty} F_A(a_c;t | x;T_2)\, \{1 - F_{A_{d2}}(x)\}\, f_A^{upd}(x;T_2)\, dx, \tag{7.21}$$
$$t > T_2.$$

Eqs. (7.10)–(7.21) assume that the crack growth process has one step memory. This assumption is exact for the random variable and the scalar Markov process models, and may be a reasonable approximation for other models.

Next, consider the case where a crack is detected during the second inspection. Again, following similar logic as presented earlier, the updated probability of failure can be written,

$$P_f^{upd}(t) = P\{A(t) \geq a_c | A(T_1) \leq A_{d1} \cap A(T_2) = A_{m2}\}$$
$$= \frac{P\{A(t) \geq a_c \cap A(T_1) \leq A_{d1} \cap A(T_2) = A_{m2}\}}{P\{A(T_1) \leq A_{d1} \cap A(T_2) = A_{m2}\}}, \tag{7.22}$$

$$P_f^{upd}(t) = 1 - F_A^{upd}(a_c;t) = 1 - \int_0^{a_c} f_A^{upd}(a;t)\, da, \qquad t \geq T_2, \tag{7.23}$$

where

$$f_A^{upd}(a;t) = \frac{1}{k_{m2}} \int_0^{\infty} f_A(a;t|x;T_2)\, f_{A_{m2}}(x)\, f_A^{upd}(x;T_2)\, dx, \qquad t > T_2, \tag{7.24}$$

$f_A^{upd}(x;T_2)$ is given in Eq. (7.10), and

$$k_{m2} = \int_0^{\infty} f_{A_{m2}}(x)\, f_A^{upd}(x;T_2)\, dx. \tag{7.25}$$

Substituting Eqs. (7.24) and (7.25) into Eq. (7.23) and integrating, the updated probability distribution becomes

$$P_f^{upd}(t) = 1 - \frac{1}{k_{m2}} \int_0^{\infty} F_A(a_c;t|x;T_1)\, f_{A_{m2}}(x)\, f_A^{upd}(x;T_2)\, dx, \qquad t > T_2. \tag{7.26}$$

As can be seen, this procedure is recursive and can be repeated for subsequent inspections and combinations of outcomes. If a repair is made, the updated probability of failure is determined based on the initial fatigue quality of the repaired structure. Further details regarding reliability updating through inspection can be found in Madsen [22], Madsen et al. [23], [58], and Yang and Chen [59].

7.2 Numerical Illustration

To illustrate the concepts presented earlier, crack growth damage accumulation in fastener holes of an F–16 lower wing skin is considered. For simplicity, a single crack at a critical location is examined, and the probability of the crack exceeding a critical length is calculated. The entire population of fastener holes should to be examined in a complete analysis.

Consider the power-law random variable crack growth model given by (see Section 2)

$$\frac{dA}{dt} = C_0 A^b, \qquad A(t_0) = A_0, \tag{7.27}$$

where C_0 and A_0 are random variables. Comparison with experimental data for fatigue cracks in fastener holes [43] indicates that C_0 is lognormally distributed with $\langle \ln C_0 \rangle = -8.91$ and $\sigma_{\ln C_0} = 0.1276$. The crack growth exponent is $b = 1.01$.

The probability distribution of the crack size $A(t)$ resulting from this model was derived in Section 2, which for the lognormal random variable model is

$$F_A(a;t|a_0;t_0) = P\{A(t) \le a\} = \begin{cases} 0, & a \le A_0 \\ \Phi\left(\dfrac{\ln h_t - \lambda}{\zeta}\right), & a > A_0, \end{cases} \tag{7.28}$$

where $h_t = (a^\beta - a_0^\beta)/\beta(t-t_0)$, λ and ζ are the mean and standard deviation of a normal random variable $\ln C_0$, respectively, and Φ is the standard normal distribution function, and $\beta = 1 - b$.

A probability distribution for $A_0(\gamma)$ that well represents the initial fatigue quality of fastener holes is given by (see Yang and Manning [44])

$$F_{A_0}(a_0) = \begin{cases} \exp\left[-\left(\dfrac{\ln(a_u/a_0)}{Q_0\beta_0}\right)^{\alpha_0}\right], & 0 \le a_0 \le a_u \\ 1.0, & a_0 > a_u, \end{cases} \tag{7.29}$$

where $a_u = 0.03$ in. is the upper bound on the initial crack size, and $\alpha_0 = 1.823$ and $Q_0\beta_0 = 1.928$ are shape parameters for the distribution.

Berens and Hovey [60] have employed regression analysis to fit seven different functional forms for the probability of detection (POD) curves to the data, which included 22,000 inspections performed on 174 cracks by 107 inspectors. The following model, termed the log-odds–log-scale model, was determined to provide the best fit of the data

$$F_{A_d}(a) = \frac{\exp(\alpha^* + \beta^* \ln a)}{1 + \exp(\alpha^* + \beta^* \ln a)}, \quad 0 \le a \tag{7.30}$$

where α^* and β^* are shape parameters for the distribution. Thus, $F_{A_d}(a)$ is the cumulative distribution of the smallest detectable crack size. Alternatively, it gives the probability of detecting (POD) a crack of length a during an inspection. The parameters in Eq. (7.30) depend upon the inspection interval and the nondestructive evaluation technique. For this example, the parameter values, taken from [59], are $\alpha^* = 13.44$ and $\beta^* = 3.95$. Figure 7.1 depicts the probability of detection curve using these parameters.

One design life for the aircraft is 8,000 flight hours. The effect of inspections for up to two lifetimes, or 16,000 flight hours, is investigated. The critical crack size a_c is taken to be 0.3 inches. Two inspections are considered — the first at $T_1 = 8{,}000$ flight hours and the second at $T_2 = 12{,}000$ flight hours.

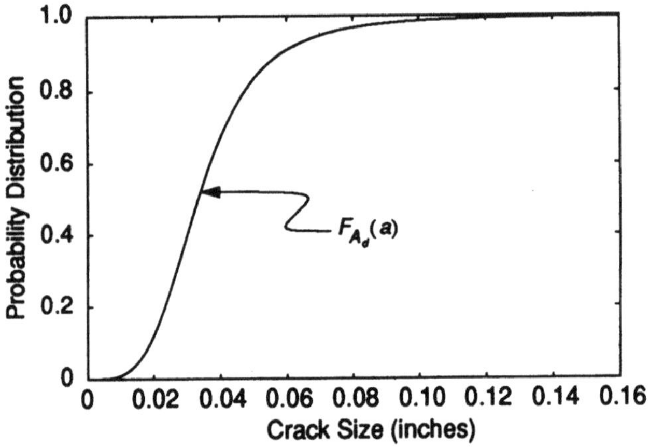

Figure 7.1. Log-odds probability of detection (POD) curve.

First, consider the case in which no cracks are found during the initial inspection. The distributions of crack size at time $t_0 = 0$ and at time $T_1 = 8{,}000$ flight hours, just before the inspection, are shown in Fig. 7.2. The results in Fig. 7.2 are calculated using Eq. (7.6), where the conditional crack size distribution $F_A(a;t|x;t_0)$ is given by Eq. (7.28). As is seen here, the mean and dispersion of the crack size have increased with time. The updated distribution of crack sizes (i.e., after the first inspection) is calculated via Eq. (7.10) and shown in Fig. 7.3, along with the distribution of crack sizes immediately prior to the inspection, which is given in Fig. 7.2. Finding no cracks during the first inspection has reduced the mean and the dispersion of the crack size, indicating that the probability of the crack exceeding the critical crack size should decrease after the inspection.

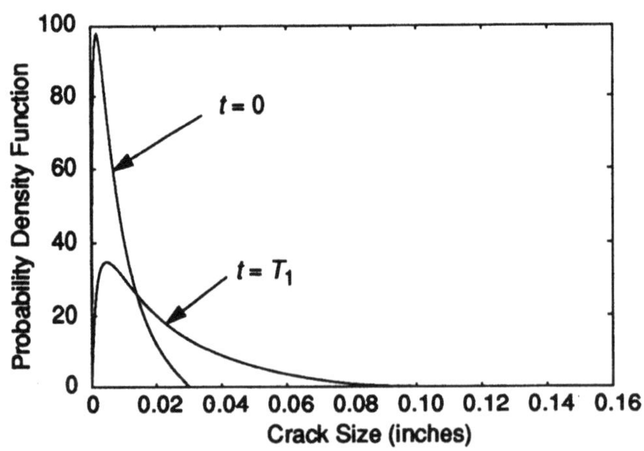

Figure 7.2. Evolution of the crack size probability distribution before the inspection at $T_1 = 8{,}000$ flight hours.

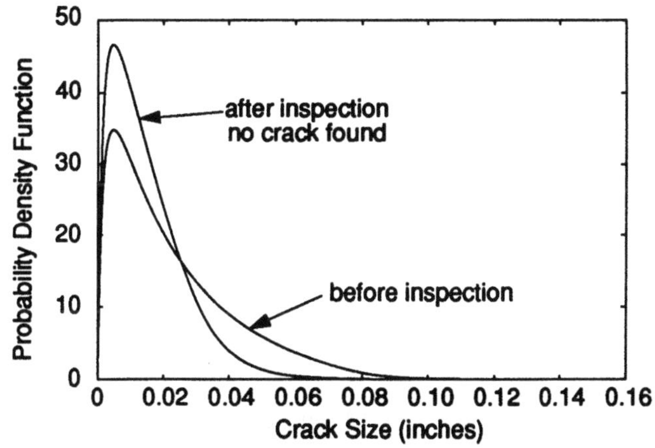

Figure 7.3. Effect on crack size distribution of not finding a crack during the first inspection at T_1 = 8,000 flight hours.

If a second inspection is conducted and no crack is found, the updated distribution of crack sizes can be calculated according to Eq. (7.10). Figure 7.4 provides a comparison of the distribution of crack sizes before and after the second inspection, finding no cracks. Again, the mean and dispersion of the crack size is decreased.

Equations (7.6), (7.12), and (7.21) are used to calculate the probability of failure as a function of time, which is shown in Fig. 7.5. Here, the probability that the crack will exceed the critical crack size decreases greatly in the short term and is lower than the probability corresponding to no inspection over the lifetime of the structure.

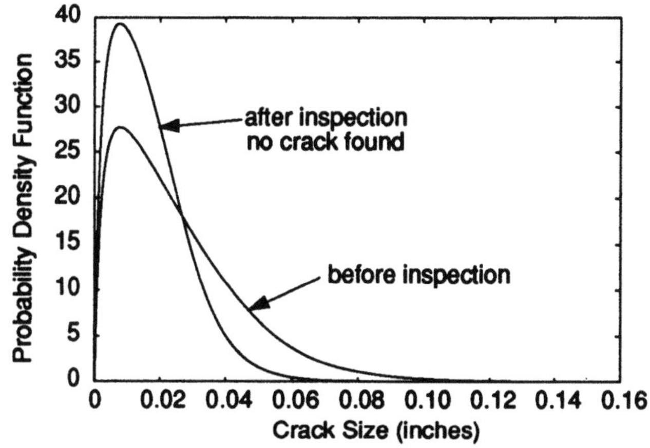

Figure 7.4. Effect on crack size distribution of not finding a crack during the second inspection at T_2 = 12,000 flight hours; no crack detected during the first inspection.

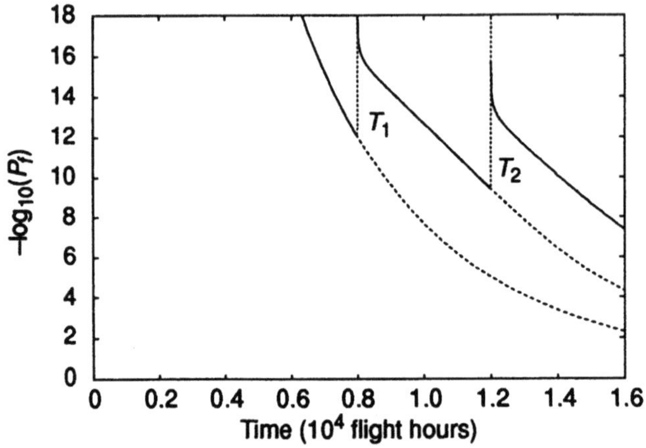

Figure 7.5. Updated probability of failure; no cracks detected during the first and second inspections.

Finally, the case is considered in which no crack is detected during the first inspection, but a crack is detected during the second inspection. The detected crack is assumed to be normally distributed with a mean size of 0.02 in. and a coefficient of variation of 10%, i.e., $F_{A_{m2}}(a)$ in Eq. (7.26) is $N(0.02,0.002)$. The distribution of crack sizes before and after finding the crack on the second inspection, along with the distribution of the measured crack, are shown in Fig. 7.6. Figure 7.7 presents the probability of failure as a function of time for this case. Here, notice that detecting a crack actually decreases the probability of failure for a short time. This is due to the fact that the inspection has confirmed that, even though a crack exists, it has

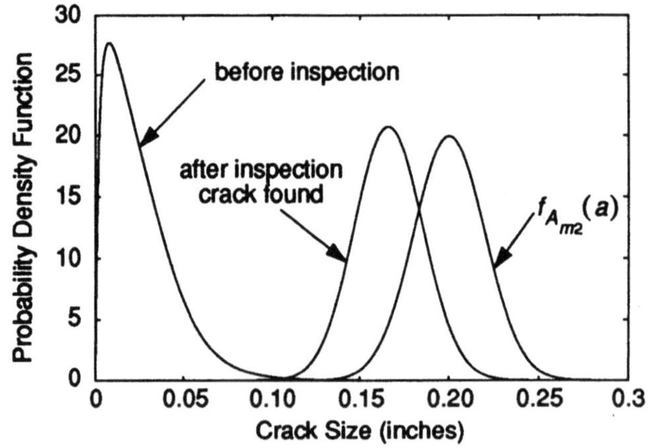

Figure 7.6. Effect on crack size distribution of finding a crack during the second inspection at T_2 = 12,000 flight hours; no crack detected during the first inspection.

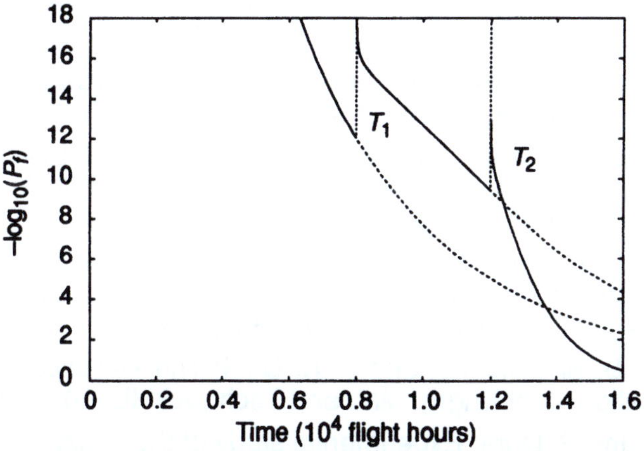

Figure 7.7. Updated probability of failure; no crack detected during first inspection; crack detected during the second inspection.

not yet reached the critical crack size. As is seen in Fig. 7.7, the period of time in which the probability of failure decreases is generally short.

Evaluation of the integrals in the preceding equations is possible in only the simplest cases, and for this example, numerical integration was required. As the complexity of the problems and the associated uncertainty increases, direct numerical integration is likely to become impossible. This is especially true when one is looking at extremely small probabilities of failure, which can be inaccurate due to finite precision of the computing hardware and limitations of the computational algorithms. If the uncertainties in the problem and crack growth model are well-represented by random variables, then a computer code such as PROBAN [45] can alleviate some of the numerical difficulties in obtaining the sought failure probabilities. (See also [23] and [58].)

8. References

1. ASCE Committee on Fatigue and Fracture Reliability, Series of articles on fatigue and fracture reliability, J. of the Struct. Div., ASCE 108 (ST1) (1982), 3-88.
2. Bannantine, J. A., J. J. Comer, and J. L. Handrock, Fundamentals of Metal Fatigue Analysis, Prentice-Hall, Englewood Cliffs, New Jersey, 1990.
3. Bluham, J. I.: Crack propagation laws, in Fracture Mechanics of Aircraft Structures, AGARD-AG-176, 95–109, 1974.
4. Miller, M. S. and G. P. Gallagher: An analysis of several fatigue crack growth rate descriptions, in Measurements and Data Analysis, ASTM STP 738, 205–251, 1981.

5. Paris, P. C. and F. Erdogan: A critical analysis of crack propagation laws, J. Basic Engrg., Trans. ASME, Series D, 85 (1963), 528–534.
6. Forman, R. G., et al.: Numerical analysis of crack propagation in cyclic load structures, J. Basic Engrg, Trans. ASME, Series D, 89 (1967), 459–465.
7. Larsen, J. M., B. J. Schwartz and C. G. Annis, Jr.: Cumulative fracture mechanics under engine spectra, Tech Report AFML–TR–79–4159, Air Force Materials Laboratory, Wright–Patterson Air Force Base, Ohio, 1980.
8. Kung, C. J., and K. Ortiz: Objective comparison of fatigue crack growth laws, in Structural Safety and Reliability, Vol. 2 (A. H-S. Ang, M. Shinozuka, and G. I. Schuëller, eds.) ASCE, New York, 1627–1630, 1990.
9. Virkler, D. A., B. M. Hillberry, and P. K. Goel: The statistical nature of fatigue crack propagation, J. of Engrg. Mat. and Tech., ASME, 101 (1979), 148–153.
10. Ghonem, H., and S. Dore: Experimental study of the constant probability crack growth curves under constant amplitude loading, Engrg. Fract. Mech., 27 (1987), 1–25.
11. Bolotin, V. V.: On safe crack size under random loading, Izvestiia Akademiia Nauk SSSR, Mekhanika Tverdogo Tela, 1 (1980) (in Russian).
12. Bolotin, V. V.: Lifetime distribution under random loading, Zhurnal Priklandnoi Mekhaniki, Tekhnicheskoi Fiziki, 5 (1980) (in Russian).
13. Ditlevsen, O.: Random Fatigue Crack Growth — A first passage problem, Engrg. Fract. Mech., 23 (1986), 467–477.
14. Ditlevsen, O., and R. Olesen: Statistical analysis of the Virkler data on fatigue crack growth, Engrg. Fract. Mech., 25 (1986), 177–195.
15. Dolinski, K.: Stochastic loading and material inhomogeneity in fatigue crack propagation, Engrg. Fract. Mech., 25 (1986), 809–818.
16. Dolinski, K.: Stochastic modelling and statistical verification of crack growth under constant amplitude loading, (manuscript) (1991).
17. Guers, F., and R. Rackwitz: Time-Variant Reliability of Structural Systems subject to Fatigue, in Proc. of ICASP-5, Vol. 1, Vancouver, Canada, 497–505, 1987.
18. Kozin, F., and J. L. Bodganoff: A Critical Analysis of some Probabilistic Models of Fatigue Crack Growth, Engrg. Fract. Mech., 14 (1981), 59–89.
19. Lin, Y. K., and J. N. Yang: On Statistical Moments of Fatigue Crack Propagation, Engrg. Fract. Mech., 18 (1983), 243–256.
20. Lin, Y. K., and J. N. Yang: A Stochastic Theory of Fatigue Crack Propagation, J. of the AIAA, 23 (1985), 117–124.
21. Madsen, H. O.: Deterministic and probabilistic models for damage accumulation due to time varying loading, DIALOG 5-82, Danish Engrg. Academy, Lyngby, Denmark, 1983.

22. Madsen, H. O.: Random fatigue crack growth and inspection, in Structural Safety and Reliability, Vol. 1, Proc. of ICOSSAR '85, Kobe, Japan (I. Konishi, A. H-S. Ang, and M. Shinozuka, eds.), Elsevier, Amsterdam, The Netherlands, 475–484, 1985.

23. Madsen, H. O., S. Krenk, and N. C. Lind, Methods of Structural Safety, Prentice-Hall, Englewood Cliffs, New Jersey, 1986.

24. Ortiz, K., and A. S. Kiremidjian: Time series analysis of fatigue crack growth data, Engrg. Fract. Mech., 24 (1986), 657–676.

25. Ortiz, K., and A. S. Kiremidjian: Stochastic modeling of fatigue crack growth, Engrg. Fract. Mech., 29 (1988), 657–676.

26. Sobczyk, K.: On the Markovian models for fatigue accumulation, J. de Mécanique Theorique et Appliqué, (Numor Special) (1982), 147–160.

27. Sobczyk, K.: Stochastic modeling of fatigue crack growth, in Proc. of the IUTAM Symposium on 'Probabilistic Methods in Mechanics of Solids and Structures,' Stockholm, Sweden, Springer-Verlag, Berlin, 111–119, 1984.

28. Sobczyk, K.: Modelling of random fatigue crack growth, Engrg. Fract. Mech., 24 (1986), 609–623.

29. Solomos, G. P.: First-passage solutions in fatigue crack propagation, Prob. Engrg. Mech., 4 (1989), 32–39.

30. Spencer, B. F., Jr., and J. Tang: A Markov process model for fatigue crack growth, J. of Engrg. Mech., ASCE, 114 (1988), 2134–2157.

31. Spencer, B. F., Jr., J. Tang, and M. E. Artley: A stochastic approach to modeling fatigue crack growth, J. of the AIAA, 27 (1989), 1628–1635.

32. Tang, J., and B. F. Spencer, Jr.: Reliability solution for the stochastic fatigue crack growth problem, Engrg. Fract. Mech., 34 (1989), 419–433.

33. Tanaka, H., and A. Tsurui: Random propagation of a semi-elliptical surface crack as a bivariate stochastic process, Engrg. Fract. Mech., 33 (1989), 787–800.

34. Tsurui, A., and H. Ishikawa: Application of Fokker–Planck equation to a stochastic fatigue crack growth model, Struct. Safety, 4 (1986), 15–29.

35. Tsurui, A., J. Nienstedt, G.i. Schuëller, and H. Tanaka: Time variant structural reliability using diffusive crack growth models, Engrg. Fract. Mech., 34 (1989), 153–167.

36. Veers, P. J.: Fatigue crack growth due to random loading, Ph.D. Dissertation, Department of Mechanical Engrg., Stanford University, Stanford, California, 1987.

37. Veers, P. J., S. R. Winterstein, D. V. Nelson, and C. A. Cornell: Variable amplitude load models for fatigue damage and crack growth, in Development of Fatigue Loading Spectra, STP-1006, ASTM, Philadelphia, 172–197, 1989.

38. Winterstein, S. R., and P. S. Veers: Diffusion models of fatigue crack growth with sequence effects due to stationary random loads, Structural Safety and Reliability, Vol. 2 (A. H-S. Ang, M. Shinozuka, and G. I. Schuëller, eds.), ASCE, New York, 1523–1530, 1990.
39. Yang, J. N., G. C. Salivar, and C. G. Annis: Statistical modeling of fatigue-crack growth in a nickel-based superalloy, Engrg. Fract. Mech., 18 (1983), 257–270.
40. Yang, J. N., and R. C. Donath: Statistical crack propagation in fastener holes under spectrum loading, J. of Aircraft, AIAA, 20 (1983), 1028–1032.
41. Yang, J. N., W. H. Hsi, and S. D. Manning: Stochastic crack propagation with application to durability and damage tolerance analyses, in Probabilistic Fracture Mechanics and Reliability (J. Provan, ed.) Martinus Nijhoff Publishers, The Netherlands, 1987.
42. Sobczyk, K. and B.F. Spencer, Jr.: Random Fatigue: From Data to Theory, Academic Press, 1992.
43. Manning, S. D., and J. N. Yang: USAF Durability Design Handbook: Guidelines for the Analysis and Design of Durable Aircraft Structures, Technical Report AFFDL-TR-84-3027, Wright–Patterson Air Force Base, Ohio, February, 1984.
44. Yang, J. N., and S. D. Manning: Distribution of equivalent initial flaw size, in Proc. of the Reliability and Maintainability Conference, 112–120, 1980.
45. PROBAN-2, A.S. Veritas Research, Det norske Veritas, Oslo, Norway, 1989.
46. Sobczyk, K., Stochastic Differential Equations with Application to Physics and Engrg., Kluwer Academic Publishers, Dordrecht, The Netherlands, 1991.
47. Cox, D. R., and H. D. Miller: The Theory of Stochastic Processes, Chapman and Hall, London, 1977.
48. Oh, K. P.: "A diffusion model for fatigue crack growth," Proc. of Royal Soc. of London, A367 (1979), 47–58.
49. Fichera, G.: On a unified theory of boundary value problems for elliptic-parabolic equations of second order, in Boundary Problems in Differential Equations (R. E. Langer, ed.), University of Wisconsin Press, Madison, Wisconsin, 97–102, 1960.
50. Enneking, T. J., B. F. Spencer, Jr., and I. P. E. Kinnmark, "Stationary two-state variable problems in stochastic mechanics," J. of Engrg. Mech., ASCE 116 (1990), 334–358.
51. Ostergaard, D. F., and B. M. Hillberry: Characterization of variability in fatigue crack propagation data, in Probabilistic Methods for Design and Maintenance of Structures, in STP-798, ASTM, Philadelphia, 97-115, 1983.
52. Ferguson, R. I.: River loads underestimated by rating curves, Water Resources Res., 22 (1986), 74–76.
53. ASTM: Appendix I, E647-86a, Recommended data reduction techniques, in Annual Book of ASTM Standards. Vol. 3.01: Metals Test Methods and Analytical Procedures, ASTM, Philadelphia, 919–920, 1987.

54. IMSL, The International Math and Statistics Subroutine Library, IMSL, Inc., Houston, Texas, 1984.

55. Johnson, W. S.: Multi-parameter yield zone model for predicting spectrum crack growth, in Methods and Models for Predicting Fatigue Crack Growth under Random Loading (J. B. Chang, and C. M. Hudson, eds.), STP-748, ASTM, Philadelphia, 85–102, 1981.

56. Enneking,T. J.: On the stochastic fatigue crack growth problem, Ph.D. Dissertation, Department of Civil Engineering, University of Notre Dame, Notre Dame, Indiana, 1991.

57. MIL-A-87221: General Specifications for Aircraft Structures, U. S. Air Force Aeronautical Systems Division, Wright-Patterson Air Force Base, Ohio, 1985.

58. Madsen, H. O., R. Skjong, A. G. Tallin, and F. Kirkemo: Probabilistic fatigue crack growth analysis of offshore structures with reliability updating through inspection, in Proc. of the Marine Structural Reliability Symposium, Arlington, Virginia, 45-55, 1987.

59. Yang, J. N., and S. Chen: Fatigue reliability of structural components under scheduled inspection and repair maintenance, Probabilistic Methods in Mechanics of Solids and Structures, Proc. of the IUTAM Symposium, Stockholm, 1984, Springer-Verlag, Berlin, 559–568, 1985.

60. Berens, A. P., and P. W. Hovey: Evaluation of nde reliability characterization, Report No. AFWAL-TR-81-4160, Volume I, University of Dayton Research Institute, Dayton, Ohio, 1981.

FATIGUE RELIABILITY OF
MARINE STRUCTURES

H.O. Madsen
Det Norske Veritas, Copenhagen, Denmark

ABSTRACT

Fatigue reliability of marine structures is an important design consideration both for ship structures, fixed and floating offshore platforms for oil and gas exploration and exploitation. The importance of performing a fatigue resistant design is increased by use of lighter materials and materials with higher static strength and the emergence of designs of structures with significant increase of response due to dynamic effects. Prediction of fatigue life for such structures is encumbered with large uncertainty due to inherent uncertainty in the loading and fatigue processes, due to uncertainty of the process parameters and due to model uncertainty. As a consequence of this, probability based methods for fatigue life prediction are well accepted. Large resources are spent in inspecting marine structures for fatigue cracks. Both visual inspection and expensive NDT methods are applied above and underwater. A probability based decision analysis for resource allocation between material spending in construction and inspection and repair is thus relevant.

1. INTRODUCTION

Typical floating and bottom supported offshore platforms are shown in Figure 1.1. These structures are made of steel except the one in Figure 1.1.e, which is a concrete platform with a steel deck.

Semi-submersibles, or column stabilized units differ radically in appearance from traditional ship vessels. They have a platform of triangular or rectangular configuration supported by columns which again are connected to large underwater displacement hulls or mounted on large vertical caissons giving the necessary buoyancy. The basic idea of this design is to reduce the wave forces by locating the major buoyancy members well beneath the surface. Sufficient stability is provided by the columns.

Floating platforms are kept on location either by a conventional spread mooring system or a dynamic positioning system. Up to now spread mooring systems consisting of 6-12 lines with anchors at the ends have been used. The tension leg platform in Figure 1.1.b is a semisubmersible type platform with buoyancy exceeding its weight thereby causing a pretension in the vertical cables. The first tension leg platform was installed at the Hutton field in the summer of 1984 and recently a very large tension leg platform has started production of oil from the Snorre field. A tension leg platform for the Heidrun field is being developed with concrete columns and pontoons and a steel deck.

The jack-up platform in Figure 1.1.d is characterized by 3 or 4 legs that can be lowered to serve as bottom support. The main deck is raised out of the water on location during drilling or production. This deck unit is watertight and has buoyancy and stability to serve as a transport unit with elevated legs during ocean or field transit. The legs are typically truss pipe structures.

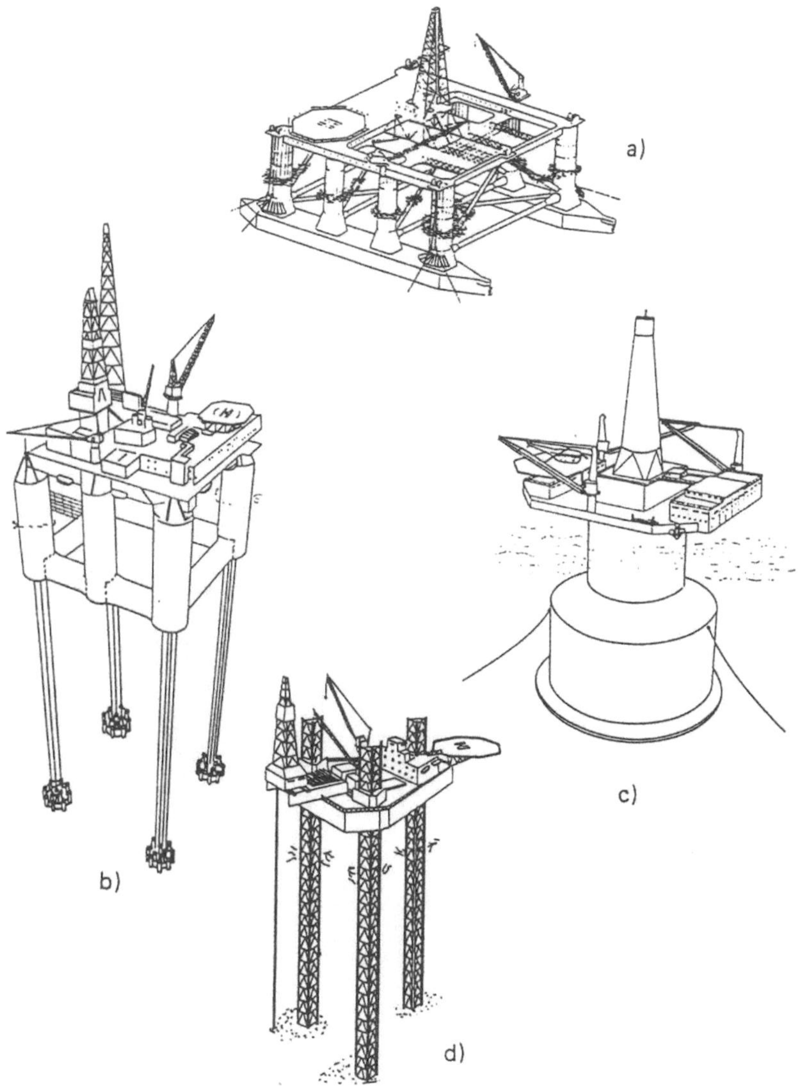

Figure 1.1 Selected offshore structures.
a) Semi-submersible platform; b) Tension leg platform;
c) Buoy-type platform; d) Jack-up

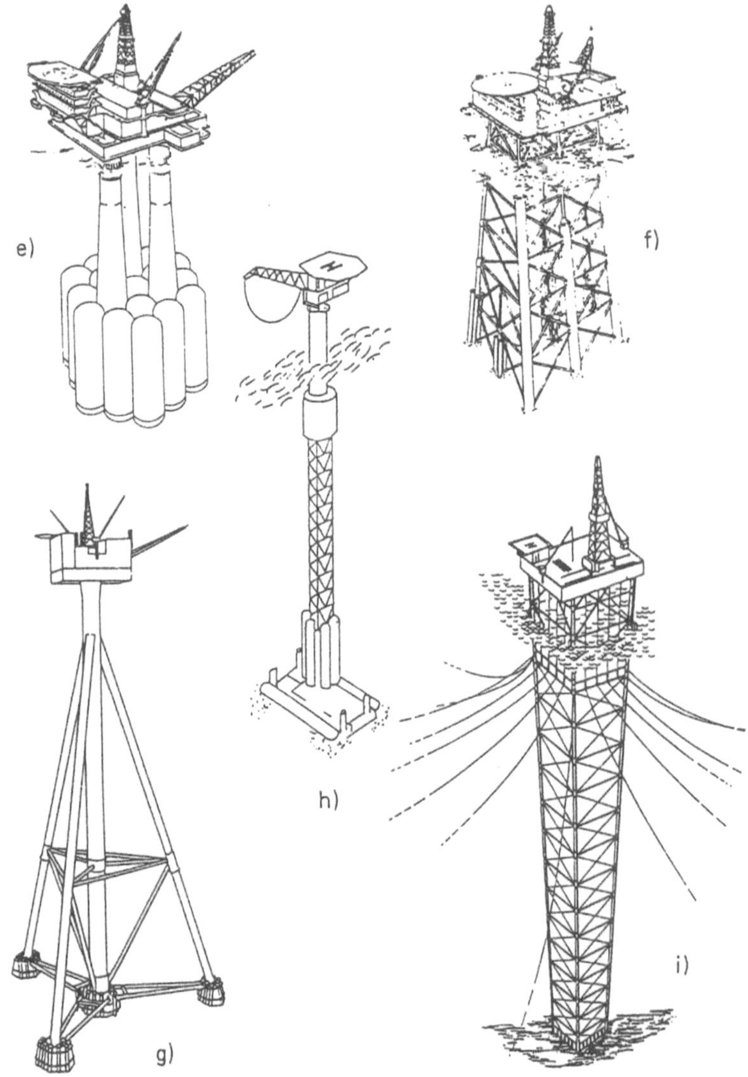

Figure 1.1 Selected offshore structures.
e) Gravity platform; f) Jacket; g) Steel tripod;
h) Articulated tower; i) Guyed tower.

A concrete gravity platform as shown in Figure 1.1.e essentially consists of a large cellular caisson supporting 3 or 4 towers with a steel deck on the top. For existing structures the upper part of the superstructure consists of a steel grillage built of plate girders or as a truss-work giving a low self-weight during tow-out to the final position. This makes it possible to install more equipment before tow-out and thus reduce the offshore installation time. The basic concept of the gravity platform is to attain stability in the permanent location without special piling. The concrete gravity structures in the North Sea are today designed for water depths in excess of 300 meter and are designed to carry a very large deck load from production and processing equipment as well as living quarters.

The jacket or template platform shown in Figure 1.1.f is a truss-work tower of tubular members with a deck on the top and with long steel piles into the sea bed. The deck and environmental loads are transferred to the foundation piles through the 4 to 16 legs. The legs are stiffened by bracing which is also able to carry resultant horizontal forces. Most of the more than 2500 existing jackets in the world are located in shallow waters with only about 50 in water depths over 100 meters.

Tripod structures as shown in Figure 1.1.g have been proposed as an alternative to the conventional truss-work platform. The small area breaking the surface reduces the wave loading and the inclined legs make the global transfer of wave loads to the seafloor very effective as this mainly introduces axial forces in the legs.

The articulated tower shown in Figure 1.1.h consists of a deck, a buoyancy chamber and a load bearing truss-work structure linked to the sea bed by a joint. So far, this type of structure has been used as loading buoys, but its potential as a production platform is being investigated.

The guyed tower shown in Figure 1.1.i is a truss-work tower resting on a spud can or with piles and moored by some 20-24 pre-tensioned cables. A buoyancy chamber located below the mean water level may be used to relieve the spud can or piles of some of the vertical load. The first guyed tower has been successfully installed in the Lena field in the Gulf of Mexico.

Except for the gravity based concrete structure most of the other structures were first developed for use in the Gulf of Mexico. Soon after oil and gas exploration began in the North Sea in the 1960's, it became apparent that the steel structure design developed for offshore activities in the Gulf of Mexico was not adequate when transferred to the more rigorous North Sea environment. In particular, fatigue cracks evolved as a result of wave action during the severe winter storms. Many repairs on the early North Sea structures had to be performed and in one serious accident fatigue played an important role.

In the late 1980's and early 1990's a new generation of fixed offshore structures were designed and introduced in the North Sea. These structures have significant dynamic amplification in certain sea conditions and also experience high frequency dynamic response due to non-linear loading and response.

Offshore steel structures are commonly composed of thin tubular members because their closed sections provide buoyancy and high torsional rigidity and minimum surface for painting and corrosion attack. Circular tubes are preferred in the submerged part of the structures because they result in smaller hydrodynamic forces than members with square cross-sections. Fabrication is, however, easier for the latter type and above the splash zone hollow rectangular sections or other sectional shapes are used.

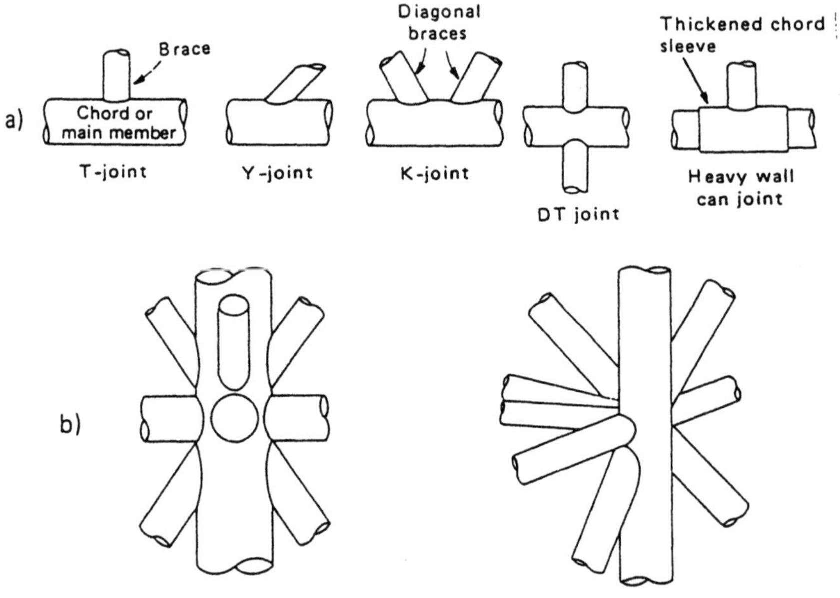

Figure 1.2 Simple tubular welded joints.
a) Simple plane joints; b) Multiplane joints.

Figure 1.2.a shows typical configurations of tubular joints for members lying in one plane. In Figure 1.2.b more complex three dimensional situations are displayed. Stiffening of a joint may be used depending on the relative diameters of the members and their plate thicknesses. The simplest form of stiffening is to increase the plate thickness in the joint area. Other possibilities include the use of internal bulkheads and stiffeners as shown in Figure 1.3. Fabricating joints with internal stiffeners obviously requires a minimum space and thus a minimum diameter of the tubes.

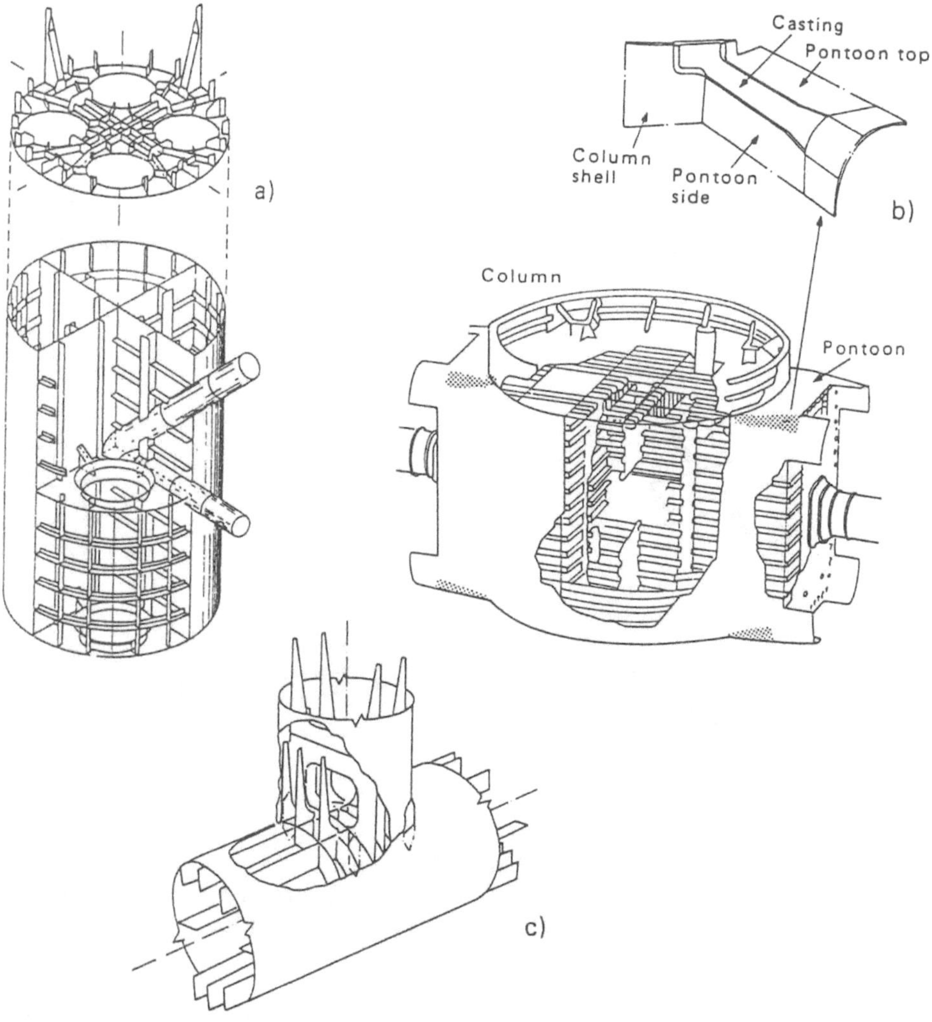

Figure 1.3 Welded joints of circular tubular members.

Rectangular members may be joined as shown in Figure 1.4 possibly with internal stiffening. Such joints are often planar joints or complex joints in perpendicular planes. Various other typical joints and fatigue prone details are also shown in Figure 1.4.

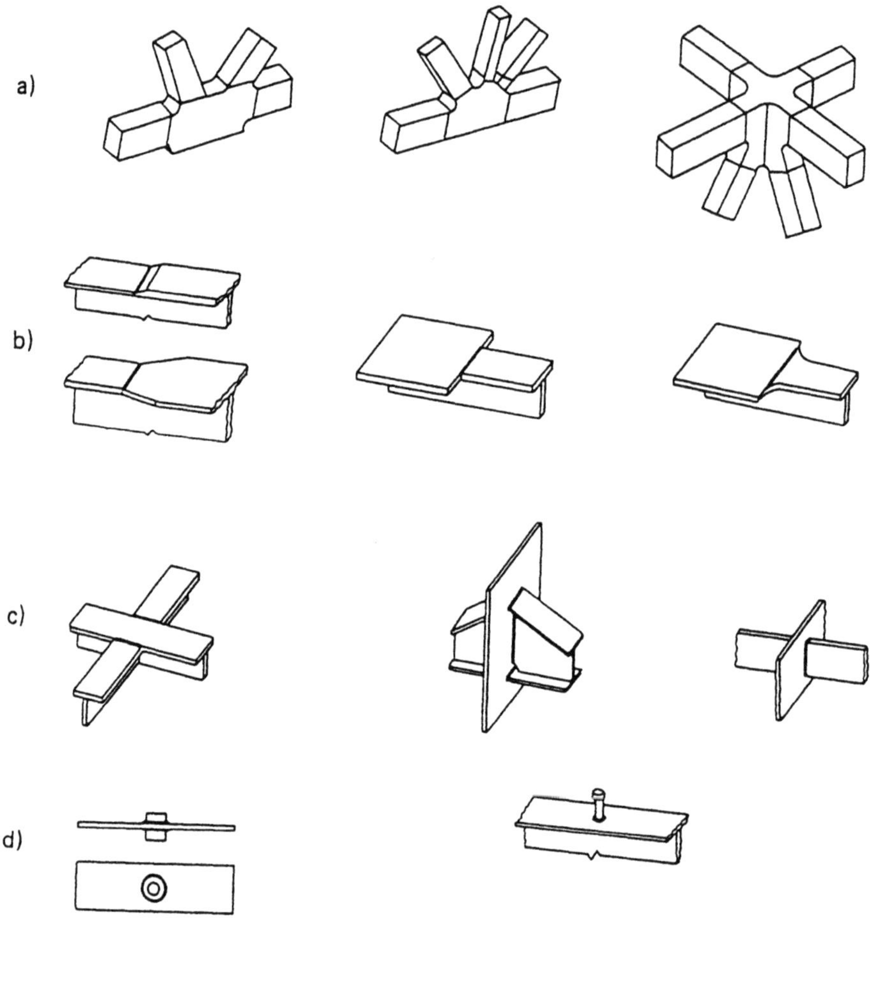

Figure 1.4 Various types of welded joints.
a) Between tubes with rectangular cross section, b) and c) between plats, d) attachment welds.

For the details shown in Figure 1.4 the areas where fatigue cracks may develop are in general not too difficult to predict. These areas are characterized by a change in geometry causing stress concentrations as well as the presence of a weld and thereby in general small initial defects. For this reason some very heavily loaded joints have been produced of cast steel and the joint connection to the brace or leg member moved outside the area of large stress concentrations.

The tension leg platform in Figure 1.1.b is connected to the sea floor by vertical elements called tethers. These are made of tubular elements which are either welded or screwed together. For these tethers all the connections are subject to almost the same load in the longitudinal and circumferential directions and it is difficult to identify single hot spots. Laboratory experiments for such tethers have shown a significant size effect when compared with fatigue strength results for a standard test specimen.

For ship structures fatigue has recently been identified as a major problem mainly because it has not been properly addressed in designs utilizing higher strength steels. Fatigue cracks in combination with extensive corrosion is believed to have been a major contributor to the recent loss of many large bulk carriers. In several very large crude oil carriers a significant amount of cracking has also been discovered in particular due to poor design of details. Although such cracks may not hamper the safety of the ship they can lead to local failure and the possibility of extensive pollution. Figure 1.5 shows a typical detail from a ship side structure and a possible crack initiation site.

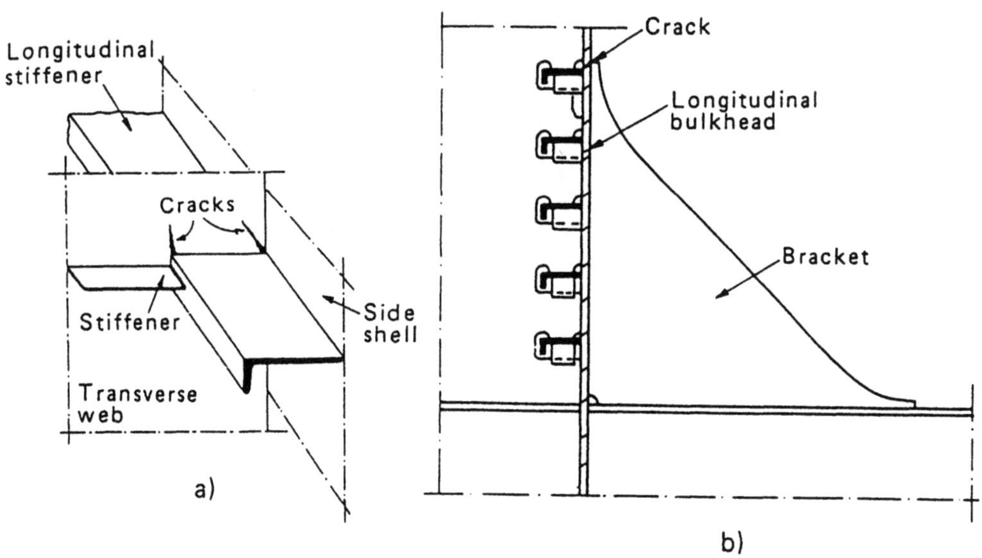

Figure 1.5 Cracks in ship details.
a) Longitudinal-transverse stiffener connection, b) bracket in transverse frame.

These notes present a general application of reliability methods to fatigue analysis of marine steel structure. A detailed description of the fatigue mechanisms is not provided, but the reader is referred to e.g.

[1] and [2]. For a comprehensive treatment of fatigue analysis of marine structures reference [3] is recommended.

Two different approaches are applied in fatigue analysis of marine structures. These are generally referred to as the S-N approach and the fracture mechanics approach. The S-N approach has until today been the standard approach for design. In the S-N approach, however, damage is not related to a measurable physical quantity and the approach can not take full account of information on observed cracks or other measures of damage. The fracture mechanics approach on the other hand describes fatigue crack growth from an initial size to a critical size and can be used in fitness-for-purpose evaluations of damaged elements. As shown in the following sections, the two approaches are, however, very similar.

A fatigue analysis requires modelling of the long term load environment, a global structural response analysis, a local strain and stress analysis, a modelling of the fatigue strength under constant amplitude loading, and a modelling of the damage accumulation under variable amplitude loading. In all modelling phases rather large uncertainties are present both in the input parameters and in the models themselves. Uncertainty modelling is thus important and reliability methods are accepted as suitable and useful. Uncertainty modelling relevant for offshore structures may be found in e.g. [4] and [5] while a more general treatment of uncertainty modelling may be found in [6].

These notes first describe an application of reliability methods for a fatigue analysis based on the S-N approach. A similar analysis based on a fracture mechanics approach is then described. Reliability updating based on inspection results is treated and finally a scheme for reliability based optimisation of design, inspection and maintenance is presented. Although the examples mainly refer to offshore structures, the methods and concepts are generally applicable. The methods have in recent years been used for inspection planning of more than 40 steel jacket platforms in the North Sea.

2. *S-N* APPROACH

The fatigue strength is expressed through the *S-N* curve (Wohler curve), which gives the number of stress cycles N under constant amplitude loading with stress range S necessary to cause failure. *S-N* curves for design are found in codes and specifications. For tubular joints in offshore structures the *S-N* curve is often selected as the T-curve from Department of Energy [7]. The T-curve is shown in Figure 2.1 and it can be written as

$$N = \begin{cases} KS^{-3}, & S>S_0 \\ K_1 S^{-5}, & S\leq S_0 \end{cases} \tag{2.1}$$

To obtain a continuous curve the constants K and K_1 are related as

$$K_1 = KS_0^2 \tag{2.2}$$

Other *S-N* curves have no change in slope in the double logarithmic plot, or they end in a horizontal branch corresponding to a fatigue threshold. The upper part of the T-curve is the result of a linear regression analysis of test results of a form shown in Figure 2.1. The distribution of $\ln N$ for a fixed value of S is assumed to be normally distributed with mean value varying linearly with $\ln S$ and with a standard deviation $\sigma_{\ln N}$ independent of S. The mean value of $\ln N$ is estimated from the test results as, [7]

$$E[\ln N] = 29.152 - 3.0 \ln S \tag{2.3}$$

and the standard deviation $\sigma_{\ln N}$ is estimated as 0.572.

The statistical analysis in [7] assumes a complete prior knowledge of the slope m ($m = 3$) and no prior knowledge of the standard deviation of $\ln N$. In [8] a more refined statistical analysis is presented. Based on a Bayesian statistical analysis with various choices for prior distribution, simple results for the posterior distribution of m, K and σ_{lnN} are derived.

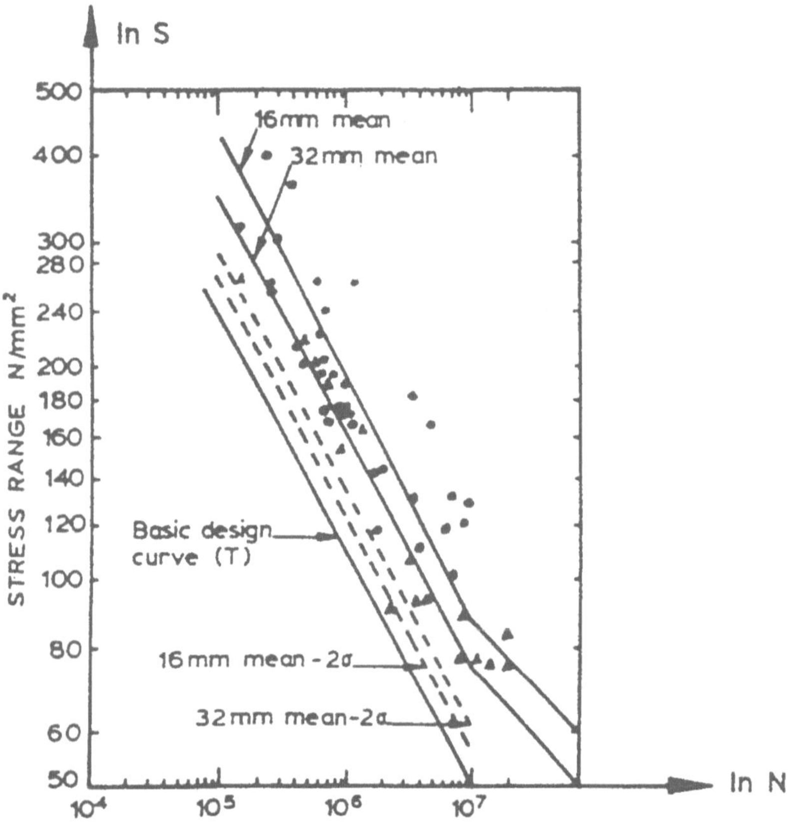

Figure 2.1 T-curve and fatigue test results.

For variable amplitude loading damage is expressed by the Miner sum [9]

$$D = \sum_i \frac{n(S_i)}{N(S_i)} \tag{2.4}$$

where $n(S_i)$ is the number of stress cycles of stress range S_i in the stress history and $N(S_i)$ is the number of stress cycles of stress range S_i necessary to cause failure. The summation is over all stress ranges. Let N_T denote the total number of stress cycles in the considered period of time. Let further $f_S(s)$ denote the probability density function of stress ranges. The damage can then be computed as

$$D = \int_0^\infty \frac{N_T f_S(s) ds}{N(s)} = N_T E[\frac{1}{N(S)}] \qquad (2.5)$$

In many applications a Weibull distribution provides a good fit to the stress range distribution

$$F_S(s) = 1 - \exp(-(\frac{s}{A})^B), \quad s>0 \qquad (2.6)$$

As an example, for the S-N curve in (2.1) with $S_0=0$, the damage becomes

$$D = N_T A^3 \Gamma(1+\frac{3}{B})/K \qquad (2.7)$$

where $\Gamma(\ )$ denotes the Gamma function.

The damage indicator D increases monotonically from an initial value of zero until failure, which is postulated to occur when D reaches the value one. Experimental results with variable amplitude loading, however, show that the value of Miner's sum at failure should be taken as a random variable Δ to account for model uncertainty. The limit state function can then be expressed as

$$g = \Delta - D \qquad (2.8)$$

Example 1 - Extended Life Time Analysis.

A problem often encountered for structures in service is whether or not to accept operation beyond the design life time. This decision must generally be taken shortly before the design life time expires and the decision should be affected by the experience gained during the service life. In this example a tubular joint of a jacket structure is considered. The design life was 20 years and the joint has successfully survived for 15 years without any repair or detectable damage. An extension of the life time to 50 years is desired. Details of the uncertainty modelling is described in [10].

First, a FORM reliability analysis is carried out for the original design specifications. Figure 2.2 shows the reliability index as a function of life time. With the requirement of $\beta \geq 2.00$ for the design life, the joint is accepted for a life time of 20 years but not for a life time of 50 years. The failure probability after 50 years is

$$P_{50} = P(g(50) \leq 0) = 2.59 \cdot 10^{-2}$$

corresponding to a reliability index

$$\beta_{50} = 1.944$$

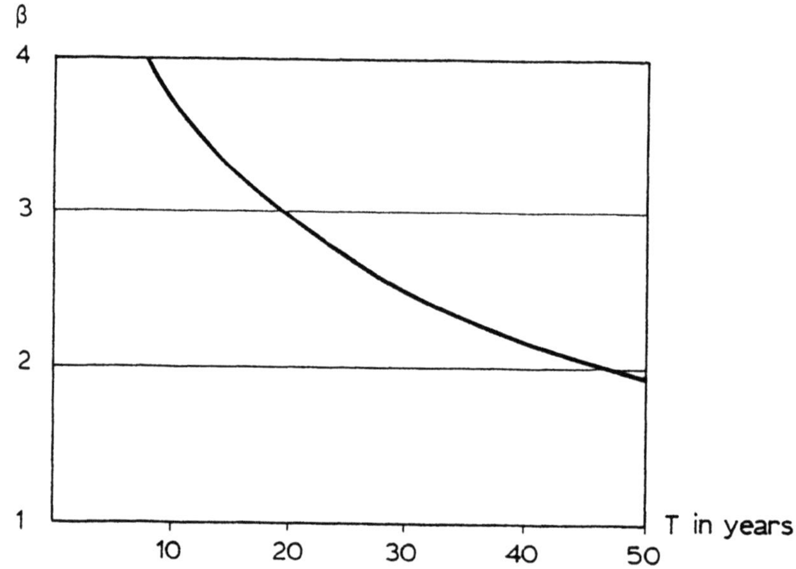

Figure 2.2 Reliability index as a function of life time, jacket structure example.

Second, it is checked whether the fact that the joint has survived 15 years of service is additional information enough to increase the reliability sufficiently. The updated failure probability after 50 years becomes

$$P(g(50) \leq 0 \,|\, g(15) > 0) = \frac{P(g(50) \leq 0) - P(g(15) \leq 0)}{1 - P(g(15) \leq 0)}$$

using that g decreases with T. An analysis for $T = 15$ years gives

$$P_{15} = P(g(15) \leq 0) = 5.65 \cdot 10^{-4}$$

The updated failure probability and the corresponding reliability index thus become

$$P_{50}' = P(g(50)\leq 0 \,|\, g(15)\geq 0) = 2.53\cdot 10^{-2}$$

$$\beta_{50}' = -\Phi^{-1}(P_{50}') = 1.955$$

The requirement $\beta\geq 2.00$ is therefore not fulfilled with this simple updating. A more efficient updating is presented in a later section.

Example 2 - Optimal Redesign for Increased Deck Mass.

A mono-tower platform has been designed with a deck mass of m_d=2000 ton. The design is illustrated in Figure 2.3. The fatigue life time distribution of the cross section at the mudline has been computed, see Figure 2.4. The anticipated life time is 50 years and the corresponding reliability index is β=3.90. An alternative design with a deck mass of 3000 tons is desired and the reliability is to be maintained at the same level.

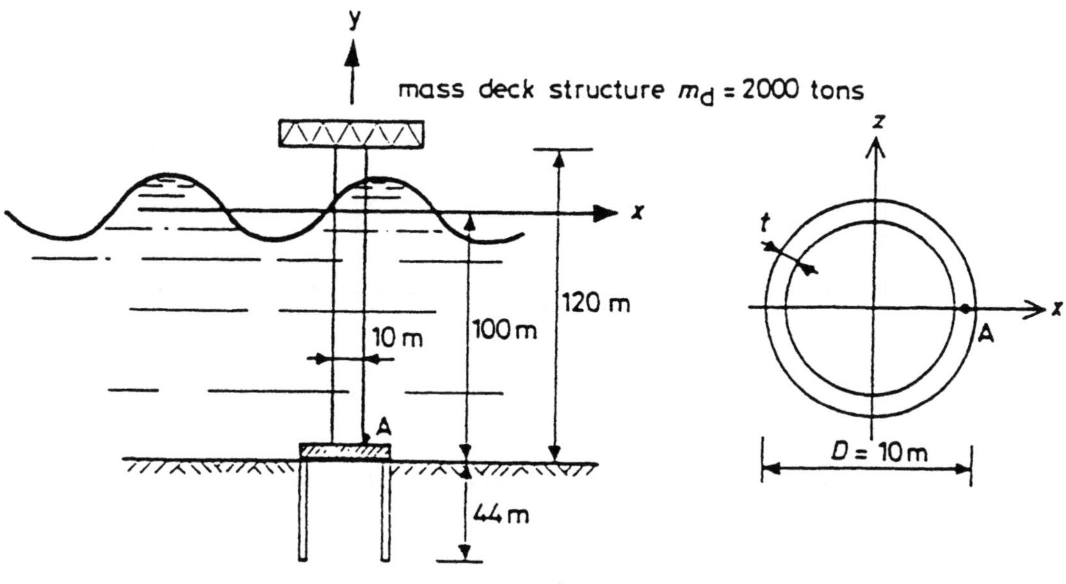

Figure 2.3 Mono-tower platform concept.

The additional deck mass increases the natural period and thus the dynamic amplification. Two alternative ways of redesign are considered and the solution which leads to the lowest weight increase of the tower is to be selected. An increased thickness t of the tower has three effects: larger sectional modulus, larger mass of the structure and lower fatigue strength due to thickness effects. The first effect increases the reliability while the two other effects lower the reliability. An increased diameter D of the tower similarly has three effects: larger sectional modulus, larger mass of the structure and larger load on the structure. Again, the first effect increases the reliability while the last two effects lower it.

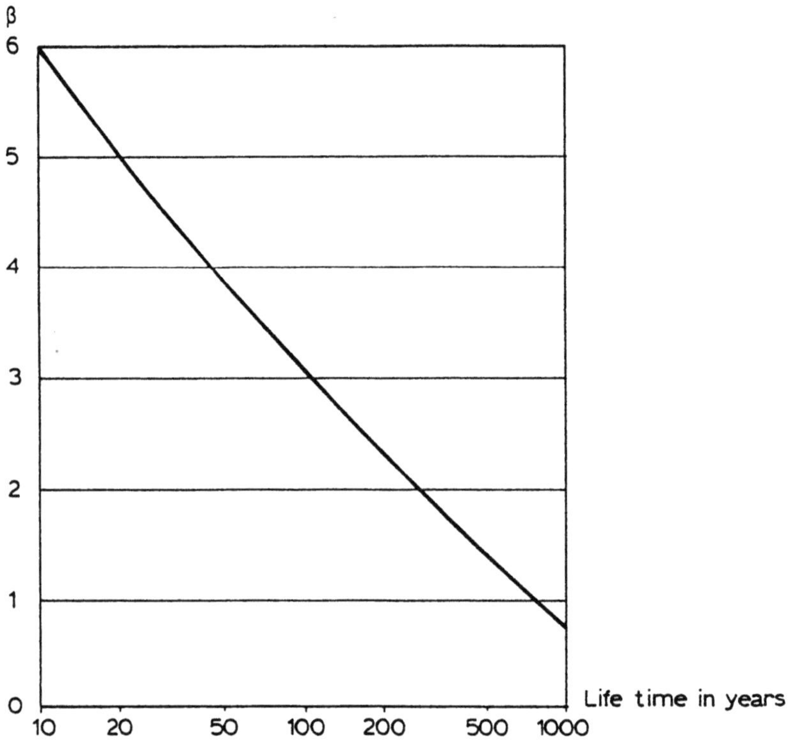

Figure 2.4 Reliability index as a function of fatigue life time, mono-tower platform example.

The derivative of β with respect to a deterministic design parameter in the limit state function is directly available from the FORM analysis, [11]. The following results are obtained for the derivatives computed at the design point for the design with a deck mass of 2000 ton.

$$\frac{\partial \beta}{\partial m_d} = -0.50 \cdot 10^{-3} \; ton^{-1}$$

$$\frac{\partial \beta}{\partial t} = 0.035 \; mm^{-1}$$

The change in the reliability index due to the increased mass is estimated as

$$\frac{\partial\beta}{\partial D} = 1.10\ m^{-1}$$

$$\Delta\beta \approx -0.50 \cdot 10^{-3}\ \text{ton}^{-1} \times 1000\ \text{ton} = -0.50$$

(the exact value is $\Delta\beta = 3.37 - 3.90 = -0.53$). The necessary increase in the diameter is then predicted as

$$\Delta D \approx 0.50/1.10\,\text{m}^{-1} = 0.45\,\text{mm}$$

leading to a weight increase of 199 ton (the exact value for $m_d = 3000$ ton and $d = 10.45$ m is $\beta = 3.95$). An alternative increase in thickness is similarly predicted to be

$$\Delta t \approx 0.50/0.035\text{mm}^{-1} = 14\,\text{mm}$$

leading to a weight increase of 412 ton. The increase in diameter is therefore preferred.

3. FRACTURE MECHANICS APPROACH

In a linear elastic fracture mechanics approach the increment in crack size, Δa, during a load cycle is related to the range of the stress intensity factor, ΔK, for the load cycle. A simple relation which is sufficient for most purposes was proposed by Paris and Erdogan,[12]

$$\Delta a = C(\Delta K)^m, \quad \Delta K > 0 \tag{3.1}$$

The crack growth equation is formulated without a positive lower threshold on ΔK below which no crack growth occurs. The equation was proposed based on experimental results, but is also the result of various mechanical and energy based models, see, e.g., [12] and [13]. C and m are material constants for which values are estimated from experimental data. The crack increment in one cycle is generally very small compared to the crack size and (3.1) is consequently written in a "kinetic" form as

$$\frac{da}{dN} = C(\Delta K)^m, \quad \Delta K > 0 \tag{3.2}$$

where N is the number of stress cycles. The stress intensity factor K is computed by linear elastic fracture mechanics and is expressed as

$$K = \sigma Y(a)\sqrt{\pi a} \tag{3.3}$$

where σ is the far-field stress and $Y(a)$ is the geometry function. The geometry function depends on the overall geometry including the geometry of the crack and the geometry of a possible weld. To explicitly account for uncertainties in the calculations of K, the geometry function is written as $Y(a) = Y(a, \mathbf{Y})$, where $\mathbf{Y}$ is a vector of random parameters. Inserting (3.3) in (3.2) and separating the variables leads to the differential equation

$$\frac{da}{Y(a,Y)^m(\sqrt{\pi a})^m} = C(\Delta\sigma)^m dN, \quad a(0) = a_0 \tag{3.4}$$

where a_0 is the initial crack size. The equation is applied both for constant and for variable amplitude loading, thus ignoring possible sequence effects.

Equations (3.1) to (3.4) describe the crack size as a scalar a, which for a through-thickness crack is the crack length. For marine structures the interest is mainly on surface breaking cracks for which a description of the crack depth, crack length and crack shape is necessary. It is common practice to assume a semi-elliptical initial shape and to assume that the shape remains semi-elliptical during the crack growth. In that case the crack depth a and the length $2c$ describe the crack. The differential equation (3.2) is replaced by a pair of coupled equations, [14]

$$\frac{da}{dN} = C_a(\Delta\sigma Y_a(a,c,Y)\sqrt{\pi a})^m \tag{3.5}$$

$$\frac{dc}{dN} = C_c(\Delta\sigma Y_c(a,c,Y)\sqrt{\pi c})^m \tag{3.6}$$

where Y_a and Y_c are the geometry functions for the deepest point and for the end point of the crack at the surface, respectively. The material constants C_a and C_c may differ due to variation in stress field tri-axiallity. The failure criterion can refer to a critical value of either a or c individually or to a function of a and c. The equations are conveniently rewritten as

$$\frac{dc}{da} = \frac{C_c}{C_a}\left[\frac{Y_c(a,c,Y)\sqrt{c}}{Y_a(a,c,Y)\sqrt{a}}\right]^m, \quad c(a_0) = c_0 \tag{3.7}$$

$$\frac{dN}{da} = [C_a(\Delta\sigma Y_a(a,c,Y)\sqrt{\pi a})^m]^{-1}, \quad N(a_0) = 0 \tag{3.8}$$

which are solved simultaneously. This can be done both for constant and for variable amplitude loading. The first equation gives c as a function of a and the initial values a_0 and c_0, but independent of the loading and number of stress cycles. The solution for c may be inserted in (3.8) which is then of the same form as (3.2). For reasons of simplicity in the presentation the following is limited to a through-thickness crack of size a.

Numerous experimental results exist for crack growth under constant amplitude loading. Figure 3.1 from [15] shows experimental results reported in [16] for 64 centre cracked specimens made of 2024-T3 aluminum. The experiments were highly controlled and performed by the same laboratory using the same equipment and the same personnel. The initial half crack length of each specimen was $a_0 = 9\ mm$ and the width of the panel was $2b = 152.4\ mm$.

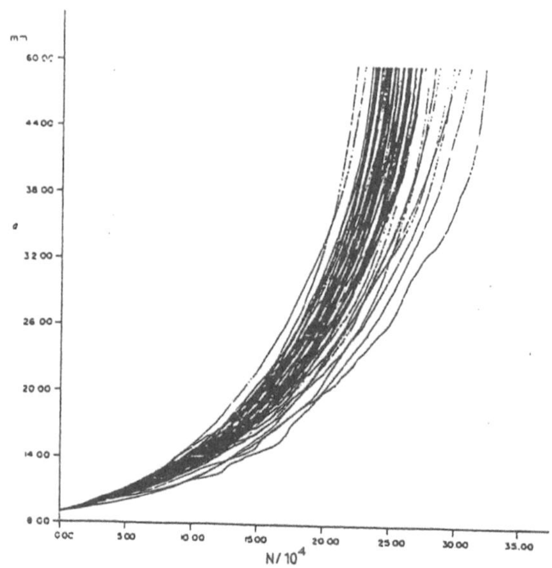

Figure 3.1 Experimental results.

The exact geometry function is known in the form of an infinite power series in a/b. The geometry function is approximated by the so-called Feddersen correction

$$Y(a) = \frac{1}{\sqrt{\cos(\pi a/b)}}, \quad \frac{a}{b}<0.7 \tag{3.9}$$

with an error less than 0.3%. For $a=0$ the geometry function is 1 corresponding to a crack in an infinite panel. Figure 3.2 shows the crack length as a function of N obtained by solving (3.4) with fixed values of C, m and a_0. Although the experimental curves resemble this curve, they are all different, they are slightly irregular and they show a good deal of intermingling. In the experiments the only non-deterministic factor in (3.4) is the material constant C. An attempt is therefore made to "randomise" C in such a way that sample functions become similar to those experimentally observed.

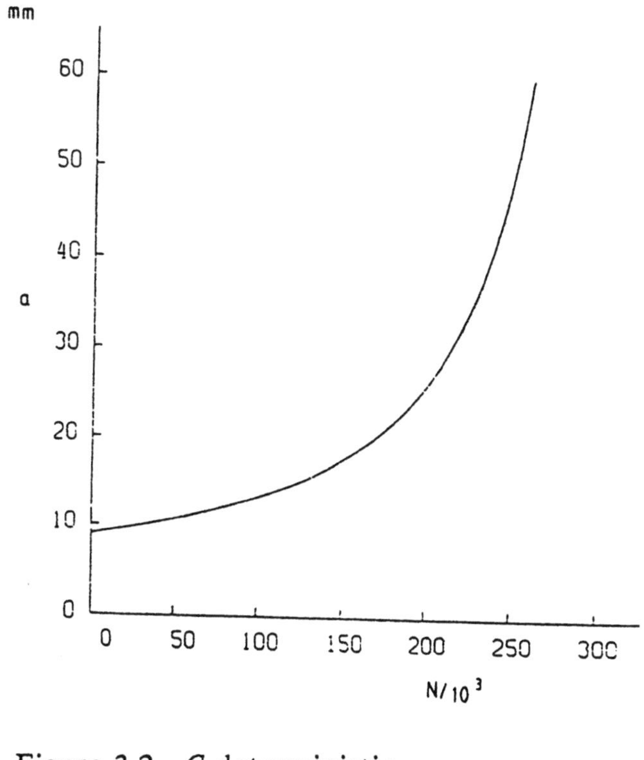

Figure 3.2 C deterministic.

First, let C be a random variable varying independently from specimen to specimen. Sample functions then appear as in Figure 3.2 from [17]. The general behaviour is still correct and different sample functions are obtained. The sample curves are, however, quite smooth and no intermingling takes place.

Second, C is written as, [18]

$$C = C(a) = \frac{C_1}{C_2(a)} \tag{3.10}$$

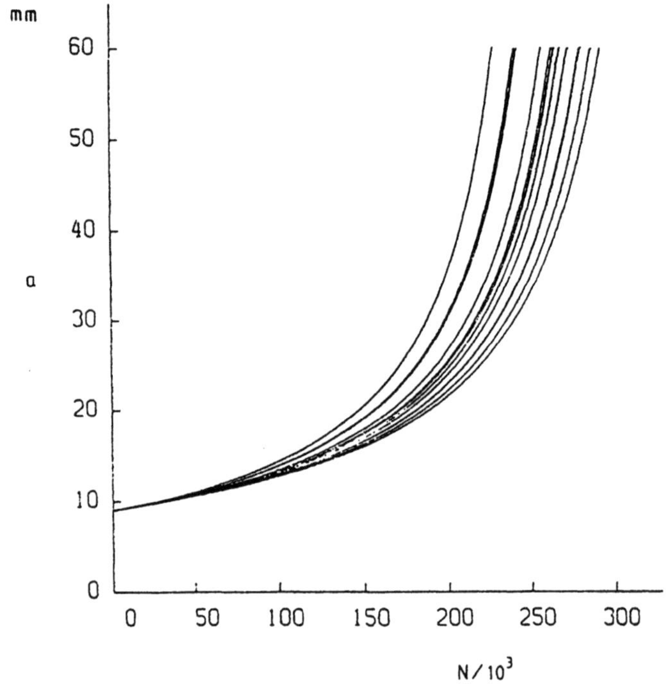

Figure 3.3 C random variable.

Here C_1 is a random variable describing random variations between mean values in different specimens, while $C_2(a)$ is a positive random process describing variations from the mean value along the crack path within each specimen. The mean value of $C_2(a)$ is one and the process is assumed to be homogeneous. The type of process is suggested as lognormal in [18]. The corresponding stochastic differential equation for a follows from (3.2) as

$$\frac{da}{dN} = \frac{C_1}{C_2(a)}(\Delta K)^m, \quad a(0) = a_0 \tag{3.11}$$

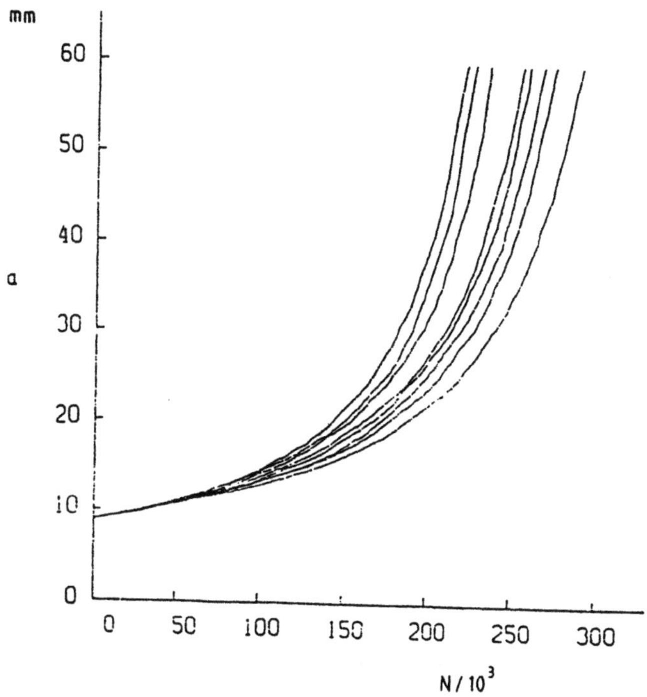

Figure 3.4 *C* random sequence.

This equation may not be readily solved but sample curves for a can be obtained by simulation of $C_2(a)$ and numerical solution of the equation. Results of such simulations as shown in Figure 3.4 resemble the results of Figure 3.1 well, and in particular the random model has three properties, which are experimentally observed in the test results:

- sample curves of a versus N are irregular and not very smooth,
- sample curves of a versus N become more smooth for larger values of a,
- sample curves of a versus N intermingle, in particular for smaller values of a.

Other stochastic models which provide the same qualitative agreement with experiments have been based on a Markov process description, [19], or have been formulated as a first-passage problem, [20].

To estimate the correlation properties of the random process $C_2(a)$ a statistical analysis of the test data from [16] has been carried out, [18]. The correlation function $\rho_2(\Delta a)$ for $C_2(a)$ is shown to decrease to zero very rapidly with Δa. The correlation radius r_C is defined as

$$r_C = \int_{-\infty}^{\infty} \rho_2(x)dx \tag{3.12}$$

and has been estimated as 0.12 *mm* for the aluminum alloy in the experiments of [16]. The variance of C_2 has been estimated as 0.062 for the same data.

A damage function $\Psi(a)$ is introduced as

$$\Psi(a) = \int_{a_0}^{a} \frac{C_2(x)}{Y(x,Y)^m(\sqrt{\pi x})^m}dx \tag{3.13}$$

The second moment statistics for the damage function conditioned upon (a_0,Y,m) are

$$E[\Psi(a)|a_0,Y,m] = \int_{a_0}^{a} \frac{E[C_2(x)]}{Y(x,Y)^m(\sqrt{\pi x})^m}dx = \int_{a_0}^{a} \frac{1}{Y(x,Y)^m(\sqrt{\pi x})^m}dx \tag{3.14}$$

$$Var[\Psi(a)|a_0,Y,m] = \int_{a_0}^{a}\int_{a_0}^{a} \frac{Cov[C_2(x_1),C_2(x_2)]}{Y(x_1,Y)^m(\pi x_1)^{m/2}Y(x_2,Y)^m(\pi x_2)^{m/2}}dx_1dx_2 \approx$$

$$\approx r_{c_2} Var[C_2]\int_{a_0}^{a} \frac{1}{Y(x,Y)^{2m}(\pi x)^m}dx \tag{3.15}$$

and

$$\rho[\Psi(a_1),\Psi(a_2)|a_0,Y,m] \approx \frac{\int\limits_{a_0}^{\min(a_1,a_2)} \frac{1}{Y(x,Y)^{2m}(\pi x)^m}dx}{\left(\int\limits_{a_0}^{a_1} \frac{dx_1}{Y(x_1,Y)^{2m}(\pi x_1)^m}\right)^{1/2}\left(\int\limits_{a_0}^{a_2} \frac{dx_2}{Y(x_2,Y)^{2m}(\pi x_2)^m}\right)^{1/2}} \tag{3.16}$$

The approximations for the variance and the correlation function are justified by the short correlation length of $C_2(a)$ compared to crack size increments of interest. The random variable $\Psi(a)|a_0,Y,m$ is essentially the sum of many independent random variables of approximately the same variance. The distribution is therefore well approximated by a normal distribution and $\Psi(a)|a_0,Y,m$ is expressed as

$$\Psi(a)|a_0,Y,m = \int\limits_{a_0}^{a} \frac{1}{Y(x,Y)^m(\sqrt{\pi x})^m}dx + U\left[r_{c_2} Var[C_2] \int\limits_{a_0}^{a} \frac{1}{Y(x,Y)^{2m}(\pi x)^m}dx\right]^{1/2} \tag{3.17}$$

where U is a standard normally distributed random variable.

The stress ranges are denoted $S_i = \Delta\sigma_i$ and solution of (3.4) gives

$$\Psi(a) = C_1\int\limits_0^N S^m dN = \begin{cases} C_1 S^m N, & \textit{constant amplitude loading} \\ C_1 \sum\limits_{r=1}^{N} S_r^m, & \textit{variable amplitude loading} \end{cases} \tag{3.18}$$

The crack growth equation (3.1) has here been directly extrapolated to variable amplitude loading where the appropriate value of S is inserted for each stress cycle. It must be emphasized that this is an extrapolation beyond experimental experience and that possible sequence effects are neglected. It is observed that the only difference between the two cases of constant and variable amplitude loading concerns the loading statistics. The crack length after N stress cycles, a_N, is obtained by solving (3.18) with respect to a.

It follows from (3.18) that if failure is defines by crack growth beyond a critical size a_C, the following equation is valid at failure under constant amplitude loading

$$NS^m = \frac{\Psi(a_C)}{C_1} = K \tag{3.19}$$

where K is a constant independent of the loading. This relation is in agreement with the S-N curves generally applied in fatigue calculations.

One way to define a damage index D in terms of the crack size is

$$D = \frac{\Psi(a)}{\Psi(a_C)} \tag{3.20}$$

Using this definition it follows from (3.18) that damage increases linearly from zero to one with the number of stress cycles. It can further be shown that the damage increment in one stress cycle of range S_i is $1/N(S_i)$, where $N(S_i)$ is the number of cycles to failure under constant amplitude loading. Damage accumulation is thus in agreement with Miner's rule and the S-N approach and fracture mechanics approach are very similar.

In the presentation is has so far been assumed that a crack is present at the time the loading is applied. With a crack initiation period before the crack reaches a size a_0 for which fracture mechanics can be applied, the solution to (3.4) is

$$\Psi(a) = \int_{a_0}^{a} \frac{C_2(x)}{Y(x,Y)^m(\sqrt{\pi x})^m}dx = \begin{cases} C_1 S^m(N-N_0), & \textit{constant amplitude loading} \\ C_1 \sum\limits_{r=N_0}^{N} S_r^m, & \textit{variable amplitude loading} \end{cases} \tag{3.21}$$

where N_0 is the (random) crack initiation period for which a separate model can be formulated.

Let the failure criterion be taken as exceedence of a critical crack size a_C in a time period with N stress cycles,

$$a_C - a_N \leq 0 \tag{3.22}$$

$\Psi(a)$ is monotonically increasing and the failure criterion can be written as

$$\Psi(a_C)-\Psi(a_N) = \int_{a_0}^{a_C} \frac{C_2(x)}{Y(x,Y)^m(\sqrt{\pi x})^m}dx - C_1 S^m N \leq 0 \tag{3.23}$$

for constant amplitude loading. The safety margin M is defined as

$$M = \int_{a_0}^{a_C} \frac{C_2(x)}{Y(x,Y)^m(\sqrt{\pi x})^m}dx - C_1 S^m N \tag{3.24}$$

For a Weibull distribution of stress ranges as in (2.6) the safety margin becomes

$$M = \int_{a_0}^{a_C} \frac{C_2(x)}{Y(x,Y)^m(\sqrt{\pi x})^m}dx - C_1 N_T A^m \Gamma(1+\frac{m}{B}) \tag{3.25}$$

For variable amplitude loading it is also of interest to study the effect of a non-zero threshold value for ΔK in (3.2), which becomes

$$\frac{da}{dN} = C(\Delta K)^m, \quad \Delta K > \Delta K_{thr} \tag{3.26}$$

Eq. (3.4) is then replaced by

$$\frac{da}{Y(a,Y)^m(\sqrt{\pi a})^m} = C(\Delta\sigma)^m 1(\Delta\sigma > \frac{\Delta K_{thr}}{Y(a,Y)\sqrt{\pi a}})\, dN, \quad a(0) = a_0 \tag{3.27}$$

where 1 denotes the indicator function. Replacing as an approximation

$$(\Delta\sigma)^m 1(\Delta\sigma > \frac{\Delta K_{thr}}{Y(a,Y)\sqrt{\pi a}})$$

by its expected value yields,[21]

$$\frac{da}{Y(a,Y)^m(\sqrt{\pi a})^m G(a)} = C\, E[(\Delta\sigma)^m]dN, \quad a(0) = a_0 \tag{3.28}$$

The reduction factor $G(a)$, $0 \leq G(a) \leq 1$ depends on the long term stress range distribution as $(S = \Delta\sigma)$

For Weibull distributed stress ranges this integration yields

$$G(a) = \frac{\int_{S_0}^{\infty} S^m f_S(S) dS}{\int_0^{\infty} S^m f_S(S) dS}, \quad S_0 = \frac{\Delta K_{thr}}{Y(a,Y)\sqrt{\pi a}} \tag{3.29}$$

$$G(a) = \frac{\Gamma\left[1+\frac{m}{B} \; ; \; \left(\frac{\Delta K_{thr}}{AY(a,Y)\sqrt{\pi a}}\right)^B\right]}{\Gamma\left[1+\frac{m}{B}\right]} \tag{3.30}$$

Example 3 - Centre Cracked Panel

Consider a panel with a centre crack as in the experiments of [16], see Figure 3.5.

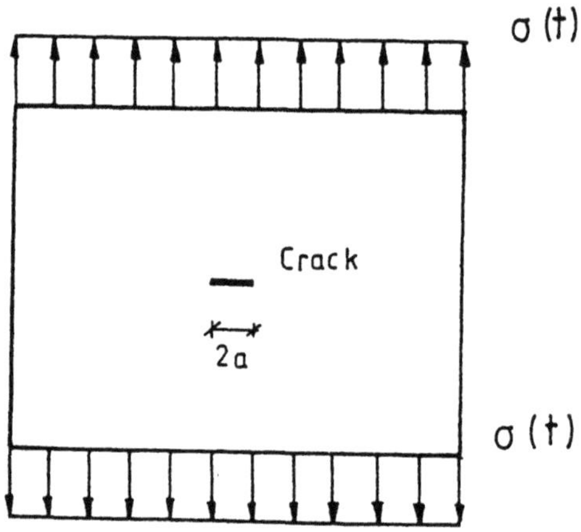

Figure 3.5 Centre cracked panel

The loading is a constant amplitude loading leading to a far-field stress range S. The geometry function is modelled as

$$Y(a,Y) = \exp(Y_1(\frac{a}{50})^{Y_2})$$

The geometry function takes the value one for $a=0$. Lengths are measured in mm and stresses in N/mm^2. The distribution of the basic variables is rather arbitrarily taken as

$$\begin{gathered} S \in N(60,10^2) \\ Y_1 \in LN(1,0.2^2) \\ Y_2 \in LN(2,0.1^2) \\ a_0 \in EX(1) \\ a_C \in N(50,10^2) \\ (lnC_1,m) \in N_2(-33.00,0.47^2,3.5,0.3^2;-0.9) \end{gathered}$$

$N(\mu,\sigma^2)$ denotes a normal distribution with mean value μ and variance σ^2. Similarly LN(μ,σ^2) denotes a log-normal distribution with mean value μ and variance σ^2. $N_2(\mu_1,\sigma_1^2,\mu_2,\sigma_2^2;\rho)$ denotes a bi-normal distribution with mean values μ_1 and μ_2, variances σ_1^2 and σ_2^2 and correlation coefficient ρ. $EX(\mu)$ denotes an exponential distribution with mean value μ. Statistics for $C_2(a)$ are taken as those reported in [18]. The example has eight basic variables which are transformed into standardised and independent normal variables.

The FORM and SORM approximations to the reliability index are shown in Figure 3.6 for various life times expressed in terms for the number of stress cycles N. The two approximations are close implying that the curvatures of the limit state surface are moderate at the design point. Statistics for the distribution of lift time T can be directly approximated from the results of Figure 3.5. For the mean life time the approximation is

$$E[T] = \int_0^\infty (1-P(T\le t))dt \approx \int_0^\infty \Phi(\beta(t))dt$$

For $N=1.5\cdot 10^6$ stress cycles the reliability index is $\beta=1.817$ and the α's are shown in Table 3.1.

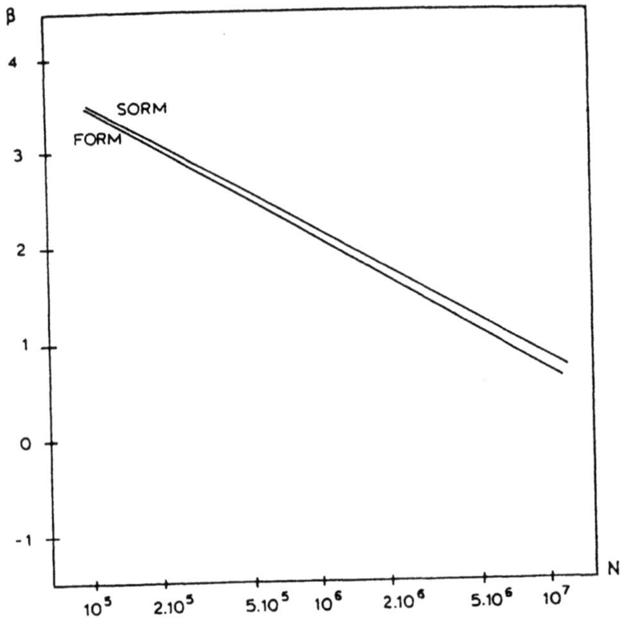

Figure 3.6 FORM and SORM reliability index design calculation

α_i^2 is interpreted as the fraction of the total uncertainty due to uncertainty arising from the ith basic variable. The major contribution to the overall uncertainty is thus in this case from the uncertainty in the material parameters. The critical crack size uncertainty is of little relative importance in this case, and the same is concluded in almost all cases where the critical crack size is significantly larger than the initial crack size. The uncertainty in the geometry function contributes very little to the total uncertainty in this case. This is because the value for $a=0$ is completely known. The uncertainty contribution from the uncertainty in the change in the geometry function from the initial value is generally found to be low. For tubular joints, where the geometry function is approximately proportional to $a^{-1/2}$ for large values of a, this statement may, however, not be true in all cases. The uncertainty contribution from the inherent spatial variability in the material is small and is generally neglected. Some experimental results for welded details are available, and these results also demonstrate that the uncertainty in C_1 is dominating over the inherent spatial uncertainty.

Table 3.1 Sensitivity Factors
$N = 1.5 \cdot 10^6$, $\beta = 1.817$

Variable	α_i	α_i^2
S	0.3577	13%
Y_1	0.0085	0%
Y_2	-0.0060	0%
a_0	0.5514	30%
a_C	-0.0001	0%
m	-0.6141	38%
$C_1 \| m$	0.4362	19%
$\Psi(a_C) \| a_0, a_C, m, Y$	-0.0248	0%

4. RELIABILITY UPDATING BASED ON INSPECTION RESULTS

Structures in service are often inspected to detect cracks before they become critical. Let a crack be detected after n_j stress cycles and its length measured as

$$a(n_j) = A_j \tag{4.1}$$

A_j is generally random due to measurement error and/or due to uncertainties in the interpretation of a measured signal as a crack length. Measurements of the type (4.1) can be envisaged for several times corresponding to different values of n_j. For each measurement (4.1) an event margin can be defined as, [22] and [23]

$$H_j = \int_{a_0}^{A_j} \frac{da}{Y(a,Y)^m(\sqrt{\pi a})^m} - C\,S^m n_j = 0, \quad j=1,2,\cdots,s \tag{4.2}$$

These event margins are zero due to (4.1).

A second type of inspection result is that no crack is detected. For an inspection at a time corresponding to n_i stress cycles this implies

$$a(n_i) \leq A_{di} \tag{4.3}$$

expressing that the crack length is smaller than the smallest detectable crack length A_{di}. A_{di} is generally random since a detectable crack is only detected with a certain probability depending on the crack length and on the inspection method. The distribution of A_{di} is the distribution of the length of non-detected cracks. This distribution is provided through the probability of detection curves (pod curves) for which experimental results exist for various inspection methods. Figure 4.1 shows experimental data and a pod curve for magnetic particle inspection (MPI).

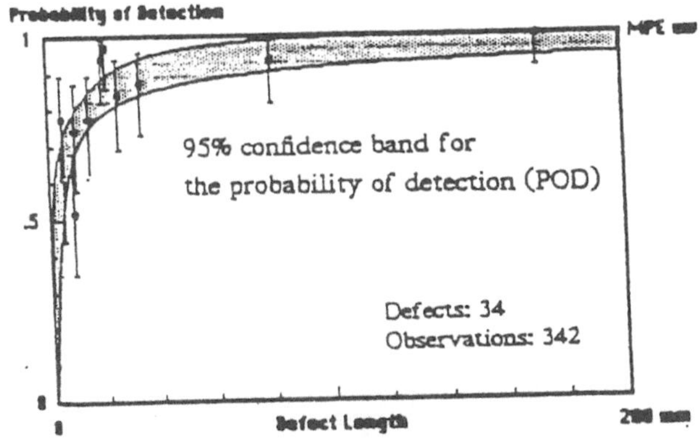

Figure 4.1 Inspection reliability for MPI

Information of the type (4.3) can also be envisaged for several times. If A_{di} is deterministic, however, and the same for all inspections, the information in the latest observation contains all the information of the previous ones. For each measurement (4.3) an event margin M_i can be defined as, [22] and [23]

$$H_j = C\, S^m n_i - \int_{a_0}^{A_{di}} \frac{da}{Y(a,Y)^m (\sqrt{\pi a})^m} \leq 0, \quad i=1,2,\cdots,r \tag{4.4}$$

These event margins are negative due to (4.3).

With one inspection result of the type (4.3) the updated failure probability is

$$P(M\leq 0 \,|\, H\leq 0) = \frac{P(M\leq 0 \cap H\leq 0)}{P(H\leq 0)} \tag{4.5}$$

Evaluation of the reliability of a parallel system (numerator) and a component (denominator) are thus required. A FORM or SORM analysis can be directly applied.

With one inspection result of the type (4.1) the updated failure probability is

$$P(M\le 0|H=0) = \frac{\frac{\partial}{\partial x}P(M\le 0 \cap H(x)\le 0)}{\frac{\partial}{\partial x}P(H(x)\le 0)} \tag{4.6}$$

where the derivatives are computed for $x=0$ and $H(x)$ is defined from (4.2) as

$$H(x) = \int_{a_0}^{A_j+x} \frac{da}{Y(a,Y)^m(\sqrt{\pi a})^m} - C\, S^m n_j \tag{4.7}$$

An evaluation of the sensitivity factor for a parallel system (numerator and a component (denominator)) are thus required and a FORM or SORM analysis can be directly applied. The analysis is easily generalised to simultaneous consideration of several inspection results.

The interest is now on updating after repair and it is assumed that a repair takes place after n_{rep} stress cycles when a crack length a_{rep} is observed. An event margin H_{rep} is defined as

$$H_{rep} = \int_{a_0}^{a_{rep}} \frac{da}{Y(a,Y)^m(\sqrt{\pi a})^m} - C\, S^m n_{rep} = 0 \tag{4.8}$$

The crack length present after repair and a possible inspection is a random variable a_{new} and the material properties after repair are m_{new} and C_{new}. The safety margin after repair is M_{new}

$$M_{new} = \int_{a_{new}}^{a_C} \frac{da}{Y(a,Y)^{m_{new}}(\sqrt{\pi x})^{m_{new}}} - C_{new} S^{m_{new}}(n-n_{rep}) \tag{4.9}$$

and the updated failure probability is

$$P(M_{new}\le 0|H_{rep}=0) = \frac{\frac{\partial}{\partial x}P(M_{new}\le 0 \cap H_{rep}(x)\le 0)}{\frac{\partial}{\partial x}P(H_{rep}(x)\le 0)} \tag{4.10}$$

where the derivatives are computed for $x=0$ and $H_{\text{rep}}(\text{x})$ is

$$H_{rep}(x) = \int_{a_0}^{a_{rep}+x} \frac{da}{Y(a,Y)^m(\sqrt{\pi a})^m} - C\, S^m n_{rep} \qquad (4.11)$$

Example 4 - Inspection of Centre Cracked Panel

The analysis in Example 3 is continued. First, the situation where a crack is found in the first inspection is considered. It is envisaged that the inspection is carried out after $N_1 = 10^5$ stress cycles and a crack length of 3.9 *mm* is measured. The measurement error is assumed to be normally distributed with standard deviation σ_A. Figure 4.2 shows the updated reliability index as a function of σ_A. The result is almost independent of σ_A in this example as the uncertainty in the initial crack size is dominating the uncertainty in A_1.

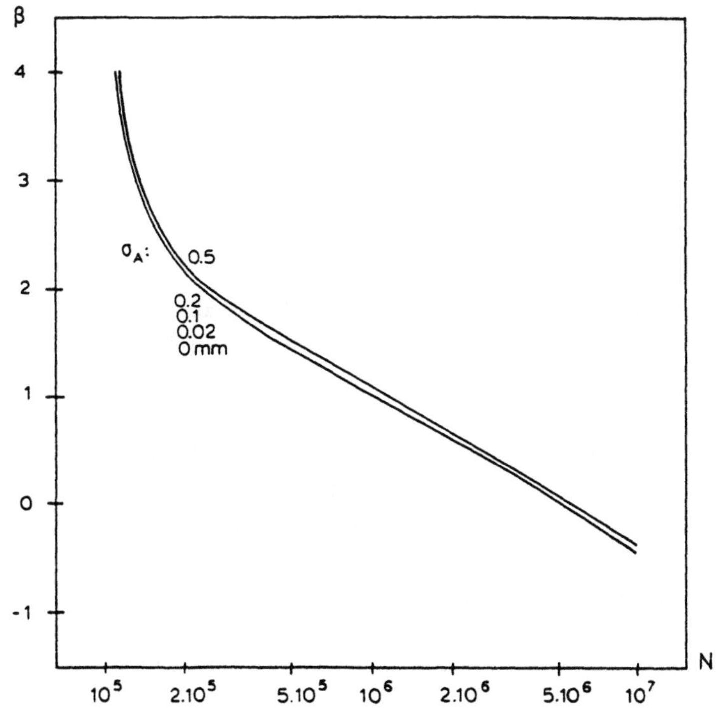

Figure 4.2 Updated first-order reliability index after first inspection with crack measurement 3.9 *mm*.

When the crack is detected, a decision has to be made and two options are present. It may be decided to repair the crack immediately or to leave the crack as it is and base a decision on repair on more inspection results. With just one inspection it is not possible to determine if the crack was initially large but grows slowly enough that repair is not needed, or the crack was initially fairly small but is growing fast and must be repaired. If a requirement on the reliability index in a period without inspections is formulated, e.g., $\beta \geq 2$, the latest time of the next inspection is determined from Figure 4.2.

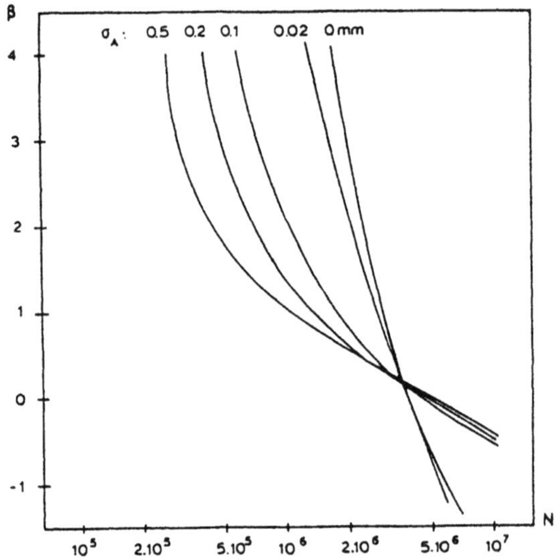

Figure 4.3 Updated first-order reliability index after second inspection with crack measurements 3.9 *mm* and 4.0 *mm*.

Assume that the crack is not repaired but a second inspection at $N=2 \cdot 10^5$ stress cycles is required. Let the inspection method be the same as in the first inspection and let the measured crack size be 4.0 *mm*. The measurement error is again assumed to be normally distributed with standard deviation σ_A and the two measurement errors are assumed to be statistically independent. Figure 4.3 shows the updated reliability index after this second inspection. Different inspection qualities now lead to very different results. With $\sigma_A=0$ the negative slope of the reliability index curve becomes very large demonstrating that the crack growth behaviour is basically determined by two combinations of the basic variables. With a large measurement uncertainty there is an immediate and large increase in reliability, but after some time the curve becomes almost identical to the curve resulting after the first inspection. Due to large uncertainty in both inspections only little information is gained on the crack growth rate. If the inspection quality is very high it may be possible to state that the crack does not grow to a critical size within the design life time. Repair and further inspections are then unnecessary. For a poorer inspection quality a time period until the next inspection can be determined and the decision on repair be further delayed.

Figure 4.4 shows the results of Figure 4.3 together with similar results for a homogeneous material. It is observed that only for very small inspection uncertainty does the material inhomogeneity significantly affect results. The estimates for material inhomogeneity used in this example are for base material and the conclusion may be somewhat different for crack growth in weld material or in base material in a heat affected zone.

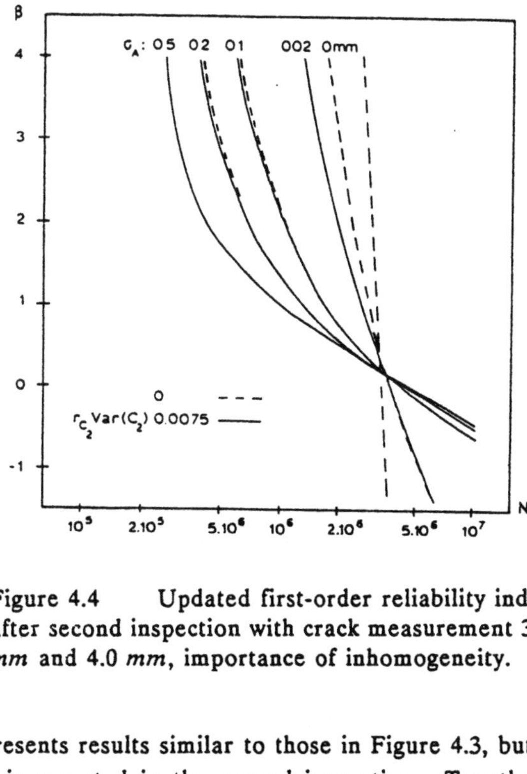

Figure 4.4 Updated first-order reliability index after second inspection with crack measurement 3.9 *mm* and 4.0 *mm*, importance of inhomogeneity.

Figure 4.5 presents results similar to those in Figure 4.3, but for the case where a crack size of 5 *mm* is reported in the second inspection. Together, the two inspection results now indicate that a large and fast growing crack is present. Repair is therefore necessary within a short period.

Consider now different situations where the inspections do not result in crack detection. An attempt is made to illustrate possible means to achieve a required reliability. Let the reliability requirement be $\beta \geq 3.0$ and let the design life time correspond to $1.5 \cdot 10^6$ stress cycles. Figure 4.6 shows the reliability index as a function of number of stress cycles for two plate thicknesses. With a plate thickness t the reliability requirement is fulfilled for the design life time and no inspections are needed. With a plate thickness of only 60% of t the reliability requirement is fulfilled for the period until $N = 2 \cdot 10^5$ stress cycles, where an inspection is needed. The quality of the inspection is reflected in the distribution of non-detected cracks. An exponential distribution is assumed with a mean value λ. Cracks initially present are cracks which have passed the inspection at the production site either because they were not detected or because they were below the acceptance level. If no cracks were accepted in fabrication, the fabrication inspection therefore corresponds to $\lambda = 1$.

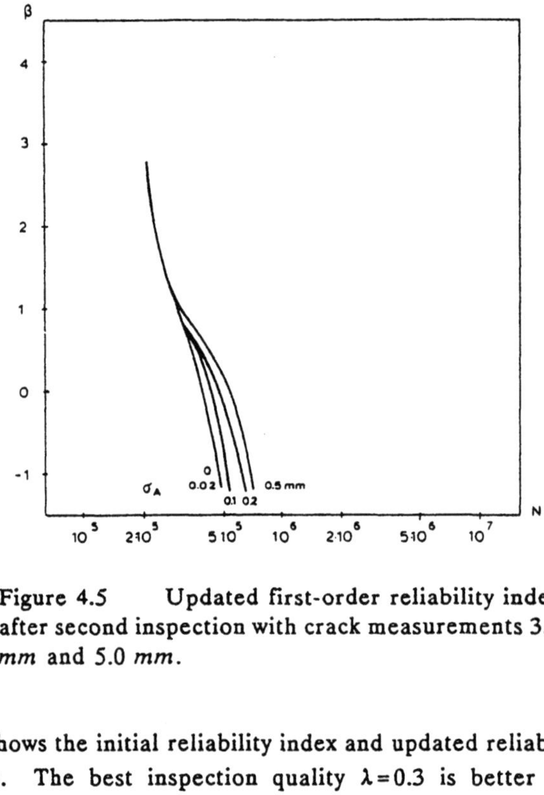

Figure 4.5 Updated first-order reliability index after second inspection with crack measurements 3.9 *mm* and 5.0 *mm*.

Figure 4.7 shows the initial reliability index and updated reliability indices for three inspection qualities. The best inspection quality $\lambda = 0.3$ is better than the fabrication inspection quality and if no crack is found with this method the increase in reliability is sufficient to make further inspections unnecessary. For the two other inspection qualities, periods are determined until the next inspection.

Figure 4.8 shows the total inspection requirement for $\lambda = 1$ when no crack is detected in any inspection. For this case two inspections are needed. Finally, Figure 4.9 shows the total inspection requirement for $\lambda = 3$ when no crack is detected in any inspection, and for this case five inspections are needed. It is thus demonstrated that different strategies on design and inspection planning can be used to achieve a required reliability. Based on costs of each strategy including expected failure costs a cost optimal solution can be determined.

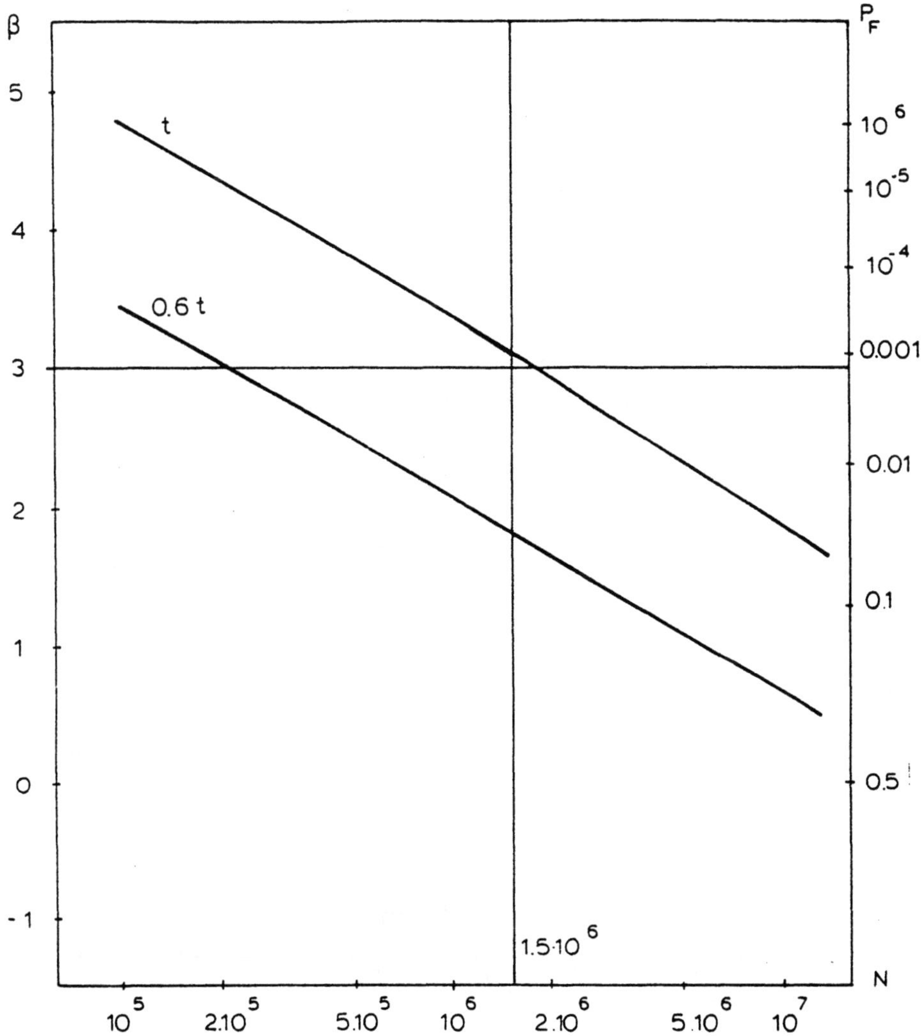

Figure 4.6 First-order reliability index for two plate thicknesses.

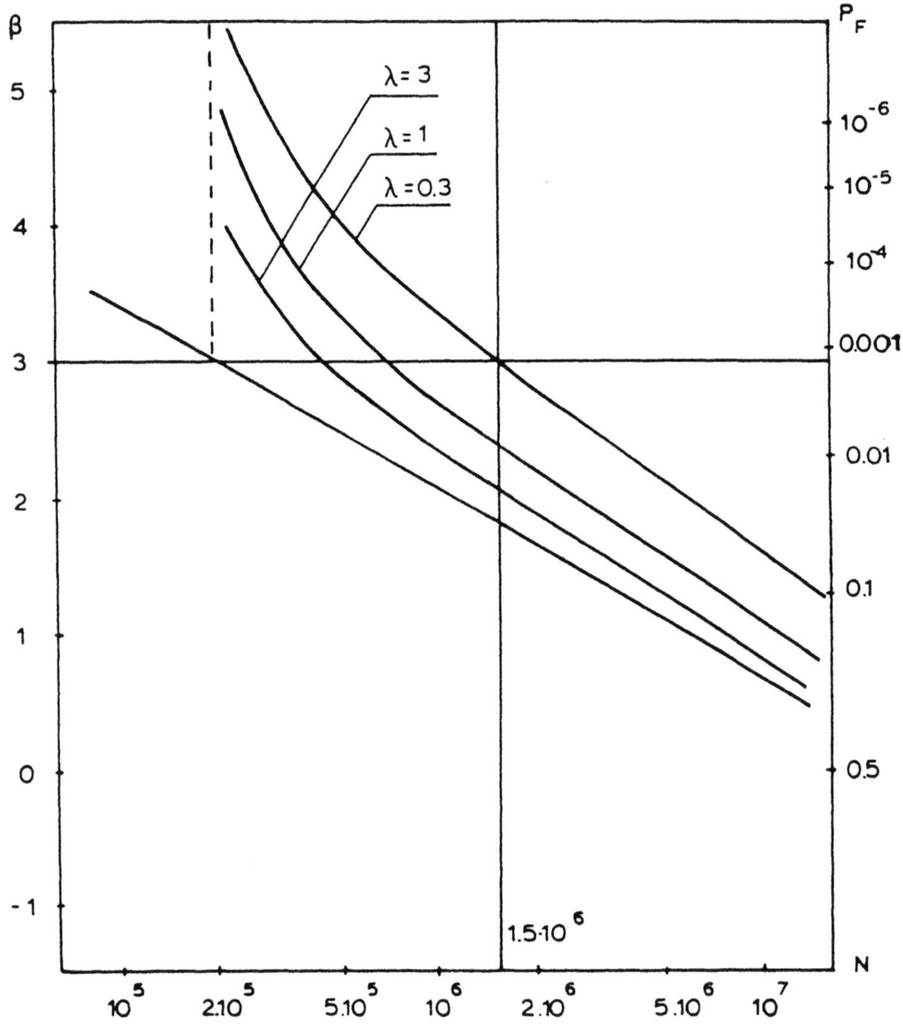

Figure 4.7 Updated first-order reliability index after first inspection with no crack detection.

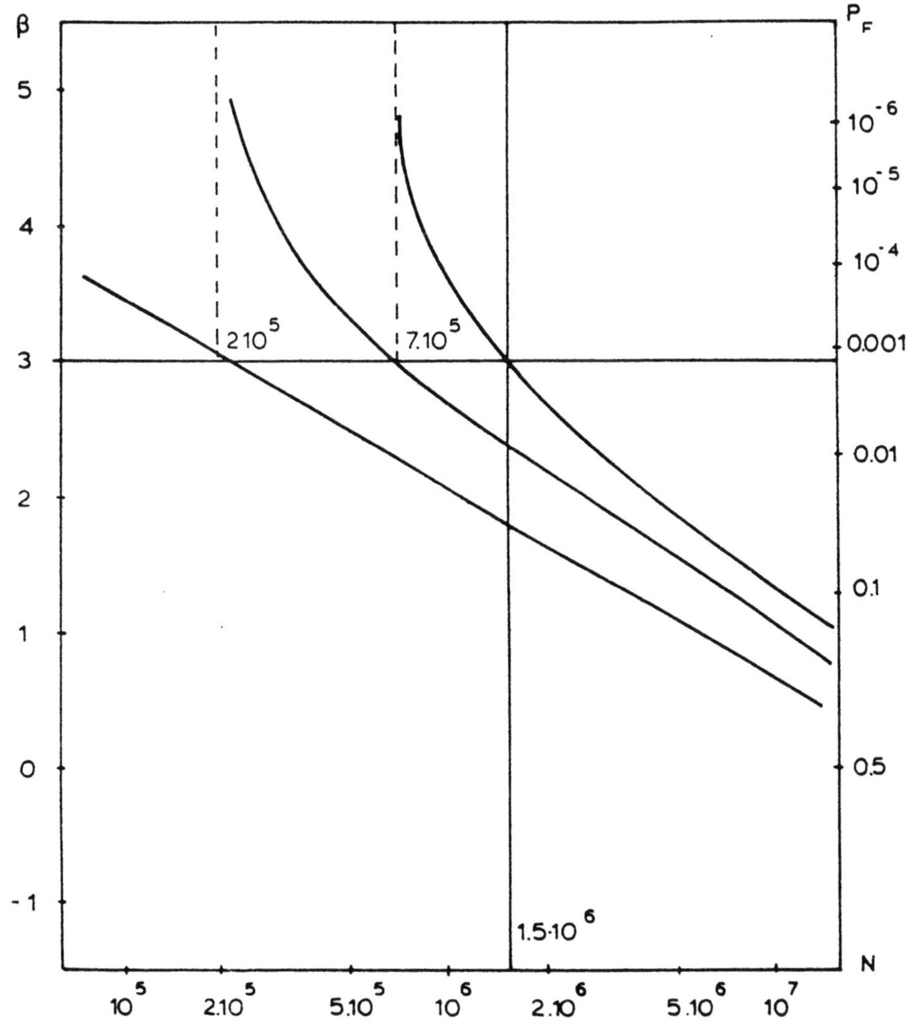

Figure 4.8 Updated first-order reliability index after inspections with no crack detection, mean size of non-detected cracks 1 *mm*.

The results of a reliability analysis following a repair of a detected crack is illustrated in Figure 4.10. It is assumed that a crack size of $a_{rep}=8$ *mm* is repaired after $N_{rep}=2{\cdot}10^5$ stress cycles. The distribution of the initial crack size after repair a_{new} is taken as an exponential distribution with a mean value of 1 *mm*, i.e., as the same initial distribution as after fabrication. Two situations are considered with either identical or independent material properties before and after repair. When independent properties are assumed the same distribution is used for the properties before and after repair. If follows from the results that there is an immediate increase in reliability after repair, but the

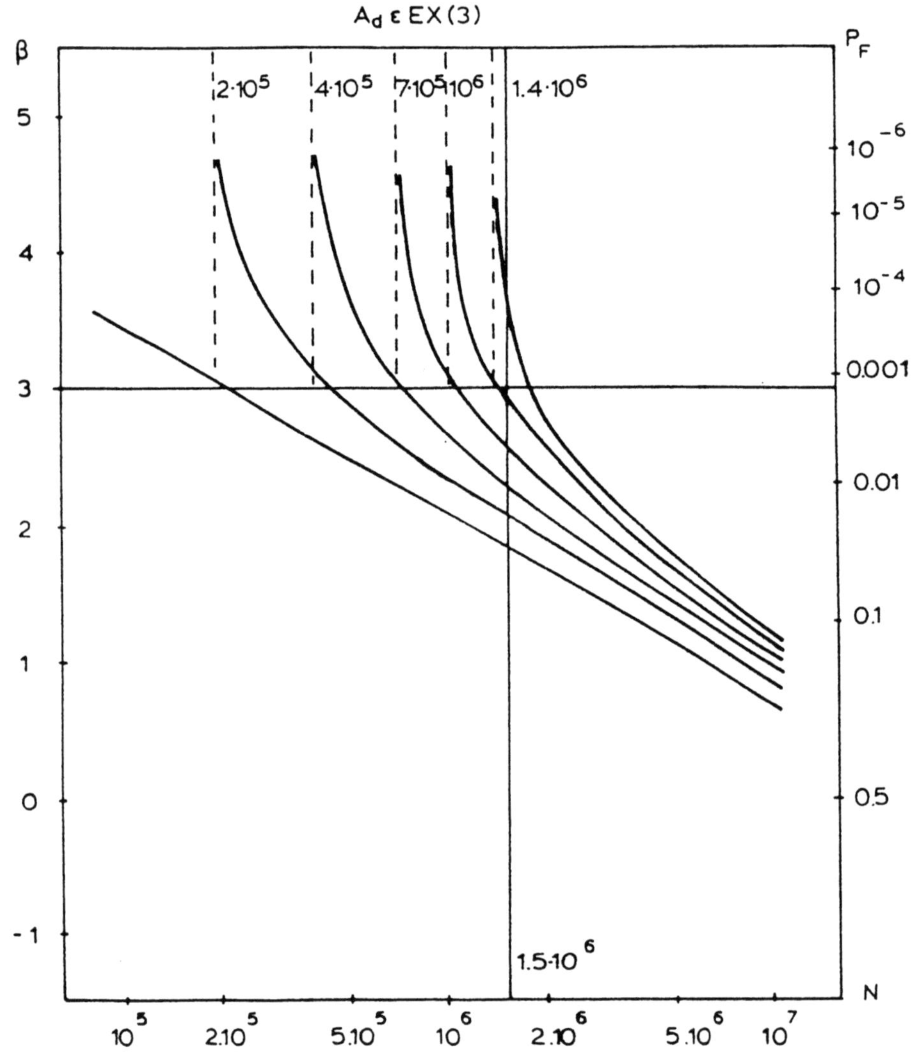

Figure 4.9 Updated first-order reliability index after inspections with no crack detection, mean size of non-detected cracks 3 *mm*.

reliability quickly drops to a level below the level obtained for the calculations before repair. This reflects the possibility that the cause for the large repaired crack size is a larger than anticipated loading of the crack tip, which is also acting after the repair.

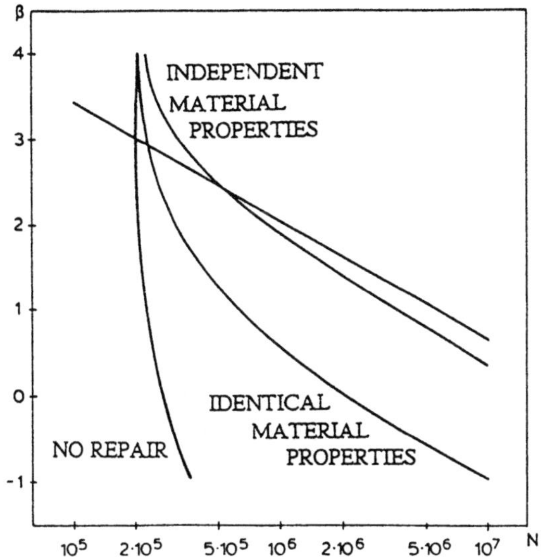

Figure 4.10 Updated first-order reliability index after repair of an 8 *mm* crack at $N=2{\cdot}10^5$ stress cycles.

Example 5 - T-joint in an offshore structure

A probabilistic fatigue analysis of a hot spot in a T-joint in an offshore steel jacket in shallow water is performed, [24]. First, a probabilistic *S-N* curve based fatigue analysis is carried out as outlined in [10]. Uncertainties in the long term environmental data, the load model coefficients, the global response analysis, the local stress analysis, the *S-N* curve, and the value of Miner's sum at failure are included. The T-curve from the Department of Energy shown in Figure 2.1 is applied, however, without a change in slope at $N=10^7$ stress cycles. The life time determined from a deterministic code is 22 years, when a design *S-N* curve (mean-2 standard deviations) is used with a Miner sum of 1 at failure. In Figure 4.11 the curve marked PROFAT gives the reliability index β as a function of years in service. The reliability index is defined in terms of the failure probability as

$$\beta = -\Phi^{-1}(P_F)$$

The reliability index is 1.4 for the deterministic life time 22 years. For this hot spot and the applied uncertainty models, the code thus implicitly gives a level of reliability corresponding to $\beta=1.4$. The same level was found for the deterministic life time of another hot spot. With an inspection strategy requiring inspections every fourth year, the reliability level at the end of the first inspection period is 3.75 ($P_F=10^{-4}$). This reliability index level agrees well with levels used in code calibrations for other failure modes.

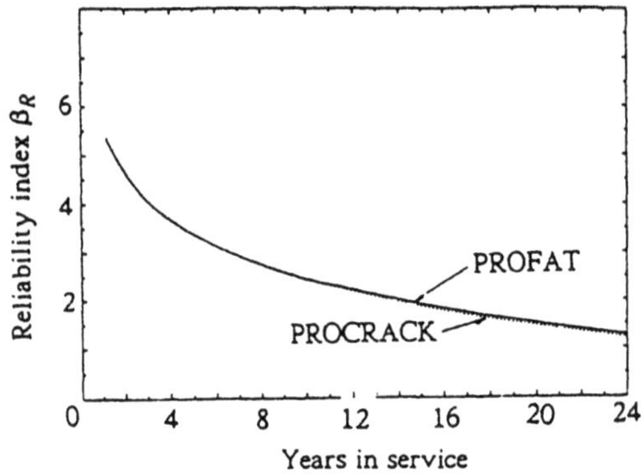

Figure 4.11 Fatigue reliability for *S-N* curve and fracture mechanics based analysis.

Next, the probabilistic fatigue analysis is performed by the fracture mechanics approach described in Section 3. A constant aspect ratio $a/c = 1/15$ is used, and crack growth is thus described in terms of the crack depth a. For the initial crack size a_0 an exponential distribution with a mean value of 0.11 *mm* is used, based on measurements reported in [25]. The critical crack size corresponds to crack growth through the thickness. A ratio of 4:1 between bending and axial stress is used, and the geometry function is shown in Figure 4.12. The geometry function is the product of the function for plates and a magnification factor due to the presence of the weld. The stress analysis is identical to the stress analysis in the *S-N* curve based fatigue analysis. Uncertainties in both the plate formulas and the magnification factor are modeled. The slope parameter in the crack growth equation (3.2) is taken as $m = 3$ and statistics for C are selected according to suggested values in the literature. The results of the fracture mechanics analysis are marked PROCRACK in Figure 4.11 and agree very well with the PROFAT results.

For a life time of 4 years the importance of the different sources of uncertainty are compared in Table 4.1.

The uncertainty modelling in the steps leading to the hot spot stress distribution involves a rather large number of random variables. To facilitate the updating calculations a two parameter long term stress range distribution is calibrated. A Weibull distribution as in (2.6) with random distribution parameters A and B is selected. Figure 4.13 shows a comparison between results for the probabilistic fracture mechanics analysis with the original stress model (marked PROCRACK) and the simplified model (marked PROINSP). A very close agreement is observed.

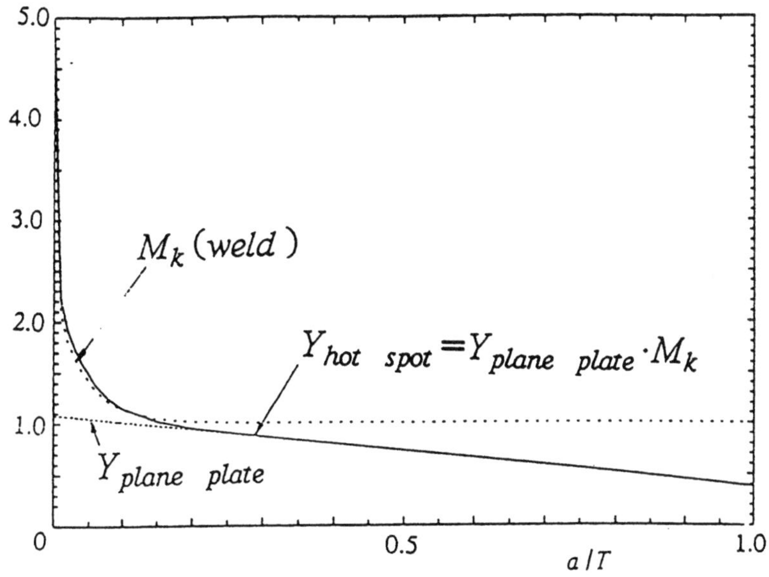

Figure 4.12 Geometry function.

Table 4.1: Importance Factors T = 4 years, β_R = 3.75	
Source of uncertainty	Importance α_i^2
Environmental description	1 %
Load model coefficients	19 %
Stress analysis	20 %
Stress intensity factor	10 %
Crack growth parameter	50 %

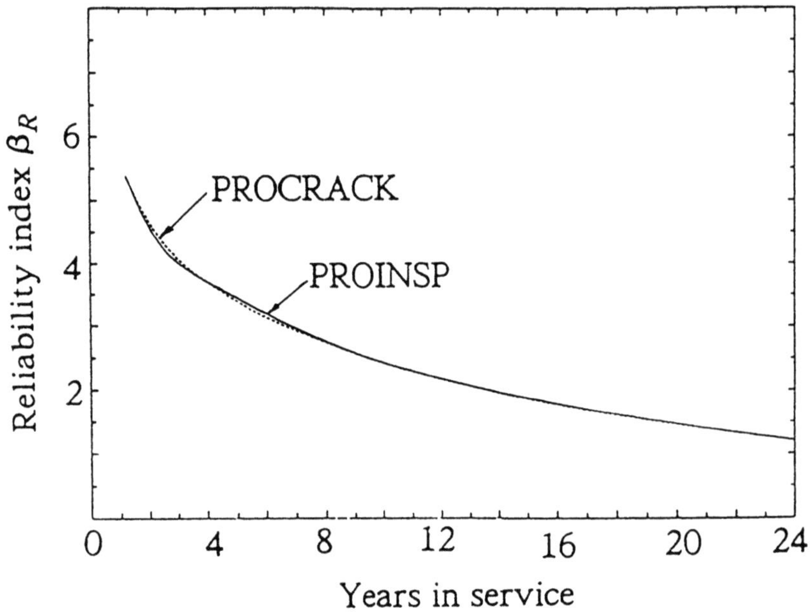

Figure 4.13 Fatigue reliability for fracture mechanics based analysis with original stress model and simplified model.

Figure 4.14 Updated fatigue reliability for fracture mechanics based analysis with MPI at regular intervals.

The effect of updating based on inspection results is now studied. Only inspections in which no cracks are detected are considered. Probability of detection (pod) curves are sparse, but the following relation has been found appropriate for visual inspection and magnetic particle inspection (MPI)

$$P(a) = \begin{cases} 1 - \exp(-a/1.3mm), & \text{Underwater MPI} \\ 1 - \exp(-a/6.5mm), & \text{Underwater visual inspection} \end{cases}$$

With the aspect ratio $a/c = 1/15$, this gives a 90% chance of detecting a 45 mm long crack by MPI and a 90% chance of detecting a 225 *mm* long crack by visual inspection. To include variations from inspector to inspector in the pod curves, the constants 1.3, respectively 6.5 mm could be replaced by random variables. The reliability analysis is not further complicated by this additional randomness.

The effect of MPI on the reliability is shown in Figure 4.14 for an inspection strategy with inspection every 4th year. It is observed that the reliability level is above $\beta = 3.75$ throughout the life time of 20 years and that the reliability level immediately before an inspection increases throughout the life time. Figure 4.15 shows the necessary amount of inspection to achieve a level above $\beta = 3.75$. It is observed that only two inspections after 4 years and 10 years are needed. A potential reduction of the amount of MPI of 50% is thus present.

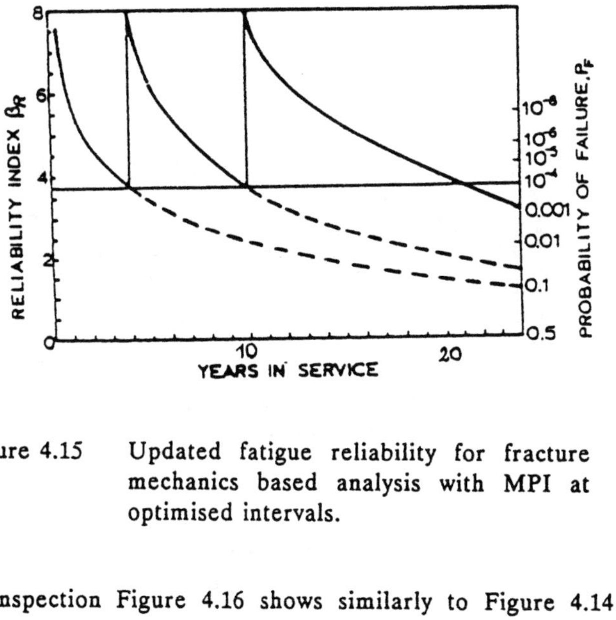

Figure 4.15 Updated fatigue reliability for fracture mechanics based analysis with MPI at optimised intervals.

For visual inspection Figure 4.16 shows similarly to Figure 4.14 for MPI the reliability for an inspection strategy with inspections every 4th year. It is again assumed that no cracks are detected. The visual inspection is not able to maintain a reliability level above $\beta = 3.75$ throughout the life time, although the reliability level immediately before an inspection increases also in this case. To maintain a reliability level larger than 3.75 thus requires more than 4 inspections, but can be obtained by 5 inspections as shown in Figure 4.17.

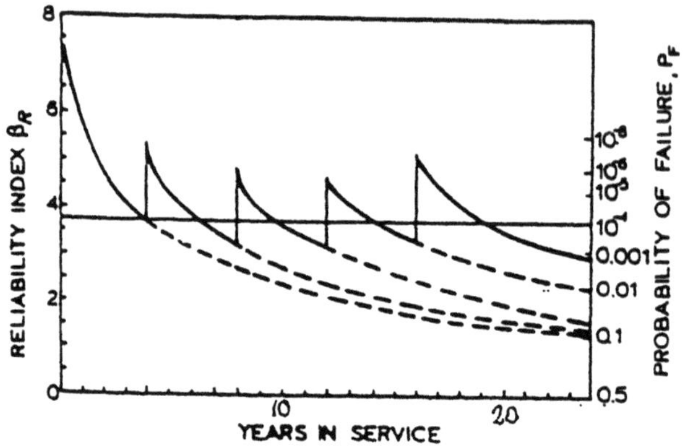

Figure 4.16 Updated fatigue reliability for fracture mechanics based analysis with visual inspection at regular intervals.

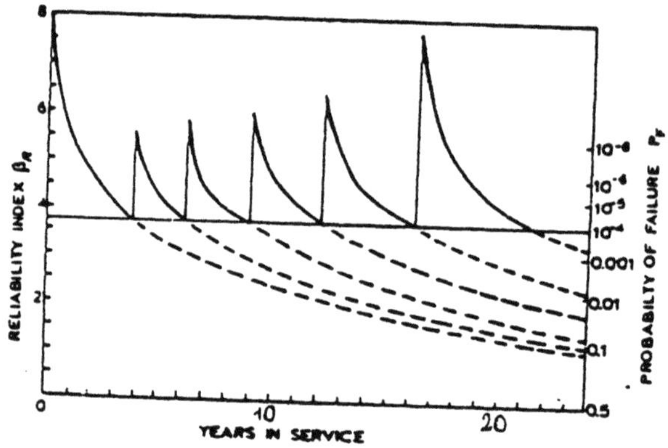

Figure 4.17 Updated fatigue reliability for fracture mechanics based analysis with visual inspection at optimised intervals.

5. PROBABILITY BASED OPTIMISATION OF DESIGN, INSPECTION AND MAINTENANCE

Reliability against fatigue crack growth is achieved through efforts in design, inspection and repair or replacement. These efforts all introduce cost and the minimum total expected cost solution is of interest. An optimisation problem can be formulated which can further include a constraint on the smallest allowable reliability, e.g. as specified by a regulatory body. Contributions to the formulation and solution of this optimisation problem can be found in e.g. [26] and [27].

This section presents a consistent formulation of cost optimal fatigue design, inspection and repair. Reliability and sensitivity calculations are performed by a first-order reliability method and the optimisation is carried out by a general non-linear optimisation algorithm.

A one-dimensional description of the crack size a is employed. A Weibull distribution with random distribution parameters A and B is used for the distribution of the stress ranges S. This is a relevant choice for the long term distribution for offshore jacket structures. The stress ranges depend on the structural optimisation parameters. The number of stress cycles per unit time is υ, and the safety margin M for failure - defined as crack growth to a critical size a_C - before time t is,

$$M(t) = \int_{a_0}^{a_c} \frac{dx}{Y(x)^m (\pi x)^{m/2}} - C \nu t A^m \Gamma(1+\frac{m}{B}) \tag{5.1}$$

The first inspection at time T_1 leads to a crack detection or not crack detection. An event margin H is defined as

$$H = \int_{a_0}^{a_{d1}} \frac{dx}{Y(x)^m (\pi x)^{m/2}} - C \nu T_1 A^m \Gamma(1+\frac{m}{B}) \tag{5.2}$$

The event margin is negative when a crack is detected and is otherwise positive. Values for the smallest detectable crack size a_d for different inspections are assumed mutually independent. When a crack is detected and repaired at time T_1, the safety margin after repair is

$$M^1(t) = \int_{a_R}^{a_c} \frac{dx}{Y(x)^m (\pi x)^{m/2}} - C\nu (t-T_1) A^m \Gamma(1+\frac{m}{B}), \quad t>T_1 \tag{5.3}$$

The geometry function is identical before and after repair. The material parameter C is either assumed fully dependent or completely independent before and after repair. Crack sizes after repair a_R are assumed mutually independent. A notation is introduced to describe the sequence of repair/no repair events. With repair at times T_1 and T_2 and no repair at T_3, the safety margin for $T_3 < t \leq T_4$ is as an example

$$M^{110}(t) = \int_{a_R}^{a_c} \frac{dx}{Y(x)^m (\pi x)^{m/2}} - C\nu (t-T_2) A^m \Gamma(1+\frac{m}{B}), \quad T_3<t\leq T_4 \tag{5.4}$$

and the event margin for crack detection at time T_4 is similarly

$$H^{110} = \int_{a_R}^{a_d} \frac{dx}{Y(x)^m (\pi x)^{m/2}} - C\nu (T_4-T_2) A^m \Gamma(1+\frac{m}{B}) \tag{5.5}$$

The crack growth formulation and subsequent optimisation has in [28] been extended to include a possible positive threshold value for the stress intensity factor range in (3.1), a constant corrosion rate, and a crack initiation period.

The following strategy for repair is selected: **All detected cracks are repaired.** In [28] also other repair and replacement strategies are included. These are

- Detected cracks smaller than a threshold value are repaired by grinding, while larger detected cracks are repaired by welding

- Only detected cracks larger than a reference size are repaired (by welding)

- All detected cracks are "repaired" by replacement of the element.

With n inspections performed at times $T_1, \ldots, T_n$, the total number of different repair courses is 2^n, see Figure 5.1.

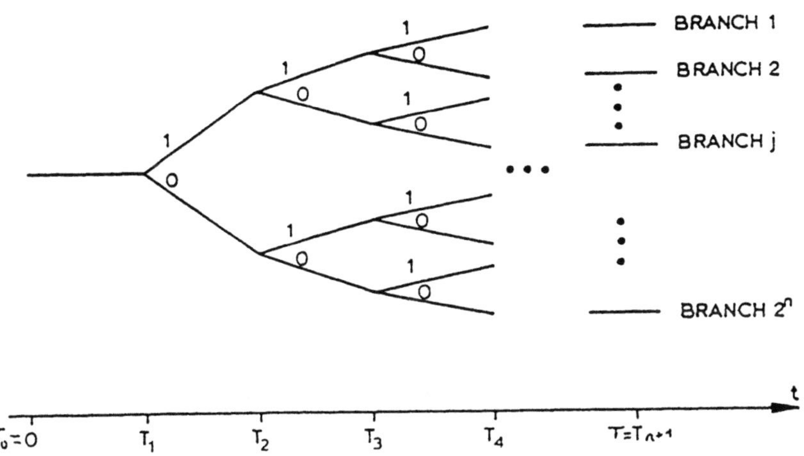

Figure 5.1 Repair realisations. 0 denotes no repair, while 1 denotes repair.

The failure probability before time t is $P_F(t)$. The corresponding reliability index is

$$\beta(t) = -\Phi^{-1}(P_F(t)) \tag{5.6}$$

In terms of the safety and event margins, failure before time t is as an example

$$\begin{aligned} P_F(t) = P(M(T_1)\leq 0) + P(M(T_1)>0 \cap H>0 \cap M^0(t)\leq 0) \\ + P(M(T_1)>0 \cap H\leq 0 \cap M^1(t)\leq 0), \quad T_1<t\leq T_2 \end{aligned} \tag{5.7}$$

and similarly for other inspection time intervals and the life time T. With n inspections between 0 and T, 2^{n+1}-1 parallel systems are analysed to compute the failure probabilities.

The expected number of repairs $E[R_i]$ at time T_i is identical to the probability of repair at time T_i. It is at time T_2

$$E[R_2] = P(M(T_1)>0 \cap H>0 \cap M^0(T_2)>0 \cap H^0 \le 0) + P(M(T_1)>0 \cap H \le 0 \cap M^1(T_2)>0 \cap H^1 \le 0) \quad (5.8)$$

and similarly for other inspection times. With n inspections between 0 and T, 2^n-1 parallel systems are analysed to compute the repair probabilities.

The inspection quality is defined by the *pod* (probability of detection) curve $p(a)$ for which an exponential form may be chosen for illustration.

$$p(a) = F_{a_d}(a) = 1 - \exp(\frac{a}{\lambda}), \quad a>0 \quad (5.9)$$

The pod curve is identical to the distribution function of the smallest detectable crack size a_d. The inspection quality is thus characterized by the parameter λ. λ can take values between 0 and ∞. In the optimisation an auxiliary measure of inspection quality q is introduced.

$$q = \frac{1}{\lambda} \quad (5.10)$$

q can take values in the interval $[0;\infty[$. $q=0$ corresponds to no inspection, while $q=\infty$ corresponds to a perfect inspection where all cracks are found.

The number of inspections n during the life time T is selected beforehand. This is done to avoid an optimisation with a mixture of integer and real valued optimisation variables. The analysis is repeated for several values of n and the resulting optimal costs are compared. The n-value with the smallest total expected cost is the optimal value. Inspection times and qualities are optimisation variables together with the structural design parameter vector z.

The following cost items are included: initial cost, $C_I = C_I(z)$, inspection cost, $C_{IN} = C_{IN}(q)$, cost of repair, C_R, and cost of failure, $C_F = C_F(t)$. Inspection and repair costs are assumed to increase with the rate of inflation. The difference between the desired rate of return and the rate of inflation is assumed constant r. The cost of failure may be a function of time.

The optimisation is formulated as a minimisation of the total expected cost, with a constraint on the reliability index for the life time and simple constraints on the optimisation parameters:

$$\min_{t,q,z} \quad C_I + \sum_{i=1}^{n} \left(C_{IN}(q_i)(1-P_F(T_i)) + C_R E[R_i]\right)\frac{1}{(1+r)^{T_i}} + \sum_{i=1}^{n+1} C_F(T_i)\,(P_F(T_i)-P_F(T_{i-1}))\frac{1}{(1+r)^{T_i}} \tag{5.11}$$

$$\beta(T) \geq \beta^{\min}$$

$$t^{\min} \leq t_i = T_i - T_{i-1} \leq t^{\max}, \quad i=1,2,\ldots,n$$

$$s.t. \quad 0 \leq T - \sum_{i=1}^{n} t_i \leq t^{\max}$$

$$q^{\min} \leq q_i \leq q^{\max}, \quad i=1,2,\ldots,n$$

$$z_i^{\min} \leq z_i \leq z_i^{\max}, \quad i=1,2,\ldots,k$$

The possibility of predetermining one or more of the inspection times and qualities as well as elements in z is available.

The optimisation problem is solved for each n using the NLPQL algorithm [29]. The value of the objective function and of the constraints are computed in a separate routine. This routine calls upon the reliability analysis program PROBAN [30] for analysis of 2^{n+1}-1 parallel systems for calculation of failure probabilities and 2^n-1 parallel systems for calculation of expected repair cost. PROBAN provides a reliability index calculated by a first-order reliability method [28] for each parallel system together with exact partial derivatives of the reliability index with respect to λ_i, T_i and z_j. From these partial derivatives, the partial derivatives with respect to q_i, t_i and z_j are easily derived. Possibilities are included to perform the optimisation with only the most important branches in the event tree in Figure 5.1 included.

A more general formulation than presented in Figure 5.1 has different inspection times and qualities in different branches. The number of optimisation variables is thereby increased drastically. To overcome this problem, a procedure can be chosen where the inspection plan is first optimized at the design stage. When the result of the first inspection is known, a new optimal inspection plan is determined applying this information in addition to the information available at the design stage.

The various failure probabilities and probabilities of repair are then conditional probabilities, conditioned upon the result of the first inspection. Actual crack measurement results can be considered at this stage. For each inspection result being available, the tree of possibilities in Figure 5.1 is reduced to one half of its size as the actual branch at the inspection time is known. With inspection results available at times T_1, ... ,T_{j-1}, the optimisation problem is formulated as

$$\min_{t,q} \cdot \sum_{i=1}^{n} (C_{IN}(q_i)(1-P_F(t_i)) + C_R E[R_i]) \frac{1}{(1+r)^{T_i}} + \sum_{i=j}^{n+1} C_F(T_i)\ (P_F(T_i)-P_F(T_{i-1})) \frac{1}{(1+r)^{T_i}} \quad (5.12)$$

$$\text{s.t.} \quad \begin{aligned} &\beta(T) \geq \beta^{\min} \\ &t^{\min} \leq t_i = T_i - T_{i-1} \leq t^{\max}, \quad i=j, \ldots, n \\ &0 \leq T - \sum_{i-1}^{n} t_i \leq t^{\max} \\ &q^{\min} \leq q_i \leq q^{\max}, \quad i=j, \ldots, n \end{aligned}$$

where failure and repair probabilities are computed conditioned upon the results of the first j-1 inspections.

Example 6 - Design and Maintenance Optimisation for Tubular Joint

The analysis of example 5 is continued. The selected input data are described in detail in [28]. Figure 5.2 shows the different cost items as a function of the number of inspections, and it shows the change in reliability index with time for the case of $n=4$ inspections. If the inspection at time T_1 does not result in a crack detection, a new optimisation is performed as described in the previous section. Figure 5.3 shows the change in reliability index with time for the new optimal solution and for the optimal solution determined with no crack detection in the first two inspections.

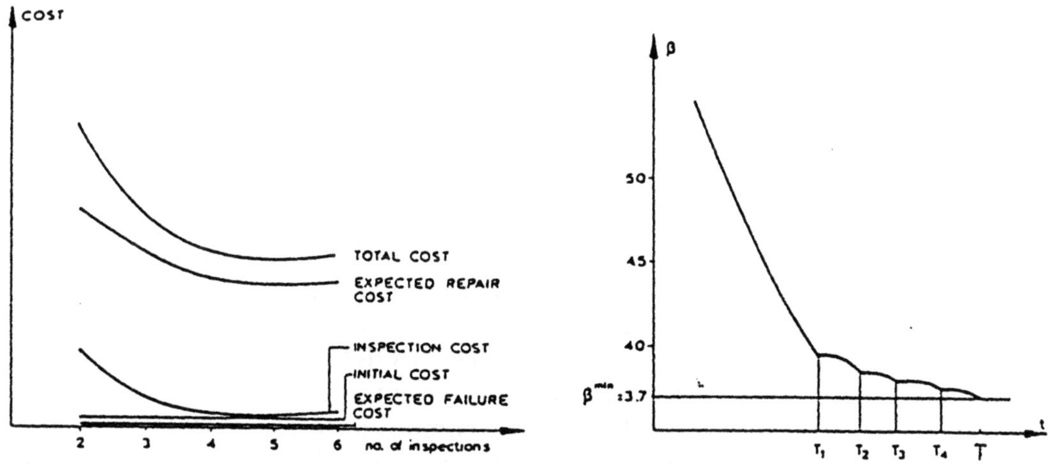

Figure 5.2 Cost functions at optimal solution; reliability index with time.

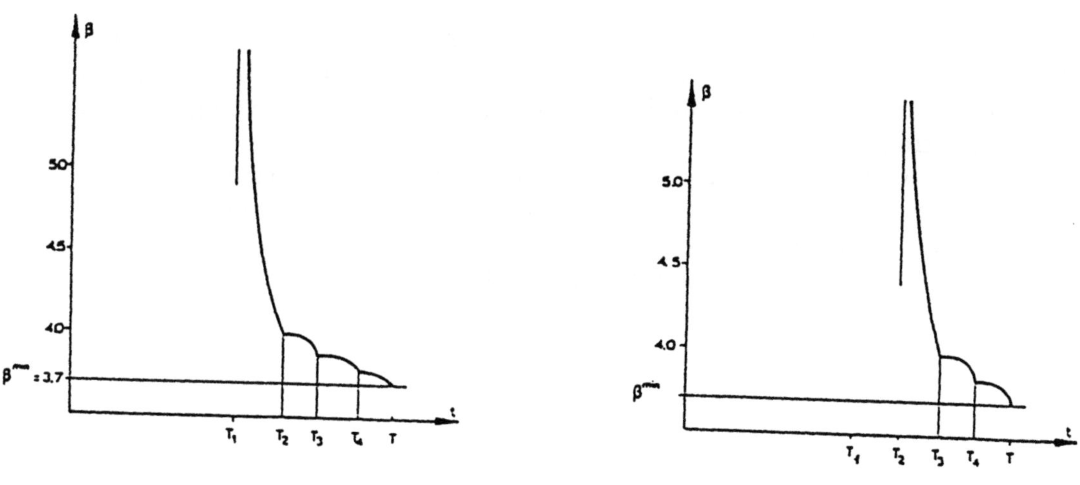

Figure 5.3 Reliability index with time in optimisation after inspections.

6. CONCLUSIONS

Reliability methods are useful and well accepted in fatigue analysis of marine structures since large uncertainties are generally present and because it is often relevant to combine uncertain design information with uncertain information obtained in service. FORM and SORM reliability methods are conveniently used to compute the necessary failure probabilities and sensitivity factors for components and parallel systems.

For fatigue design both an *S-N* approach and a fracture mechanics approach are applicable. The two approaches are very similar when applied for design calculations. Inclusion of information from inspections on crack sizes can, however, only be done within the fracture mechanics approach.

A scheme for probability-based optimisation of design, inspection planning and repair strategies is presented. A cost optimal solution is determined at design and is updated based on in service information from e.g. inspections.

Reliability methods for fatigue analysis are already being used to a certain extent in daily engineering practice in some industries. This use is expected to increase rapidly with the latest conceptual developments and development of associated commercial software.

7. REFERENCES

1. Broek,D., *Elementary Fracture Mechanics*, Martinus Nijhoff Publishers, Fourth Edition, 1986.

2. Sobczyk,K. and Spencer Jr.,B.F.: *Random Fatigue - From Data to Theory*, Academic Press 1992.

3. Almar-Naess,A. (ed.), *Fatigue Handbook*, Tapir, 1985.

4. Wirshing,P., "Probability-Based Fatigue Design for Offshore Structures," Final Project Report, API-PRAC Project #81-85, University of Arizona, Arizona, 1983.

5. Wirshing,P., "Fatigue Reliability of Offshore Structures," *Journal of Structural Engineering*, ASCE, Vol. 110, 1984, pp. 2340-2356.

6. Series of Articles by the ASCE Committee on Fatigue and Fracture Reliability, *Journal of the Structural Division*, ASCE, Vol. 108, 1982, pp. 3-88.

7. Department of Energy: "New Fatigue Design Guidance for Steel Welded Joints in Offshore Structures," Recommendations, 1982.

8. Madsen,H.O., "Bayesian Fatigue Life Prediction," in *Probabilistic Methods in the Mechanics of Solids and Structures*, Julius Springer Verlag, Berlin, 1985.

9. Miner,M.A., "Cumulative Damage in Fatigue," *Journal of Applied Mechanics*, ASME, VOL. 12, 1945.

10. Madsen,H.O., Skjong,R. and Moghtaderi-Zadeh,M., "Experience on Probabilistic Fatigue Analysis of Offshore Structures," in *Proceedings*, OMAE Conference, Tokyo, Japan, 1986, Vol. II, pp. 1-8.

11. Madsen,H.O., Krenk,S. and Lind.N.C., *Methods of Structural Safety*, Prentice-Hall Inc., Englewood Cliffs, New Jersey, 1985.

12. Paris,P. and Erdogan,F., "A Critical Analysis of Crack Propagation Laws," *Journal of Basic Engineering*, Trans. ASME, **85**, 1963, pp. 528-534.

13. Irving,P.E. and McCartney,L.N., "Prediction of Fatigue Crack Growth rates: Theory, Mechanisms and Experimental Results," Fatigue 77 Conference, University of Cambridge, in *Metal Science*, Aug./Sept. 1977, pp. 351-561.

14. W.Shang-Xian: "Shape Change of Surface Crack During Fatigue Growth", *Engineering Fracture Mechanics*, **22**, 1985, pp. 897-913.

15. Kozin,F. and Bogdanoff,J.L, "A Critical Analysis of Some Probabilistic Models for Fatigue Crack Growth," *Engineering Fracture Mechanics*, **14**, 1981, pp. 59-89.

16. D.A.Virkler, B.M.Hilberry and P.K.Goel: "The Statistical Nature of Fatigue Crack Propagation," *Journal of Materials and Technology*, **101**, 1979, pp. 148-153.

17. Madsen, H.O., "Probabilistic and Deterministic Models for predicting Damage Accumulation due to Time Varying Loading," *DIALOG 5-82*, Danish Engineering Academy, Lyngby, Denmark, 1983.

18. Ortiz.K., "On the Stochastic Modelling of Fatigue Crack Growth," Ph.D. Dissertation, Stanford University, 1985.

19. Lin,Y.K. and Yang,J.N., "On Statistical Moments of Fatigue Crack Propagation," *Engineering Fracture Mechanics*, **18**, 1983, pp. 243-256.

20. Ditlevsen,O., "Random Fatigue Crack Growth - A First Passage Problem," *Engineering Fracture Mechanics*, **23**, 1986, pp. 467-477.

21. Wirshing,P.H., Ortiz,K. and Chen,Y.N., "Fracture Mechanics Fatigue Model in a reliability Format," in Proceedings OMAE '87, Houston, Texas, May 1987, pp. 331-337.

22. Madsen,H.O., "Random Fatigue Crack Growth and Inspection," in *Structural Safety and Reliability*, Proceedings of ICOSSAR '85, Kobe, Japan, IASSAR, 1985, **1**, pp. 475-484.

23. Madsen,H.O., "Model Updating in First-Order Reliability Theory with Application to Fatigue Crack Growth," in *Proceedings*, Second International Workshop on Stochastic Methods in Structural Mechanics, University of Pavia, Pavia, Italy, 1985.

24. Madsen,H.O., Skjong,R. and Kirkemo,F., "Probabilistic Fatigue Analysis of Offshore Structures - Reliability Updating through Inspection Results," *Integrity of Offshore Structures -3*, D. Faulkner et al. (eds.), Elsevier, 1988.

25. Bokalrud,T. and Karlsen,A., "Probabilistic Fracture Mechanics Evaluation of Fatigue Failure from Weld Defects in Butt Welded Joints", in *Proceedings*, International Conference on Fitness for Purpose Validation of Welded Constructions, The Welding Institute, London, 1981.

26. Skjong,R., "Reliability-Based Optimisation of Inspection Strategies," in *Proceedings*, ICOSSAR '85, IASSAR, Vol. III,1985, pp. 614-618.

27. Thoft-Christensen,P. and Sørensen,J.D., "Optimal Strategies for Inspection and Repair of Structural System," *Civil Engineering Systems*, Vol. 4, 1987, pp. 94-100.

28. Madsen,H.O., "Theoretical Manual PRODIM - PRObability-based Design, Inspection and Maintenance," A.S. Veritas Research Report No. 88-2019, Høvik, Norway, 1988.

29. Schittkowski,K., "NLPQL: A FORTRAN Sub-Routine Solving Constrained Non-Linear Programming Problems," *Annals of Operations Research*, 1986.

30. Tvedt,L., "User's Manual PROBAN - PROBabilistic ANalysis," AS Veritas Research Report No. 86-2037, Høvik, Norway, 1987.

MIX
Papier aus verantwortungsvollen Quellen
Paper from responsible sources
FSC® C105338

If you have any concerns about our products,
you can contact us on
ProductSafety@springernature.com

In case Publisher is established outside the EU,
the EU authorized representative is:
Springer Nature Customer Service Center GmbH
Europaplatz 3, 69115 Heidelberg, Germany

Printed by Libri Plureos GmbH
in Hamburg, Germany